Analytical Characterization of Biotherapeutics

Analytical Characterization of Biotherapeutics

Edited by Jennie R. Lill and Wendy Sandoval

Department of Microchemistry, Proteomics and Lipidomics,
Genentech Inc.
South San Francisco, CA, USA

WILEY

This edition first published 2017

Registered Office
John Wiley & Sons, Inc., 111 River Street, Hoboken, NJ 07030, USA

Editorial Office
111 River Street, Hoboken, NJ 07030, USA

For details of our global editorial offices, customer services, and more information about Wiley products visit us at www.wiley.com.

Wiley also publishes its books in a variety of electronic formats and by print-on-demand. Some content that appears in standard print versions of this book may not be available in other formats.

Library of Congress Cataloging-in-Publication Data

Names: Lill, Jennie R., editor. | Sandoval, Wendy, editor.
Title: Analytical characterization of biotherapeutics / edited by Jennie R. Lill, Wendy Sandoval.
Description: Hoboken, NJ : Wiley, 2017. | Includes bibliographical references and index. |
Identifiers: LCCN 2017013986 (print) | LCCN 2017022128 (ebook) | ISBN 9781119384427 (pdf) | ISBN 9781119384403 (epub) | ISBN 9781119053101 (hardback)
Subjects: LCSH: Proteins–Therapeutic use. | Proteins–Analysis. | BISAC: SCIENCE / Chemistry / Analytic.
Classification: LCC RM666.P87 (ebook) | LCC RM666.P87 A53 2017 (print) | DDC 615.7–dc23
LC record available at https://lccn.loc.gov/2017013986

Cover image: (Background) © Zffoto/Gettyimages;(Illustration) Courtesy of Allison Bruce
Cover design by Wiley

Set in 10/12pt Warnock by SPi Global, Pondicherry, India

10 9 8 7 6 5 4 3 2 1

"To Joe and Charlie, thank you for your wonderful smiles and boundless energy"

Jennie R. Lill

"To my daughters, Nicolina and Olivia, who are my daily inspiration"

Wendy Sandoval

Contents

List of Contributors

Corey E. Bakalarski
Departments of Microchemistry, Proteomics & Lipidomics and Bioinformatics & Computational Biology
Genentech Inc.
South San Francisco, CA, USA

Anne Baldwin
Departments of Biomolecular Engineering and Proteomics & Biological Resources
Genentech Inc.
South San Francisco, CA, USA

Karen Billeci
Department of Proteomics & Biological Resources
Genentech Inc.
South San Francisco, CA, USA

John B. Briggs
Department of Protein Analytical Chemistry
Genentech Inc.
South San Francisco, CA, USA

Natalie Castellana
Digital Proteomics LLC
La Jolla
and
Mapp Biopharmaceutical, Inc.
San Diego, CA, USA

Paola Di Lello
Department of Structural Biology
Genentech Inc.
South San Francisco, CA, USA

Diego Ellerman
Department of Protein Chemistry and Structural Biology
Genentech Inc.
South San Francisco, CA, USA

Adrian Guthals
Digital Proteomics LLC
La Jolla
and
Mapp Biopharmaceutical, Inc.
San Diego, CA, USA

Robert F. Kelley
Drug Delivery Department
Genentech Inc.
South San Francisco, CA, USA

Denise Krawitz
Department of Analytical Operations
Genentech Inc.
South San Francisco, CA, USA

Jennie R. Lill
Department of Microchemistry, Proteomics & Lipidomics
Genentech Inc.
South San Francisco, CA, USA

Yichin Liu
Department of Biochemical and Cellular Pharmacology
Genentech Inc.
South San Francisco, CA, USA

T. Noelle Lombana
Department of Antibody Engineering
Genentech Research and Early Development
South San Francisco, CA, USA

Patrick Lupardus
Department of Structural Biology
Genentech Inc.
South San Francisco, CA, USA

Till Maurer
Department of Protein Chemistry and Structural Biology
Genentech Inc.
South San Francisco, CA, USA

Christopher M. Overall
Center for Blood Research
University of British Columbia
Vancouver, British Columbia, Canada

Wilson Phung
Department of Microchemistry, Proteomics & Lipidomics
Genentech Inc.
South San Francisco, CA, USA

Jason C. Rouse
Department of Analytical Research and Development
Pfizer
Cambridge, MA, USA

Wendy Sandoval
Department of Microchemistry, Proteomics & Lipidomics
Genentech Inc.
South San Francisco, CA, USA

Justin M. Scheer
Antibody Engineering Department
Boerhinger Ingelheim, Ridgefield, Connecticut; Department of Protein Chemistry and Structural Biology
Genentech Inc.
South San Francisco, CA, USA

Kurt Schroeder
Departments of Biomolecular Engineering and Proteomics & Biological Resources
Genentech Inc.
South San Francisco, CA, USA

Lovejit Singh
Departments of Biomolecular Engineering and Proteomics & Biological Resources
Genentech Inc.
South San Francisco, CA, USA

Nestor Solis
Center for Blood Research
University of British Columbia
Vancouver, British Columbia,
Canada

Justin B. Sperry
Department of Analytical Research and Development
Pfizer
Cambridge, MA, USA

Christoph Spiess
Department of Antibody Engineering
Genentech Research and Early Development
South San Francisco, CA, USA

Martin Vanderlaan
Department of Analytical Operations
Genentech Inc.
South San Francisco, CA, USA

1

Introduction to Biotherapeutics

Jennie R. Lill

Department of Microchemistry, Proteomics & Lipidomics, Genentech Inc., South San Francisco, CA, USA

Abbreviations

ADAs	antidrug antibodies
ADC	antibody–drug conjugate
ADCC	antibody-dependent cell-mediated cytotoxicity
CDR	complementary-determining region
Fab	antigen binding fragment
Fc	cystallizable fragment
NMR	nuclear magnetic resonance
PEG	polyethyleneglycol
PTM	posttranslational modification

1.1 Introduction

Biotherapeutics, also known as biologics, include protein-based and nucleic acid-based drugs that are commonly derived by recombinant expression in living organisms although a few are made by chemical synthesis. This book focuses on the characterization of protein-based biotherapeutics, exploring the various analytical technologies that have enabled in-depth molecular characterization while discussing current triumphs and limitations.

The first human protein therapeutic derived from recombinant DNA technology was human insulin (Humulin®) created at Genentech, developed by Eli Lilly, and approved by the US Food and Drug Administration (FDA) in 1982. Since that time, major advancements in both recombinant DNA technology

Analytical Characterization of Biotherapeutics, First Edition. Edited by Jennie R. Lill and Wendy Sandoval.

and recombinant protein production have contributed to the development of several hundred biotherapeutics [1] including relatively simple molecules such as interferons, insulin, and the human growth hormone to more complexly engineered moieties including ADCs such as trastuzumab emtansine [2] and brentuximab vedotin [3].

Unlike conventional small molecule (chemical) drugs such as aspirin, antibiotics, and various chemo-therapeutics, the manufacturing process for biotherapeutics is typically far more cumbersome as they are larger compounds with more complex structures and their production can be extremely sensitive to changes in fermentation and environmental conditions. In addition, biotherapeutics are often less stable than many small molecules and can be prone to aggregation [4] or deamidation, oxidation, and other modifications [5]. Since the manufacturing of biotherapeutics is often dependent upon the host cells of living organisms, complex process development is required to ensure reproducible fermentations, isolation, and characterization [6].

1.2 Types of Biotherapeutics and Manufacturing Systems

There are several different types of marketed biotherapeutics including antibody-based drugs, anticoagulants, blood factors, bone morphogenetic proteins, engineered protein scaffolds, enzymes, Fc (cystallizable fragment) fusion proteins, growth factors, hormones, interferons, interleukins, and thrombolytics (Figure 1.1).

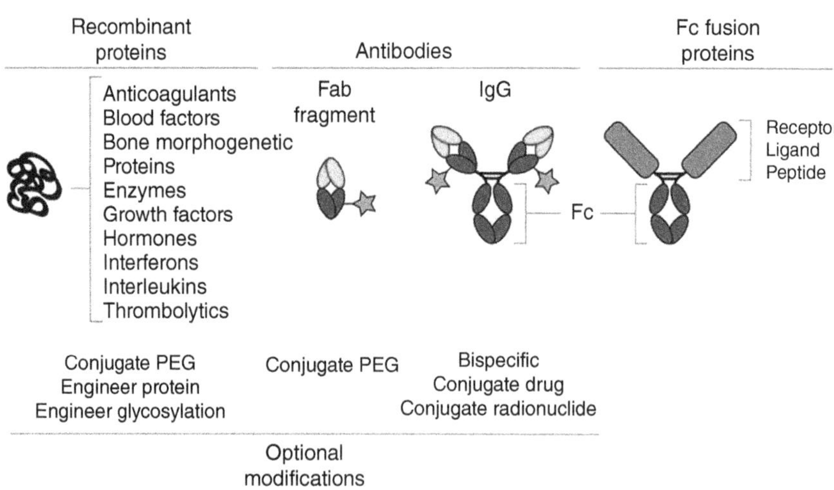

Figure 1.1 Various categories of the main types of biotherapeutics currently marketed. Source: Carter [7]. Reproduced with permission of Elsevier.

Antibody-based drugs represent the largest and most rapidly expanding class of biotherapeutics [1]. Figure 1.2 shows the diverse mechanisms by which the antibody structure can be modified to increase its biotherapeutic potential.

Humanized and other chimeric versions of these antibodies now dominate the market [11] and in the past 5 years have accounted for nearly 30% of all approvals. Various antibody isotypes are now being explored to provide a wealth of functional diversity that is present through the various IgG subclasses that can be exploited to improve clinical safety and performance by increasing stability, reducing adverse events, modulating effector functions, and by the engagement of two antigens by a single antibody [8]. Several variants that have been Fc engineered for reduced effector function have entered the clinic, for example, Eculizumab, a novel engineered IgG isotype, IgG2m4, with reduced Fc functionality. IgG2m4 is engineered based on the IgG2 isotype with four key amino acid residue changes derived from IgG4 (H268Q, V309L, A330S, and P331S). This antibody was demonstrated to have an overall reduction in complement and Fc gamma receptor binding in *in vitro* binding analyses while maintaining the normal *in vivo* serum half-life in rhesus [12].

Biosimilars (biologically identical antibodies, for example) and so-called biobetters (moieties with improved properties such as pharmacodynamic (PD) and pharmacokinetic (PK) readouts, higher potency, longer half-lives, and less immunogenicity, for example) are also starting to emerge, which presents new challenges in terms of testing for the presence of liabilities such as degradative properties, changes in immunogenicity through addition of novel contaminant proteins from new manufacturing processes, and so on. New formats such as glucagon-like peptide 1GLP fused proteins, for example, Eperzan (albiglutide) [13], and PEGylated proteins such as Plegridy (e.g., peginterferon beta-1a) [14] offer improved PK or PD properties but also increased analytical challenges due to their larger masses and increased heterogeneity.

Typically, expression of non-mAb biotherapeutics has been performed in *Escherichia coli* or a noneukaryotic system. This has many advantages for biotherapeutics that are not reliant on PTMs for their optimal activity. Over the years, however, there has been a gradual increase in the prevalence of mammalian expression systems. Of the mammalian expression systems, the Chinese hamster ovary (CHO) cell-based model (reviewed by Krawitz and Sandoval in Ref. [11]) remains the most employed expression system with a smaller percentage of therapeutics manufactured in other mammalian cell lines such as the murine myeloma line, NSO, and baby hamster kidney cells [15, 16]. Nonmammalian eukaryotic expression systems such as yeast [17] are also utilized, each again presenting their own challenges with regard to the correct PTM of the protein, occasionally adding to adverse properties [18, 19].

More recently transgenic animal production systems (e.g., expression of recombinant products in the animals' milk [19, 20], rabbits, and goats) have been explored as a means of biopharmaceutical production although to date

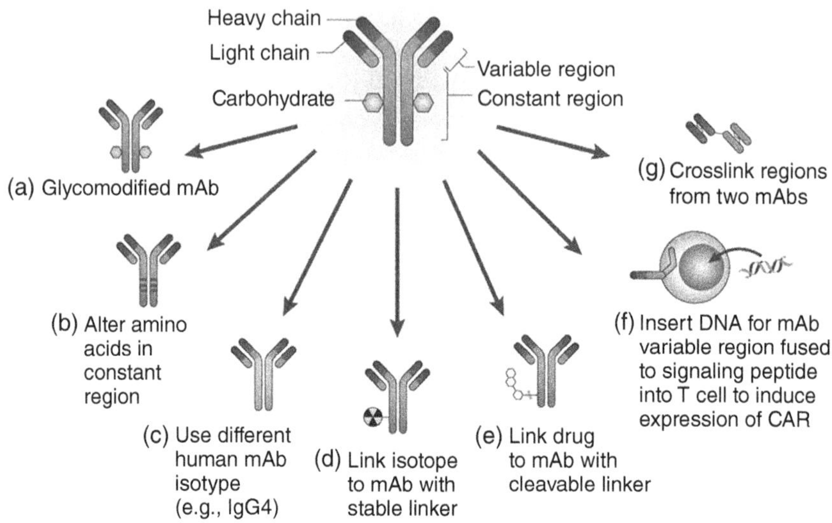

Figure 1.2 Monoclonal antibody (mAb) structure can be modified on the basis of the desired mechanism of action. Immunoglobulin G1 (IgG1) is the most effective naturally occurring human IgG isotype at mediating antibody-dependent cell-mediated cytotoxicity (ADCC). Glycomodified afucosylated mAbs (part a) (such as Obinutuzumab) demonstrate enhanced binding to IgG Fc receptors (FcγRs) and enhanced ADCC. In addition antibody-dependent cellular phagocytosis, a process mediated by macrophages, can also occur [8]. Afucosylated mAbs are produced using cell lines that lack the enzymes responsible for fucosylation. Modifying the amino acid sequence of mAb Fc (part b), as was done to produce ocaratuzumab [9], can also result in enhanced binding to FcγRs and enhanced ADCC. For mechanisms of action in which ADCC is not desirable, IgG4 may be a more appropriate isotype, as IgG4 mAbs do not mediate ADCC to the same degree as IgG1 (part c) although this isotype can still engage macrophage effector function *via* nanomolar affinity binding to FcγRI. Nivolumab, an IgG4 mAb that blocks programmed cell death protein 1 (PD1) on T cells, is one such example. Producing radioimmunoconjugates involves linking the radioisotope to the mAb. A stable linker is most desirable (part d) to limit the leakage of the free radioactive isotope. Conversely, optimal antibody–drug conjugates (ADCs) use a cleavable linker (part e). To avoid nonspecific toxicity, it is desirable for drugs used in ADCs to be cytotoxic once inside the target cell but nontoxic when bound to the mAb in the circulation. Linkers that are pH-sensitive or enzymatically cleaved are now a standard component of ADCs. Chimeric antigen receptor (CAR) T cells get their specificity from mAb variable regions but are a form of gene, not protein, therapy. They are produced by inserting DNA coding for the mAb variable region fused to DNA coding for signaling peptides into T cells (part f). Some bispecific antibodies lack a functional constant region so that they do not nonspecifically crosslink activating receptors and activate T cells (part g). The lack of a constant region on such constructs results in a short half-life, thus requiring continuous infusion to achieve the desired exposure. Source: Weiner [10]. Reproduced with permission of Nature Publishing Group.

there are many challenges associated with this type of biotherapeutic production with few benefits. Throughout this book the challenges of characterizing both the biotherapeutic moiety itself and the contaminant proteins such as CHO-derived proteins are discussed.

1.3 Types of Analyses Performed

Throughout this book a variety of analytical procedures are described. Many of them have been implemented for characterizing biotherapeutic molecules for as long as these moieties have existed. Others have evolved as the need arises. One such example of developing such sets of tools to answer a newly arisen problem is for the *de novo* sequencing of antibodies [21, 22]. Occasionally antibodies are discovered that are of great interest for preclinical testing, for which the cDNA or any genetic information is not available. In these scenarios, researchers have to sequence antibodies at the protein level, one amino acid at a time, and then reverse engineer the antibodies to the nucleotide level. In Chapter 6 Castellana and Guthals provide technical details and review the innovative approaches employed to quickly gain sequence information through a *de novo* approach.

As well as sequence information at the amino acid level, PTM profiling is also an important element in characterizing biotherapeutics [23]. There are a plethora of cotranslational modifications and PTMs that play key roles in the folding of proteins, in their secretion, and in their ultimate stability and effector functionality *in vivo*.

Glycosylation is important both for antibody secretion by B-cells and for *in vivo* antibody effector function. Glyco-engineering is a rapidly growing field, whereby glycosylation sites and composites are engineered to produce antibodies with specific glycoforms which may have an effect on therapeutic efficacy. Obinutuzumab (Gazyva®) [24], for example, is a humanized therapeutic monoclonal antibody that binds to an epitope on the B cell antigen, CD20. This antibody is engineered in a platform that allows control of the proteins' glycosylation, in this case the platform enforces the overexpression of two glycosylation enzymes MGAT3 and the golgi mannosidase 2. This results in the generation of antibodies with bisected nonfucosylated sugars, thereby increasing the antibodies' ability to activate natural killer cells. This means that Obinutuzumab can induce cell death through a dual mechanism of action, both by the antibody directly binding to B cells and by antibody-mediated cytotoxicity by recruiting the immune system to attack B cells. Some types of glycosylation are sometimes not beneficial. For example, Cetuximab, a chimeric mouse–human IgG1 monoclonal antibody against the epidermal growth factor receptor (EGFR) approved for use in colorectal cancer and squamous-cell carcinoma of the head and neck has a high prevalence of hypersensitivity reactions which has been attributed to cross reactivity to a V domain glycosylation site. In some patients severe adverse events have been observed including anaphylaxis, which was found to be due to the generation of patient-specific antibodies to the galactose-alpha-1,3-galactose modification [25].

Protein terminal modifications have the effect of modifying a protein's function, half-life, or cellular localization. Pyroglutamate formation, for example, is a highly prevalent modification whereby glutamine and glutamate at the N termini of recombinant monoclonal antibodies can cyclize spontaneously to pyroglutamate (pE) *in vitro* [26]. Proteolytic processing is also an irreversible modification that affects the vast majority of proteins, often with great functional consequences. Intracellular proteolytic processing has distinct effects on the functionality of proteins and can either abrogate or antagonize function, modify half-life, or also determine cellular localization. During protein synthesis, manufacturing, purification, and storage proteolysis events can occur, thereby changing a protein's functionality or stability [27]. Either through direct mass spectrometric analysis as reviewed in Chapter 2, through a variety of historic analytical techniques such as gel electrophoresis or Edman degradation, or by employing a variety of new biochemical-based methodologies for determining the termini of recombinant proteins as reviewed in Chapter 3, the determination of proteolytic processing remains a key analytical need for the characterization of biotherapeutic moieties.

Beyond linear sequence determination, structural analyses are also instrumental in the overall characterization of biotherapeutics. The biomolecular architecture is a vital component in dictating the specificity and overall efficacy of therapeutic proteins. The higher order structure (HOS) of a protein includes the secondary, tertiary, and quaternary structures of a protein that are required for its function. There is a diverse range of biophysical methods including circular dichroism, isothermal calorimetry, which are available for the characterization of a protein HOS, each of them with associated benefits and limitations. Related to conformational analysis is structural analysis as it pertains to epitope and paratope mapping. Again, several well-established techniques such as nuclear magnetic resonance (NMR) [28] and X-ray crystallography as well as some newer techniques such as mass spectrometric-based structural tools [29] including hydrogen deuterium exchange are described in Chapters 4 and 5.

1.4 Future perspectives

Nature has provided us with various types of protein scaffolds to explore as frameworks for building new types of biotherapeutics and there is a growing field of using these scaffolds as alternatives to antibodies [30]. Each of these types of engineered molecular structures offers new advantages in terms of stability and specificity. One example of this is the cystine knot mini proteins/peptides (knottins); these are peptide-based alternative molecules to monoclonal antibodies which are raised/designed against tumor-associated receptors and other antigens of interest. Knottins contains a disulfide-bonded core that exhibits a high level of resistance to proteolysis and increased thermal stability. Knottins emerged as an attractive molecular candidate for drug development as they fill the niche between small molecule drug design and protein biologics. Knottins have the potential to bind clinical targets with both high selectivity and affinity [31]. There are several naturally occurring knottins that have been approved as biotherapeutics for the treatment of pain [32] and irritable bowel syndrome and for tumor imaging purposes [33].

Elucidating disulfide bonding patterns of any biomolecules, but in particular a structure which relies on disulfide bonding patterns for their folding, stability, and activity, is an important part of molecular characterization. A variety of techniques can be employed for doing this from simple intact molecular weight measurement to more complex top-down proteomic protocols [34] and these methodologies continue to mature as more of these types of molecules emerge on the market.

Another growth area for biotherapeutics is increasing the molecules' *in vivo* half-life. For therapeutics that involves frequent or uncomfortable delivery, for example, injectable ocular therapeutics, or to make drugs that have poor PD properties more tolerable, there is a strong drive to create molecules that have increased *in vivo* stability (and potentially decreased immunogenicity). Some common mechanisms of molecular half-life extension include the generation of Fc fusion proteins, the formulation of biomolecules into various different nanoparticle systems, or the addition of stabilizing peptides. For a comprehensive list of these efforts please refer to Table 1.1.

The addition of albumin to stabilize the half-life of proteins has also been explored. Albumin is the most abundant plasma protein in humans and mice and is highly soluble, extremely stable, and has a circulatory half-life of ~20 days in man [36]. By fusing albumin to therapeutic proteins, these molecules become less susceptible to renal filtration and circulatory clearance and one of the main reasons albumim has a long half-life is due to its ability to bind to FcRn. Association, conjugation, or fusion of therapeutic drugs to albumin has been shown to correlate with superior PK.

In addition to albumin addition, PEGylation, the process of adding polyethyleneglycol (PEG) chains to a molecule by incubating a reactive derivative of PEG with the biomolecule of interest, also appears to work universally for improving therapeutic protein *in vivo* half-life [37]. PEGylation also increases the molecule's hydrodynamic size, thereby prolonging its time in circulation by reducing renal clearance. In addition, PEGylation can also make hydrophobic drugs with poor PK properties more water-soluble. There are a number of PEGylated molecules on the market and many companies are exploring this concept on a wider range of biotherapeutics [38, 39]. Characterizing these modified proteins, such as albumin-conjugated or PEGylated species, brings along increased analytical challenges, some of which are covered by Bakalarski *et al.* in Chapter 2 and by Ellerman *et al.* in Chapter 4.

Table 1.1 The circulatory half-life of a therapeutic protein can be extended by several strategies depending upon the endogenous clearance mechanism of the drug.

Protease degradation	**Novel/alternative delivery strategies avoiding need to extend half-life**
N-terminal acetylation or C-terminal amidation	Controlled-release depot (subcutaneous, intramuscular, intravenous); e.g. Bydureon–poly(lactic-*co*-glycolic acid) (PLGA) microspheres (Alkermes)
Nonnatural amino acids at labile sites	
Cyclization using disulfide bonds	
Microspheres or nanoparticles	
Increase size/hydrodynamic volume to prevent clearance by kidneys	**Increase size/hydrodynamic volume to prevent clearance by kidneys and attach to protein with a long half-life**
Di- or multimers	Attach to Fc (the natural antibody constant region)
Attach PEG—conjugation	Genetic fusion; e.g. Enbrel (Amgen), Mimetibody™ (Centocor), SynFusion (Biogen Idec/Syntonix)
Advanced PEGylation—modification of peptides (and prodrugs) by attaching with specific polymer chains	Covalent attachment (site-specific) (CovX/Pfizer)
Site-specific PEGylation (polytherics)—enables more selective PEGylation, reducing likelihood of protein deactivation upon conjugation and reducing immunogenicity	Attach to human serum albumin
Glyco-PEGylation (Neose)	Genetic fusion, albufuse® (Novozymes Biopharma/GSK/Teva/CSL)
Releasable PEGylation (Enzon)	Conjugation (Novozymes Biopharma, ConjuChem, Cardiovax)
Protein "rPEG"—genetic fusion	Albumin binding peptides/proteins/affinity tags (Genentech/Roche, Ablynx, Philochem, Affibody, Adnexus/BMS)
Poly-glycine	Albumin binding single domain antibodies fused to bioactive peptides (GSK/Domantis, Ablynx, Haptogen/Wyeth/Pfizer)
PASylation (XL-protein)	Albumin binding fatty acids (Novo Nordisk)
XTEN (Amunix)	
Other	
Hyaluronic acid (Novozymes)	

Table 1.1 (Continued)

Increase size/hydrodynamic volume to prevent clearance by kidneys	Increase size/hydrodynamic volume to prevent clearance by kidneys and attach to protein with a long half-life
Hydroxyethyl Starch (HESylation®—Fresenius Kabi)	
Polysialic acid (PolyXen®—Xenetic Biosciences)	
Elastin-like polypeptide (ELP) technology (Phase Bio Pharmaceuticals Inc.)	

Source: Sleep [35]. Reproduced with permission of Elsevier.
These include the reduction in the endogenous degradation of the drug; slow release/depot formulations; increasing the hydrodynamic volume of the drug by attachment of a large bulky polymer or extension of the therapeutic protein by addition of a linear but unstructured protein; or the addition of a large structured protein which additionally can take advantage of the FcRn-mediated recycling.

PEG polymers themselves are often highly heterogeneous and the increase in molecular weight that occurs when biomolecules are PEGylated means that more specialized mass spectrometric techniques or other analytical methods need to be employed [40]. With an increase in these types of complex molecules being developed for improved drug stability, the analytical challenges posed with their characterization also increase in complexity.

In addition to analyzing the intact biomolecules both *in vitro* and *in vivo* for assessing molecular stability, PK and PD properties, additional types of analyses to assess traits such as immunogenicity are starting to emerge. All biotherapeutics, including monoclonal antibodies and their derivatives, are immunogenic to varying degrees in various patients, with chimeric antibodies representing more of an immunogenic risk than humanized antibodies. These mouse human chimeric antibodies can induce some patients to develop antidrug antibody (ADA) responses [41]. Certain factors are known to influence biotherapeutic immunogenicity including structural homology with respect to human amino acid sequences and various PTMs. Although frameworks can be designed to minimize the potential for immunogenicity, the complementarity-determining regions (CDRs) of antibodies and other variable domains of biotherapeutics can be highly sequence variable making it difficult to predict how immunogenic a regent maybe [42]. Several companies have adopted strategies to assess immunogenicity retrospectively for therapeutics that have shown an ADA response [43]. Prospectively, a T cell activation assay, whereby CD4 + T cells are monitored for activation by antigen-presenting cells (APCs) loaded with biotherapeutics, can be employed along side a major histocompatibility complex associated peptide proteomics (MAPPs) assay. The MAPPs assay involves the *in silico* prediction of DR4 or other MHC class II peptide presentation using one of several programs such as SYPETHI, NETMHC, and so on. These algorithms predict potential T cell epitopes derived from the therapeutic protein. Additional analyses whereby MHC class II complexes are immune-precipitated from cells, peptides isolated and analyzed by mass spectrometry are also sometimes performed to complement the results from *in silico* predictions. These types of analyses are starting to be incorporated into early molecular assessment workflows to minimize the potential risk of adverse immunogenicity of new molecules prior to ADAs being reported.

In addition to these types of analyses, additional complexities arise due to chemically modified/conjugated biotherapeutics. Antibody–drug conjugates and antibody–antibiotic conjugates are gaining momentum in the clinic and these hybrid molecules that are composed of an antibody, a cleavable linker, and a chemotherapeutic or potent antibiotic molecule raise their own set of analytical challenges. In Chapter 7, Liu describes the importance of measuring drug-to-antibody ratios (DARs) [44] and the types of chromatographies and mass spectrometric techniques employed for characterizing these complex molecules. Liu describes the applications of increasingly diversified mass spectrometric techniques employed for characterizing the integrity of ADCs from the perspective of production to *in vitro* and *in vivo* testing as well as in terms of elucidating the mechanisms of ADC biotransformations.

So, now we have a variety of different biotherapeutic material in the research environment, manufacturing, and the clinic. How do we organize such materials? Prior to official lot validation that occurs once a biomolecule hits our pipeline, how do we ensure that the antigens we are using for immunization are from the same batch of cells, have been purified in the same manner and already characterized for correct protein sequence, folding, and structure? Biorepositories [45, 46] are growing in popularity and necessity in biotechnology companies. Official curation, stable storage, and easy retrieval of various lots of cDNAs, antigens, antibodies, ADCs and all the associated modifications are now becoming commonplace to allow researchers more consistency in the quality and reproducibility of their assay and other results.

As analytical technologies increase in sensitivity and complexity, low-level heterogeneity of therapeutic biomolecules that were previously unknown becomes revealed. Many of these modifications will be irrelevant in terms of the protein's stability or function; however, when reporting such heterogeneity to the FDA and other regulatory bodies one has to take care to not provide information that might be misconstrued. So how sensitive should the analysis be? At what percentile of the overall nonmodified protein should a mutation or modification be before it is reported or deemed important for functionality?

Some of the techniques involved in the molecular assessment of therapeutic biomolecules are described by Phung *et al.* in Chapter 11. These are all questions that analytical chemists and biochemists face as we embark on the design of new biotherapeutic moieties and see the emergence of biosimilars.

Acknowledgments

Thank you to Wendy Sandoval and Paul J. Carter for informative discussions and editing.

References

1 Walsh G (2014) Biopharmaceutical benchmarks 2014. *Nature biotechnology* 32(10):992–1000.
2 Niculescu-Duvaz I (2010) Trastuzumab emtansine, an antibody-drug conjugate for the treatment of HER2+ metastatic breast cancer. *Current opinion in molecular therapeutics* 12(3):350–360.
3 Borate U, *et al.* (2016) Treatment of CD30-positive systemic mastocytosis with brentuximab vedotin. *Leukemia research* 44:25–31.
4 Ratanji KD, Derrick JP, Dearman RJ, & Kimber I (2014) Immunogenicity of therapeutic proteins: influence of aggregation. *Journal of immunotoxicology* 11(2):99–109.
5 Jostock T & Knopf HP (2012) Mammalian stable expression of biotherapeutics. *Methods in molecular biology* 899:227–238.
6 Li F, Vijayasankaran N, Shen AY, Kiss R, & Amanullah A (2010) Cell culture processes for monoclonal antibody production. *mAbs* 2(5):466–479.
7 Carter PJ (2011) Introduction to current and future protein therapeutics: a protein engineering perspective. *Experimental cell research* 317(9):1261–1269.
8 Brezski RJ & Georgiou G (2016) Immunoglobulin isotype knowledge and application to Fc engineering. *Current opinion in immunology* 40:62–69.
9 Bowles JA, *et al.* (2006) Anti-CD20 monoclonal antibody with enhanced affinity for CD16 activates NK cells at lower concentrations and more effectively than rituximab. *Blood* 108:2648–2654.
10 Weiner GJ (2015) Building better monoclonal antibody-based therapeutics. *Nature reviews. Cancer* 15(6):361–370.
11 Henricks LM, Schellens JH, Huitema AD, & Beijnen JH (2015) The use of combinations of monoclonal antibodies in clinical oncology. *Cancer treatment reviews* 41(10):859–867.
12 An Z, *et al.* (2009) IgG2m4, an engineered antibody isotype with reduced Fc function. *mAbs* 1(6):572–579.

13 Trujillo JM & Nuffer W (2014) GLP-1 receptor agonists for type 2 diabetes mellitus: recent developments and emerging agents. *Pharmacotherapy* 34(11):1174–1186.
14 Khan UT, Tanasescu R, & Constantinescu CS (2015) PEGylated IFNbeta-1a in the treatment of multiple sclerosis. *Expert opinion on biological therapy* 15(7):1077–1084.
15 Eldering JA, Felten C, Veilleux CA, & Potts BJ (2004) Development of a PCR method for mycoplasma testing of Chinese hamster ovary cell cultures used in the manufacture of recombinant therapeutic proteins. *Biologicals: journal of the International Association of Biological Standardization* 32(4):183–193.
16 Thomsen C, *et al.* (1993) A pharmacological characterization of the mGluR1 alpha subtype of the metabotropic glutamate receptor expressed in a cloned baby hamster kidney cell line. *Brain research* 619(1–2):22–28.
17 Lee YJ & Jeong KJ (2015) Challenges to production of antibodies in bacteria and yeast. *Journal of bioscience and bioengineering* 120(5):483–490.
18 Frenzel A, Hust M, & Schirrmann T (2013) Expression of recombinant antibodies. *Frontiers in immunology* 4:217.
19 Spiesberger K, *et al.* (2015) Large-Scale Purification of r28M: A Bispecific scFv Antibody Targeting Human Melanoma Produced in Transgenic Cattle. *PLoS one* 10(10):e0140471.
20 Redwan el RM (2009) Animal-derived pharmaceutical proteins. *Journal of immunoassay & immunochemistry* 30(3):262–290.
21 Castellana NE, *et al.* (2011) Resurrection of a clinical antibody: template proteogenomic de novo proteomic sequencing and reverse engineering of an anti-lymphotoxin-alpha antibody. *Proteomics* 11(3):395–405.
22 Pham V, *et al.* (2006) De novo proteomic sequencing of a monoclonal antibody raised against OX40 ligand. *Analytical biochemistry* 352(1):77–86.
23 Heffner KM, *et al.* (2014) Exploiting the proteomics revolution in biotechnology: from disease and antibody targets to optimizing bioprocess development. *Current opinion in biotechnology* 30:80–86.
24 Ratner M (2014) Genentech's glyco-engineered antibody to succeed Rituxan. *Nature biotechnology* 32(1):6–7.
25 Chung CH, *et al.* (2008) Cetuximab-induced anaphylaxis and IgE specific for galactose-alpha-1,3-galactose. *The New England journal of medicine* 358(11):1109–1117.
26 Liu H, *et al.* (2014) In vitro and in vivo modifications of recombinant and human IgG antibodies. *mAbs* 6(5):1145–1154.
27 Ryan BJ & Henehan GT (2013) Overview of approaches to preventing and avoiding proteolysis during expression and purification of proteins. *Current protocols in protein science* Chapter 5:Unit5 25.
28 Stamos J, *et al.* (2004) Convergent recognition of the IgE binding site on the high-affinity IgE receptor. *Structure* 12(7):1289–1301.

29 Beck A, Sanglier-Cianferani S, & Van Dorsselaer A (2012) Biosimilar, biobetter, and next generation antibody characterization by mass spectrometry. *Analytical chemistry* 84(11):4637–4646.

30 Skrlec K, Strukelj B, & Berlec A (2015) Non-immunoglobulin scaffolds: a focus on their targets. *Trends in biotechnology* 33(7):408–418.

31 Moore SJ, *et al.* (2013) Engineered knottin peptide enables noninvasive optical imaging of intracranial medulloblastoma. *Proceedings of the national academy of sciences of the United States of America* 110(36):14598–14603.

32 Murray JK, *et al.* (2015) Engineering potent and selective analogues of GpTx-1, a tarantula venom peptide antagonist of the Na(V)1.7 sodium channel. *Journal of medicinal chemistry* 58(5):2299–2314.

33 Jiang L, *et al.* (2014) A radiofluorinated divalent cystine knot peptide for tumor PET imaging. *Molecular pharmaceutics* 11(11):3885–3892.

34 Reinwarth M, Avrutina O, Fabritz S, & Kolmar H (2014) Fragmentation follows structure: top-down mass spectrometry elucidates the topology of engineered cystine-knot miniproteins. *PLoS one* 9(10):e108626.

35 Sleep D, Cameron J, & Evans LR (2013) Albumin as a versatile platform for drug half-life extension. *Biochimica et Biophysica acta* 1830(12): 5526–5534.

36 Sleep D (2015) Albumin and its application in drug delivery. *Expert opinion on drug delivery* 12(5):793–812.

37 Schmidt PG, Campbell KM, Hinds KD, & Cook GP (2007) PEGylated bioactive molecules in biodegradable polymer microparticles. *Expert opinion on biological therapy* 7(9):1427–1436.

38 Moore DJ, Adi Y, Connock MJ, & Bayliss S (2009) Clinical effectiveness and cost-effectiveness of pegvisomant for the treatment of acromegaly: a systematic review and economic evaluation. *BMC endocrine disorders* 9:20.

39 Goel N & Stephens S (2010) Certolizumab pegol. *mAbs* 2(2):137–147.

40 Gong C, *et al.* (2014) Development and validation of an LC-MS/MS assay for the quantitation of a PEGylated anti-CD28 domain antibody in human serum: overcoming interference from antidrug antibodies and soluble target. *Bioanalysis* 6(18):2371–2383.

41 Tatarewicz SM, *et al.* (2014) Strategic characterization of anti-drug antibody responses for the assessment of clinical relevance and impact. *Bioanalysis* 6(11):1509–1523.

42 van Schie KA, Wolbink GJ, & Rispens T (2015) Cross-reactive and pre-existing antibodies to therapeutic antibodies—Effects on treatment and immunogenicity. *mAbs* 7(4):662–671.

43 Karle A, Spindeldreher S, & Kolbinger F (2016) Secukinumab, a novel anti-IL-17A antibody, shows low immunogenicity potential in human in vitro assays comparable to other marketed biotherapeutics with low clinical immunogenicity. *mAbs* 8(3):536–550.

44 Bryant P, *et al.* (2015) In vitro and in vivo evaluation of cysteine rebridged trastuzumab-MMAE antibody drug conjugates with defined drug-to-antibody ratios. *Molecular pharmaceutics* 12(6):1872–1879.

45 Baker M (2012) Biorepositories: Building better biobanks. *Nature* 486(7401):141–146.

46 Siwek M (2015) An overview of biorepositories-past, present, and future. *Military medicine* 180(10 Suppl):57–66.

2

Mass Spectrometric Characterization of Recombinant Proteins

Corey E. Bakalarski[1,2], *Wendy Sandoval*[1], *and Jennie R. Lill*[1]

[1] *Department of Microchemistry, Proteomics & Lipidomics, Genentech Inc., South San Francisco, CA, USA*
[2] *Department of Bioinformatics & Computational Biology, Genentech Inc., South San Francisco, CA, USA*

Abbreviations

CCS	collisional cross section
CHCA	alpha-cyano-4-hydroxycinnamic acid
DC	direct current
DHB	2,5-dihydroxybenzoic acid
EMR	extended mass range
ESI	electrospray ionization
FAIMS	field asymmetric ion mobility spectrometry
HPLC	high-performance liquid chromatography
IEC	ion exchange chromatography
IMS	ion mobility spectrometry
ISD	in-source decay
MALDI	matrix assisted laser desorption ionization
MS	mass spectrometry
MS/MS	tandem mass spectrometry
m/z	mass-to-charge ratio
PSM	peptide–spectral match
QTOF	quadrupole time of flight
RF	radio frequency
SEC	size exclusion chromatography

Analytical Characterization of Biotherapeutics, First Edition. Edited by Jennie R. Lill and Wendy Sandoval.

2.1 Introduction

A mass spectrometer in its most simplistic terms is an instrument capable of determining the mass of individual atoms or molecules. In reality one measures the mass (m)-to-charge (z) ratio (m/z) of an ion and since atoms of each element have different masses, the accumulative molecular composition of a molecule (e.g., a protein) can be determined by its m/z.

Antibodies and other recombinant proteins are one of the leading types of therapeutic moieties for the treatment of human disease and often have advantages over traditional small molecule-based drugs in that they are highly specific and typically exhibit fewer off-target side effects. All therapeutic related biomolecules, whether they are antigens to generate immunization campaigns or the actual therapeutic protein itself, may exhibit heterogeneity due to various modifications, either synthetically induced or naturally occurring during the manufacturing process [1].

N-terminal heterogeneity, generated from either truncated or elongated termini from exposure to aberrant proteolysis events during the secretion process, is a common observation on recombinant proteins [2, 3]. Co- or posttranslational modifications (PTMs) on proteins due to expression in and purification from eukaryotic expression systems can occur and often their presence or absence can be indicative of suboptimal manufacturing processes and potentially affect the molecule's activity or stability. Kinases in their active conformation, for example, are often auto-phosphorylated [4]; for IgG antibodies, glycosylation status is important for dictating correct Fc effector function [5]. Due to these inherent heterogeneities extensive analytical characterization is imperative to ensure reproducible, reliable, and safe production of these therapeutic proteins. The fastest, easiest, most accurate, and arguably most sophisticated analytical method to measure each of these events is through intact molecular weight mass measurements using mass spectrometry (MS).

2.1.1 Ionization

The first step in the mass spectrometric analysis of proteins is ionization. After purification and often dialysis into an appropriate MS-compatible buffer (e.g., a composition of an organic solvent such as methanol or acetonitrile in combination with water and an ion pairing reagent), the samples are introduced into the mass spectrometer [6]. To facilitate entry into the mass analyzer, samples must be ionized using one of several techniques; the most common types of ionization for large biomolecules are (i) matrix assisted laser desorption ionization (MALDI) [7], where a sample is spotted onto a MALDI plate for subsequent analysis, and (ii) electrospray ionization (ESI) [8, 9] whereby the sample is injected *via* either direct infusion or elution from an orthogonal analytical device such as a high-performance liquid chromatograph (HPLC) into an ESI source. Ionization is a critical step in the mass spectrometric analysis of any molecule, as it is this process that converts the molecule(s) of interest from a liquid or solid phase into a gaseous state, which is then amenable to introduction into and detection by the mass analyzer. The methods described herein are considered "soft" ionization techniques, as the actual process of ionization is amenable to keeping the biomolecule of interest intact (and thus retaining molecular weight information) during its conversion to a gaseous ion and entry into the mass spectrometer.

2.1.1.1 Matrix Assisted Laser Desorption Ionization

MALDI is a soft ionization technique typically coupled to MS that was first described by Hillenkamp and Karas [7] and a schematic of a typical MALDI ionization process is depicted in Figure 2.1. During MALDI analysis, the molecule of interest is cocrystallized with an organic matrix which is chosen based on the nature and size of the analyte. For example, intact mass measurement on a MALDI instrument is most often performed using 3,5-dimethoxy-4-hydroxycinnamic acid (a.k.a sinapinic acid) to minimize fragmentation, whereas peptides are commonly analyzed in the presence of alpha-cyano-4-hydroxycinnamic acid (CHCA) or 2,5-dihydroxybenzoic acid (DHB) matrices. The matrix absorbs ultraviolet light generated from a laser (most typically a nitrogen laser light at a wavelength

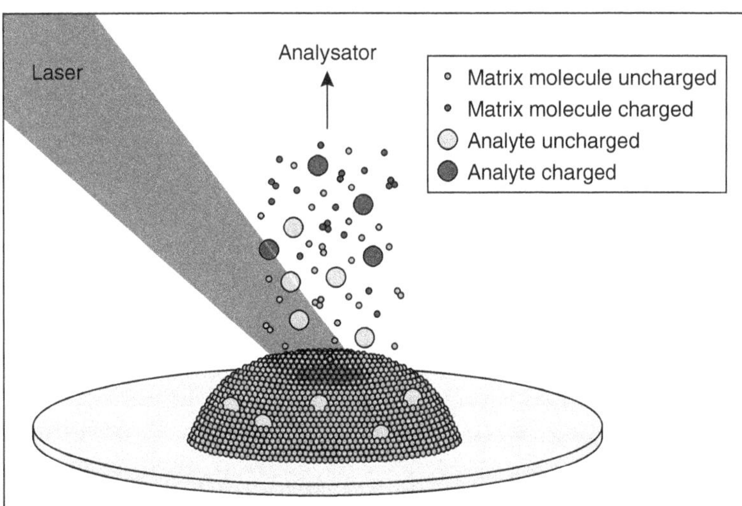

Figure 2.1 *MALDI ionization.* The uncharged analyte is mixed with an uncharged matrix molecule. After receiving energy *via* the laser pulse the now charged matrix donates a proton to the analyte *via* desorption, therefore allowing the analyte to become charged. The charged analyte is guided through the MALDI mass spectrometer for *m/z* analysis.

of 337 nm) and converts it to heat energy. A small part of the matrix heats in a matter of nanoseconds causing it to vaporize, together with the sample. There are many hypotheses to the mechanism of ionization, but, in general, MALDI produces only singly to low number of molecular charges [10]. This produces a charged gaseous ion (neutral ions are pumped out of the system and not detected) that is now amenable to mass spectrometric analysis. MALDI is a quick and easy tool for rapid analyses of noncomplex mixtures and is routinely employed in many laboratories.

MALDI ionization sources are generally coupled to Time of Flight mass analyzers (described in Section 2.1.2.1). MALDI-MS is rapid technique, providing spectra in seconds (thousands of scans are averaged in this time frame). MALDI is more tolerant to the presence of salts in sample buffers compared to other ionization modes. At a typical protein mass range (>10 kDa), the linear mode of detection is used and mass resolution is sacrificed for sensitivity and ease of use and predominantly single charge states up to 100 kDa affords immediate verification of mass. Figure 2.2 shows a high-mass MALDI spectrum of secretory IgA (sIgA). The sIgA complex was purified over a sizing column and

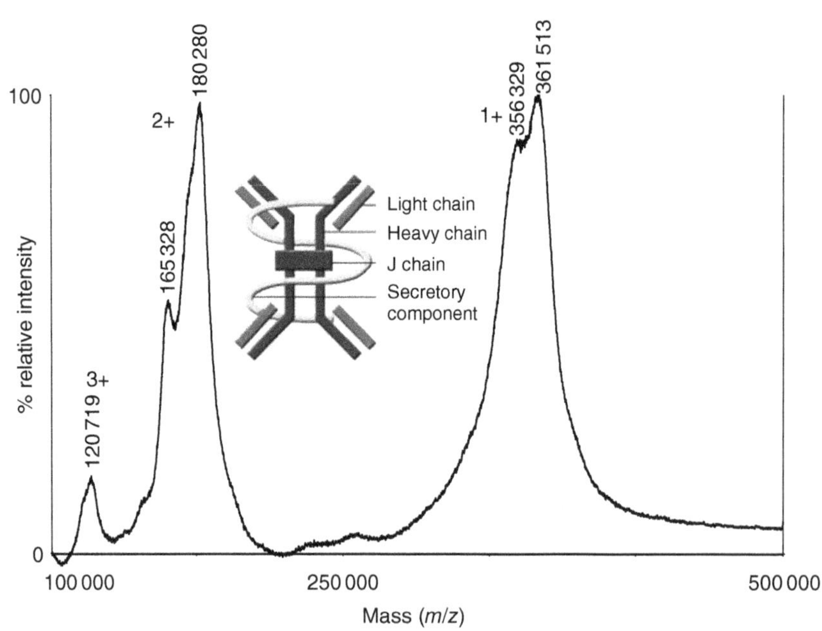

Figure 2.2 *High-mass MALDI-TOF spectrum of IgA dimer.* Secretory immunoglobulin A (sIgA) exists as a dimer and is the main immunoglobulin found in mucus secretion. A high-mass linear detector (CovalX, Saugus, MA) attached to a 4800 MALDI-TOFTOF (Sciex, Redwood City, CA) enabled the detection of a 362 kDa glycosylated IgA dimer and demonstrated that the ratio of antibody to J-chain was 2 : 1. Doubly (181 kDa) and triply (120 kDa) charged species are also observed.

was confirmed to exist in dimeric form using MALDI. Although other higher-resolution mass spectrometric techniques were attempted, heterogeneity of the individual components in the complex (including extensive glycosylation) prevented deconvolution of highly charged mass envelopes.

2.1.1.2 Electrospray Ionization

ESI is also a soft ionization technique where ions are produced using an electrospray which in the presence of a high voltage creates gaseous ions from a liquid aerosol (Figure 2.3). Unlike MALDI and other ionization techniques, ESI typically produces multiply charged ion species, often allowing higher mass species to be analyzed because of generating lower m/z for a species of identical molecular weight. Yamashita and Fenn first described ESI [8, 9], and were awarded the Nobel Prize for the development of ESI for the analysis of biological macromolecules in 2002. There are now many flavors of ESI sources on the market and for sources that operate at lower flow rates and associated voltages, the term "nanospray" ionization is more typically adopted [11]. ESI is ideally suited to analyses where a separation device such as capillary electrophoresis

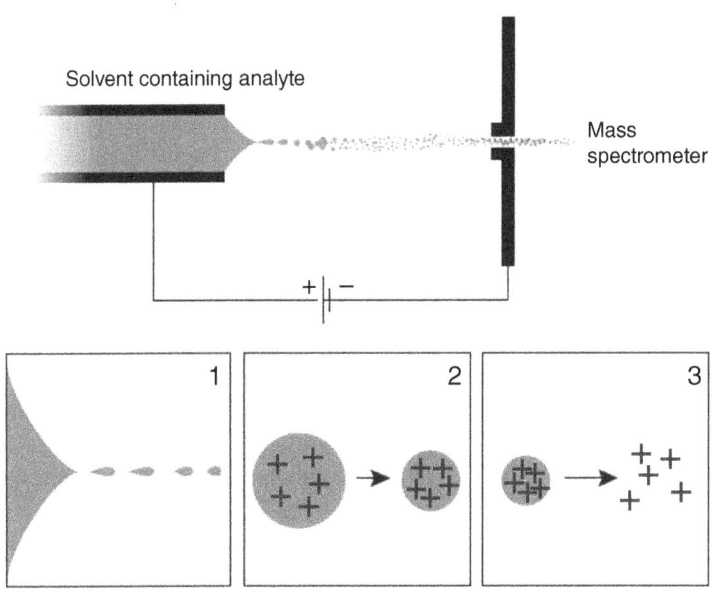

Figure 2.3 *Schematic of electrospray ionization.* (1) Under high voltage, the eluent from a syringe, tip, or HPLC creates a cone shape (a Taylor cone) and represents the initiation of the transfer of ions from the liquid to the gas phase. (2) Evaporation of the droplets occurs leaving them increasingly highly charged. (3) When the charge exceeds the Rayleigh limit (the maximum amount of charge a liquid droplet can carry before throwing out fine jets of liquid), the droplet completely dissociates leaving a stream of charged gaseous ions that now enter the mass spectrometer for separation and detection.

(CE) or high-pressure liquid chromatography is coupled in-line with a mass spectrometer. Desolvation of ions from these systems can be performed online, by eluting liquid from the chromatography systems directly into the electrospray source [12].

2.1.2 Mass Analyzers for Intact Molecular Weight Measurement of Biotherapeutics

There are a variety of mass spectrometers on the market, some of which are more specifically capable of performing high-mass accuracy analysis of large biotherapeutic entities such as antibodies and other recombinant proteins. The choice of mass spectrometer is an important one when considering what types of analysis one wishes to perform, the necessary throughput for the analysis in question, and the robustness of the system. Two examples of popular mass spectrometers for the intact molecular weight analysis of proteins are described here, although there are many instruments on the market that are capable of such analyses.

2.1.2.1 Time of Flight and Quadrupole Time of Flight Mass Spectrometers

Since the introduction of MALDI and ESI, two techniques that revolutionized the ability to study proteins and peptides by MS, the mass spectrometer of choice for determining intact molecular weight mass measurements for recombinant proteins has been the Time of Flight (TOF) mass spectrometer and its more sophisticated counterpart the Quadrupole Time of Flight (Q-TOF) [13]. These are robust instruments that typically have high-resolution capabilities and allow accurate intact mass measurements of larger biomolecules. In the most simplistic terms, the instrument measures the m/z of an ion by a time-based measurement. After ionization, ions are introduced into the mass analyzer, accelerated by application of an electric field into to a flight tube, and the velocity of the ion as it passes along the flight tube depends on its m/z. Ions with smaller m/z values (lighter ions) and more highly charged ions move faster through the drift space until they reach the detector, whereby the time of ion flight differs according to the m/z value of the ion. Figure 2.4 shows a schematic of a typical Q-TOF instrument.

To increase specificity, many TOF instruments are now coupled to a quadrupole (Q-TOF). A quadrupole consists of four parallel metal rods with oscillating direct current (DC) and Radio Frequency (RF) voltages on each pair of rods. Ions accelerating through the quadrupole between the rods and ions with a particular m/z have a certain trajectory. Ions can therefore be selected based upon m/z and the quadrupole can act as a mass filter, or, at broadband RF mode (allowing a wide mass transfer), increase signal-to-noise ratio by decreasing the transmission of superfluous chemical noise or neutral ions before they enter the

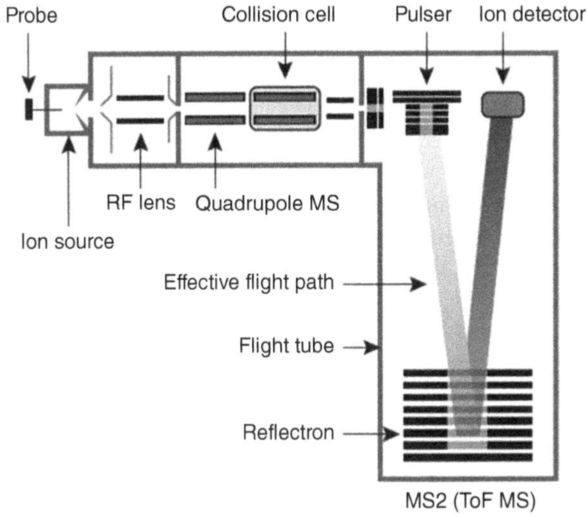

Figure 2.4 *Schematic of a quadrupole time of flight mass spectrometer.* An analyte, for example, a protein of interest undergoes ionization at the ion source and gaseous ions are transmitted into the mass spectrometer. After ions enter the source, the pressure is raised to increase transmission efficiency. Ions then pass through the quadrupole (where mass selection can take place) before being accelerated into the collision cell. The collision cell is filled with an inert gas. Analytes are separated through the flight tube where they are resolved based upon the time it takes them to traverse across a predefined distance. To enable higher resolution using reflectron mode, ions are "reflected" back along the flight tube and detected by the second detector.

TOF tube. For additional information on the various techniques, for example, top-down proteomic methodologies using a Q-TOF, please refer to Section 4.3.1.

2.1.2.2 High-Resolution Intact Mass Measurement and Native MS

Many analyses of biotherapeutic molecules are typically performed under denaturing conditions; however, a subset of proteins, for example, bispecific antibodies, drug/protein conjugates, or noncovalent multimeric complexes, require the analysis of proteins in their native conformational state. In 2014 Thermo Fisher Scientific introduced the Exactive Plus Extended Mass Range (EMR) Orbitrap [14]. This instrument combines high-resolution, accurate mass measurements over an EMR (up to 20 000 m/z), thereby allowing researchers to gain highly accurate m/z information for large biomolecules such as antibodies and their derivatized counterparts. In addition, the instrument is capable of interrogating noncovalent interactions between biomolecules and other molecular entities such as small molecules, metabolites, and lipids as well as allowing characterization of the structure and topology of native-like tertiary and quaternary protein structures.

In this process known as native MS, the original buffer solution is typically exchanged with a volatile solution such as ammonium acetate that mimics a native buffer but is also amenable to solvent evaporation during ESI. Ionized proteins under native conditions typically have lower charge state distributions than the counterparts generated by typical ESI. Ions generated by native MS therefore present at a higher *m*/*z* window than their nonnative counterparts. By having native MS capabilities, the EMR Orbitrap Plus not only provides high mass accuracy and high-resolution data, but also proves to be a high-performing mass spectrometer for screening complex biomolecules [15, 16] such as antibody–drug conjugates, PEGylated species, and complex glycoforms of proteins [17].

It has previously been reported that IgG antibodies can coordinate into ordered hexamers on cell surfaces after binding their antigen. These hexameric complexes bind the first component of complement C1 inducing complement-dependent target cell killing. There are various efforts under way both in academic and in the biopharmaceutical industries to exploit these antibody complexes as therapeutic moieties. Given the large size of these structures, and the noncovalent interactions between the monomers making up the hexameric species, these are particularly challenging to study [18]. In order to investigate the suitability of the IgG mutants for their potential development as hexameric complex forming biopharmaceuticals, de Jong *et al.* analyzed their biophysical characteristics with an emphasis on methods that could detect solution-phase multimers. Despite trying to detect hexameric formation by conducting high-performance size exclusion chromatography (HP-SEC) and multiangle laser light scattering analysis of hollow fiber flow field-flow fractionation, the most successful analytical tool for detecting multimerization was by analyzing the mutants by native MS [19]. The EMR Orbitrap instrument is predicted to be a game-changing analytic in the characterization of complex biomolecular entities. Figure 2.5 shows the mass spectrometric charge envelope of a small amount of monomer and mutant antibody hexamer, illustrating the purity of the mutant antibody preparation. Native MS is being employed to measure the efficiencies of such mutants in forming hexameric antibody complexes.

Thompson *et al.* [16] have also employed native MS to look at composite antibody mixtures. It is understood that a combination of multiple antibodies can lead to an increased effector response and can minimize escape mechanisms that can propagate diseases. To characterize a mixture of 15 antibodies, native MS on an EMR instrument unambiguously identified each antibody to within 7 ppm mass accuracy.

2.1.2.3 Ion Mobility Spectrometry

Ion mobility spectrometry (IMS) is a complementary analytical technique to MS that is used to separate and characterize ionized molecules in the gas phase based upon their mobility by a carried buffer gas. Such gas-phase ion mobility

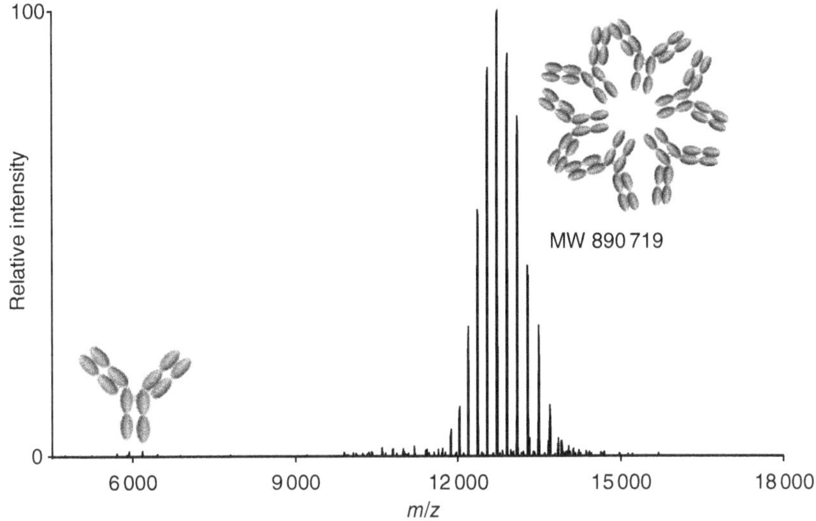

Figure 2.5 *Native MS analysis of and RGY–antibody hexamer.* Under native conditions (10 mM ammonium acetate) on an EMR Exactive Orbitrap, using high C-trap pressure the RGY-IgG mutant is observed to be a hexameric species. The charge envelope of the IgG monomer appears around *m/z* 6000 and that of the IgG hexamer around *m/z* 13 000.

provides a collisional cross section (CCS) of a given ion as it drifts through an inert buffer gas in the presence of a low electrical field. During this process, the ion collides with inert buffer gas molecules and provides readout in terms of the time it takes for the ion to pass through the drift region. Ions that have a large cross-section drift through the inert gas more slowly than those with smaller cross sections. This analytical tool can therefore be employed to separate proteins and protein complexes based on their native conformational states. Indeed, mobility cross sections can be applied to differentiate between the structure of two proteins sharing the same *m/z*, but with differing conformations in the gas phase. There are several types of IMS including Drift-Tube Ion Mobility Spectrometry (DT-IMS), Traveling-Wave Ion Mobility Spectrometry (TW-IMS), and Field Asymmetric Ion Mobility Spectrometry (FAIMS)/Differential Mobility Spectrometry (DMS). For a comprehensive review please refer to Cumeras *et al.* [20, 21]. Figure 2.6 shows a schematic of a typical ion mobility cell.

Ion mobility has been employed to allow the rapid characterization of disulfide variants in intact IgG2 monoclonal antibodies [22], characterizing the thermal stability of monoclonal antibodies [23], and for measuring drug-to-antibody ratios (DARs) for antibody–drug conjugates [24]. It is also a powerful tool for distinguishing differential analytical features that may not be evident by MS alone for analysis of biosimilars. For example, in a head-to-head

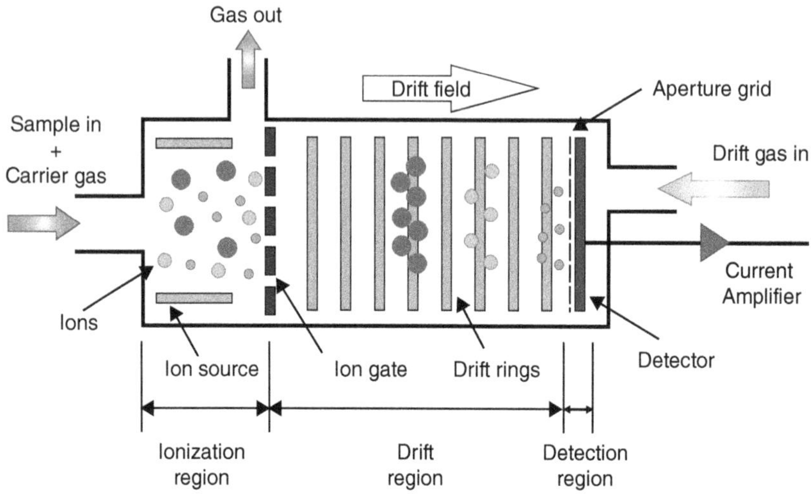

Figure 2.6 *Schematic of a conventional drift time IMS (DTIMS)* system showing three ions of different sizes in the reaction region and then migrating at different velocities in the drift region.

comparison of trastuzumab and cetuximab with corresponding biosimilars and biobetter candidates, various analytical techniques including MS, native MS, and ion mobility MS were employed to look for variations in PTMs, disulfide bond pairing, and other features [25].

For additional information on IMS in this book please refer to Chapter 8 where Yichin Liu describes the use of IMS for the characterization of antibody–drug conjugates (ADCs) and Section 8.2.1 describes how IMS is employed for studying conformational differences in various monoclonal antibody isoforms.

2.1.3 Software for the Analysis of Intact Molecular Weight Measurements

As mass spectrometers measure not the overall mass of a molecule but rather its m/z, the observed signal is dictated by a function of both its mass and an apparent charge state. This signal can be difficult to reverse engineer from the observed spectrum, owing to the accumulation of chemical and electronic noise. In practice, several methods have been proposed for the deconvolution of the charge information from the uncharged protein mass. Perhaps the most popular choice is the application of the probabilistic approach of maximum entropy [26], which seeks to find the most probable parent mass spectrum which would yield the observed, convoluted signal. This approach relies on Bayes' Rule to yield the appropriate parent spectrum given the signal observed.

While implementations may vary in how they model the observed noise, such approaches are commonplace in vendor software and produce acceptable results in a wide variety of analyses. Nevertheless, alternative implementations for specific applications are also available [27–29].

2.1.4 Separation Devices for the Characterization of Biotherapeutics

During the initial identification of a recombinant protein to test its ability as a tool compound or first generation therapeutic, a single round of chromatographic separations after isolation from the cellular supernatant is often performed followed by mass spectrometric analysis. For final drug products, the purification and testing process is much more arduous and often follows a set of analytical characterization steps including determining quality (appearance, pH, osmolality), concentration (typically measuring the concentration at A280), identity (human IgG and enzyme Western blotting), purity (reducing and nonreducing sodium dodecyl sulfate polyacrylamide gel electrophoresis (SDS-PAGE), size exclusion HPLC, anion exchange HPLC), potency (human insulin receptor (HIR) binding affinity, enzyme specific activity) as well as other purity determining measurements. After these purification and separation steps, mass spectrometric characterization of the biomolecule of interest occurs either by intact molecular weight mass measurement as previously described in Section 2.1.2 or by peptide mass fingerprinting (PMF) or peptide sequencing using LC/MS/MS as mentioned in Sections 2.1 and 2.2.

Several of the analytical separation procedures mentioned earlier are performed in-line with a mass spectrometer. The IMS system is just one example of such an orthogonal separation strategy that can be coupled to a mass spectrometer; however, for the majority of cases, a liquid chromatographic separation such as size exclusion chromatography (SEC), ion exchange chromatography (IEC), or more commonly high-performance liquid chromatography (HPLC) is employed.

2.1.4.1 High-performance Liquid Chromatography

HPLC is an analytical technique employed to separate, identify, and quantify biological (or other) components from a mixture. HPLC works by pumping a pressurized liquid or solvent containing the analyte(s) through a column filled with a solid matrix, typically a silica-based stationary phase (C18 resins being more compatible with peptidic analyses and C4 or C8 for proteins, although many variations on these matrices exist). Peptides and proteins which are present in the injected sample flow within the solvent being delivered by the HPLC pumps are adsorbed onto the hydrophobic surface of the column, remaining there until the concentration of the organic modifier (typically methanol or acetonitrile) exceeds the threshold to elute the molecules from the hydrophobic surface. The elution order is directly related to the increasing hydrophobic properties of the

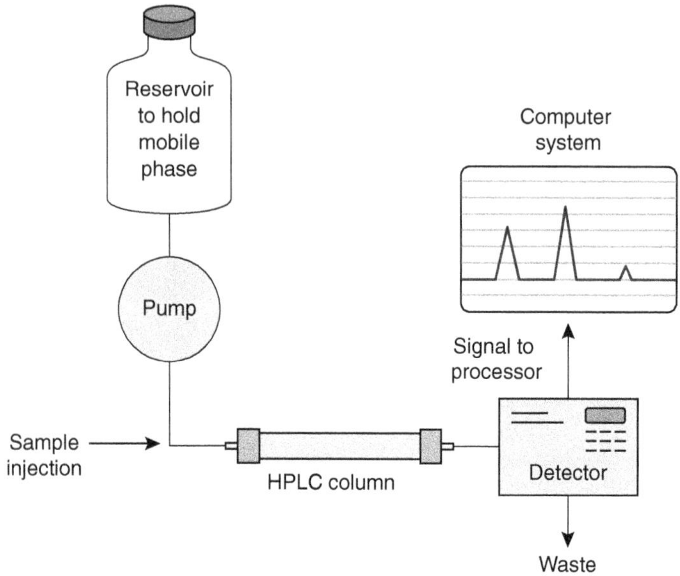

Figure 2.7 Schematic of a high-performance liquid chromatography (HPLC) system: Analytes are separated on the solid stationary phase by displacement with a liquid mobile phase. The stationary phase is the packing material contained in the HPLC column (hollow tube containing particles). The mobile phase is the solvent or mixture of solvents (and sometimes other ingredients) that is passed through the HPLC column to flush the compounds through the column. Different compounds have different affinities that are attracted to the particles in the HPLC column and different affinities to being flushed through the column using the mobile phase. The competition between a compound's affinity to bind to the stationary phase or elute away in the liquid mobile phase is what enables the separation. A mixture of different compounds can be separated by chromatography and then detected individually by a detector (e.g., UV absorbance). The absorbance response for the separated compound will be proportional to the concentration of the compound in solution. Comparison of the detector response for an unknown to that of a known standard (known concentration of a pure compound) can determine the concentration present in the unknown. The analytes (compound you are testing the sample for) of interest must be dissolved in a solvent to perform an HPLC separation. Source: Reproduced from http://www.actalabs.com/ac/HPLC.php

solute; the more soluble a solute is in water or the more hydrophilic the solute, the faster it will be eluted. Figure 2.7 shows a schematic of a typical HPLC and for detailed information about HPLCs, the background theory to chromatography and advice on their operation please refer to Stoll [30].

2.1.4.2 Capillary Electrophoresis

CE is an electrokinetic separation technique utilizing small (submillimeter) diameter capillaries. In CE, analytes of interest migrate through electrolyte solutions in the presence of an electric field. Analytes are separated according to their ionic mobility and may be focused based on gradients of conductivity

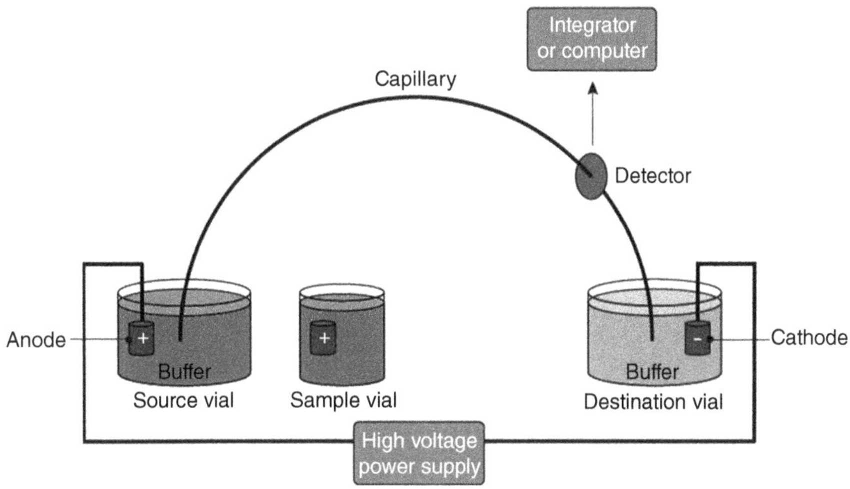

Figure 2.8 *Schematic of a capillary electrophoretic system.* The analyte of interest is introduced into the CE system by placing the capillary inlet into the samples vial where it traverses through the capillary by capillary action, pressurization, siphoning, or through electrokinetic mechanisms, where finally the sample exits at the source vial. Migration of analytes occurs when an electric field (supplied by the electrodes) is applied between the source and destination vials and in most CE systems all ions, positive or negative, navigate through the capillary in the same general direction by electroosmotic flow. Analytes are detected near the outlet end of the capillary and data is displayed as an electropherogram.

and pH. The main components of a simple CE system include the sample vial, the source and destination vials, the electrophoretic capillary, electrodes, and a detector. A schematic of a simple CE system is shown in Figure 2.8.

CE was first used in parallel with MS by Smith *et al.* and since then this coupled set of devices has proven an invaluable tool for the characterization of biotherapeutic moieties [31]. This is because CE-MS is a very sensitive technique where only a very small amount (nanoliters) of analyte is needed for analysis. Several injection strategies have been described (either by pressure or by electrokinetic injection) that allow the concentration of analytes into a narrow zone near the inlet of the capillary. To increase throughput CE instruments now exist in arrayed formats where multiple samples can be analyzed simultaneously. Such capillary array electrophoretic (CAE) devices come with either 16 or 96 capillaries and allow for analytes to be sampled from 96 well plates [32]. CE-MS has been used extensively for the characterization of biotherpeutics including monitoring the stability and pharmacokinetic properties of complex materials such as pegylated interferon-beta-1b [33], measuring deamidation and isomerization of biomolecules [34], and characterizing dimer formation for therapeutic antibodies [35]. The field continues to evolve with new types of multiplexed formats of capillaries and miniaturized system being constantly introduced [36].

2.1.4.3 Microfluidic Chromatographic Devices

Over the past decade the bioanalytical chemistry world has seen an increase in the number of microfluidic chip-based devices that allow online liquid chromatography (LC)-based separations with mass spectrometric analysis. One example of this is the Agilent Chip Cube, a device that is coupled directly to the mass spectrometer (typically a Q-TOF) to allow low nanoliter scale separations on a variety of resins including reverse phase and polymeric reverse phase (PLRP) resins, a polystyrene divinylbenzene copolymer for peptide and protein analyses. In addition, Agilent published their work on the utilization of a novel polymeric microfluidic device with an on-chip enzyme reactor that is used for the characterization of recombinant glycoproteins. The enzyme reactor chip is packed with PNGaseF-modified solid support material in combination with a microfluidic glycan "clean-up" reverse phase chip and a commercially available HPLC chip. This microfluidic device therefore allows researchers to perform deglycosylation, protein removal, glycan capture, and analysis all in line with a TOF-MS. Such analyses are high-throughput, sensitive, and reproducible. These microfluidic devices are typically robust and highly sensitive and allow a high-throughput analysis of biotherapeutic molecules. Figure 2.9 shows a microfluidic device and how it operates for the analysis of glycosylated moieties. Figure 2.10 shows the spectra for an antibody after on-chip deglycosylation over three time points.

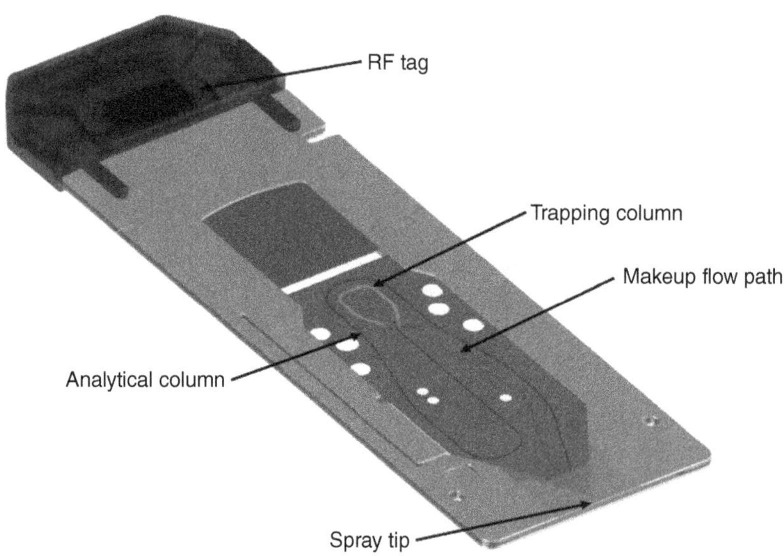

Figure 2.9 Agilent microfluidic device. Source: Reproduced with permission of Agilent (Santa Clara, CA).

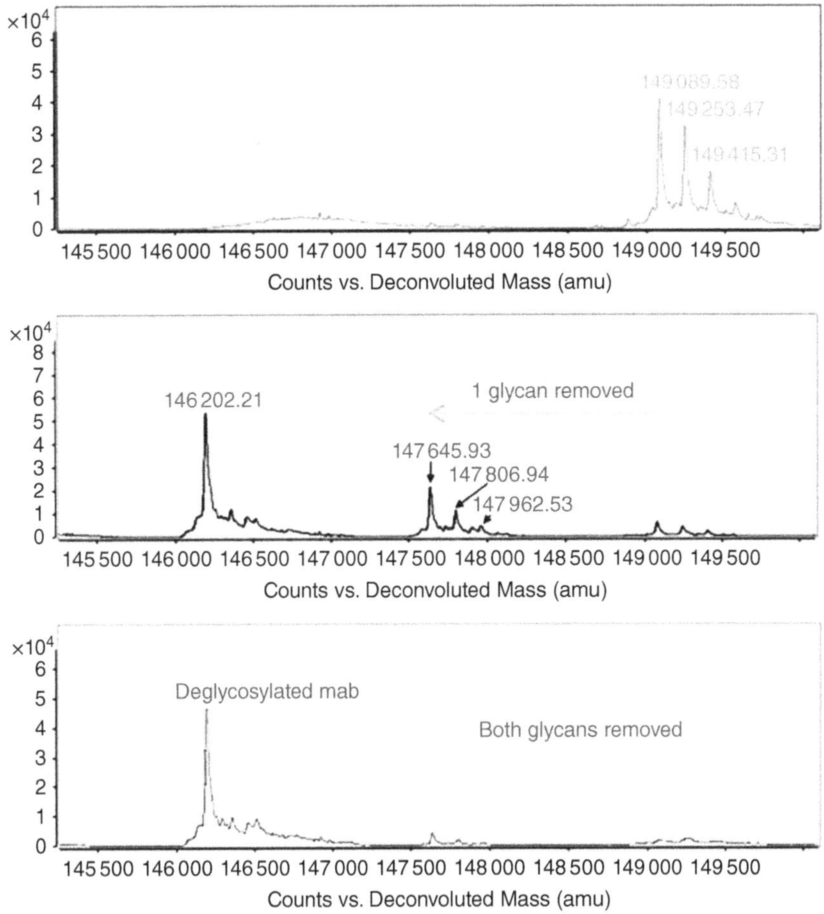

Figure 2.10 *Analysis of deglycosylated antibody with on-chip deglycosylation.* Three main peaks are observed prior to deglycosylation. These peaks are suggestive of the antibodies with the combination of G0, G1, and G2 glycans. The partially deglycosylated antibody follows a 3 s residence time. Antibodies with one glycan are observed. The deglycosylated antibody peak after a 6 s residence time in the PNGase F enzyme reactor chip. Source: Bynum *et al.* [37]. Reproduced with permission of American Chemical Society.

2.2 Peptide Mass Fingerprinting

PMF is a mass spectrometric method in which a protein is cleaved into smaller peptides using a proteolytic enzyme with predictable cleavage sites (e.g., trypsin cuts predictably to the C-terminus of Arginine and Lysine). Typically PMF is performed on a MALDI-TOF or ESI-TOF instrument and the method was first described over two decades ago [38]. Once peptide masses are generated these are compared to a database containing known protein sequences. Results are statistically analyzed to find the best match of the spectrum to the theoretical peptide fingerprint and there are several algorithms available for deciphering protein matches based upon PMF data.

PMF is perhaps one of the earliest examples of proteomic analysis using MS that employs computational tools for data analysis. First described in 1989, the approach relied upon a computer algorithm, FRAGFIT, which dynamically generated peptide masses based on the cleavage patterns from theoretical protein sequences [38]. This approach allowed for rapid analysis of any protein and became one of the earliest examples of high-throughput MS.

Today, the procedure remains largely unchanged. PMF analysis tools can be found in a number of software suites, including Protein Prospector [39] and Mascot [40]. Each tool takes as input a deconvoluted peptide mass fingerprint spectrum and matches it against the list of generated peptide masses for the protein in question, scoring the match based on the appearance of the required masses. While this approach works well to confirm the presence of a known protein, it does not identify the presence of additional proteins or proteoforms within the sample; detection of these forms requires more in-depth analysis using many advanced techniques which have since been defined within the field. Nevertheless, PMF remains a useful tool for the rapid verification of protein sequence.

This method is a robust, quick, and easy analytical technique for determining that a single protein has been correctly generated, processed, and purified. Figure 2.11 shows a typical workflow involving PMF. Although mass spectrometric instrumentation has evolved to have high mass accuracy and sensitivity, this method is still challenging for identifying mixtures of proteins. Also, upon analysis of the sample, if a peak that does not match the predicted sequence is identified, the researcher will have to decipher the identity of the erroneous peak by performing MS/MS or through additional analyses. Despite these limitations, PMF is popularly employed for characterizing biotherapeutics and biosimilars [41].

2.3 Tandem Mass Spectrometric Characterization of Biomolecules

Tandem mass spectrometry (MS/MS or MS^2) involves the selection of an ion or ions in the mass spectrometer, followed by fragmentation, resulting in the generation of product ions (also referred to as daughter ions). There are various types of MS/MS fragmentation that can be applied including collision-induced dissociation (CID) [42], photo dissociation (PID) [43], electron transfer dissociation (ETD) [44], and electron capture dissociation (ECD) [45].

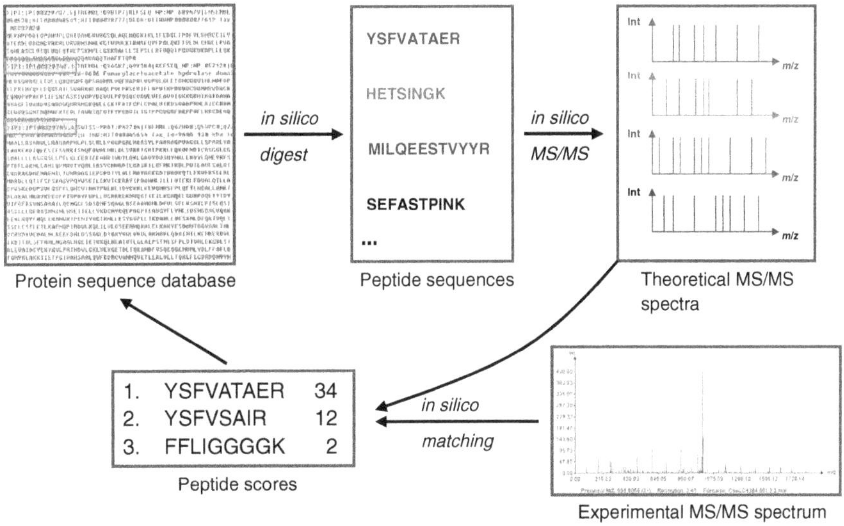

Figure 2.11 *Schematic of a peptide mass fingerprinting workflow.* The sequence of a protein of interest is selected from a database. An *in silico* digestion is performed (e.g., in this case trypsin is employed, assuming cleavage at the C-terminus of Arginine and Lysine). Theoretical masses are then assigned to these peptides and, either manually or using an algorithm, predicted masses are correlated with experimental masses, allowing researchers to determine if the protein being analyzed matches the protein in the database. Source: Lennart Martens. https://creativecommons.org/licenses/by-sa/4.0/.

When researchers refer to protein characterization they often categorize analysis as being "bottom-up" or "top-down" analyses. Bottom-up analysis refers to the analysis of a protein by its proteolytic cleavage into peptides, MS/MS on these peptides, and then the accumulative information for these peptides is used to present an overall sequence coverage of a protein. In top-down analyses, the converse is true and the entire protein (or large polypeptides from said protein) is fragmented. There are pros and cons to both methods with bottom-up proteomics often losing some contextual information, and some peptides being generated that are too small and hydrophilic to be retained on the chromatography column or be selected for MS/MS fragmentation in the mass spectrometer.

For top-down analyses, although becoming more popular in recent years, there are still many challenges in dealing with the various isoforms of a protein and separating them at the protein level and, for large highly charged proteins, obtaining spectra that can be adequately deconvoluted. Figure 2.12 shows the various types of top-down and bottom-up analyses that are employed in the characterization of biotherapeutics.

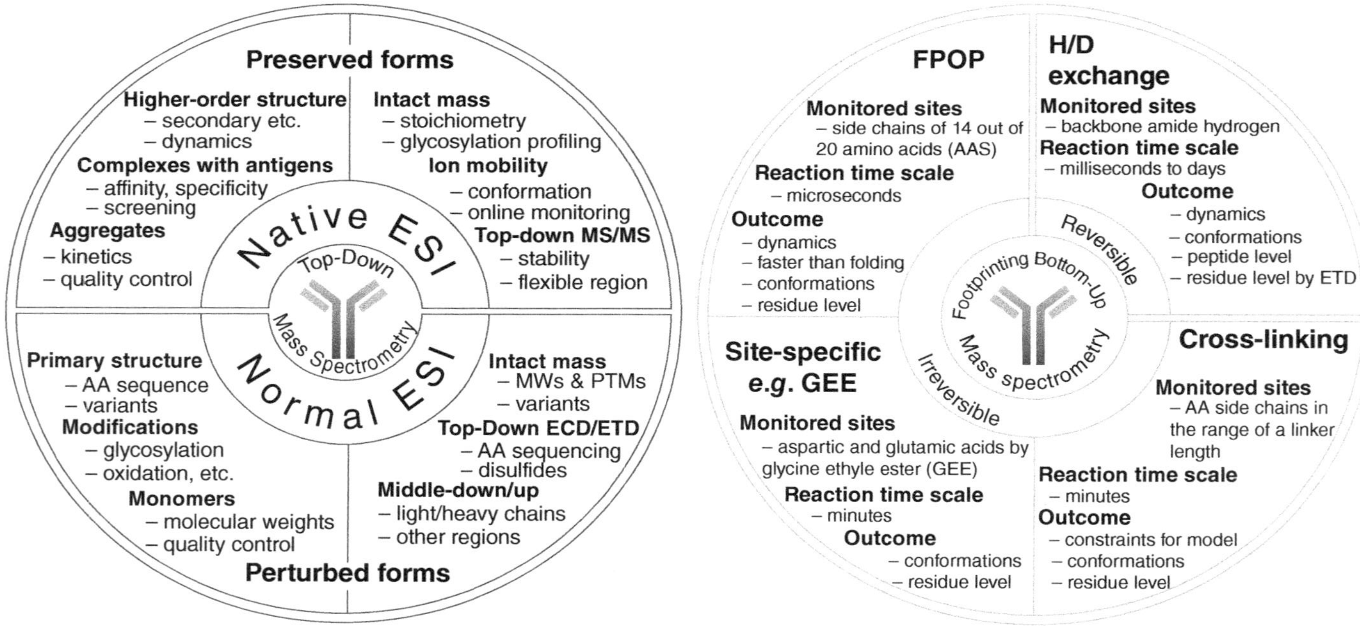

Figure 2.12 *Overview of top-down and bottom-up MS*-based protein biophysical studies (using antibody as example). The left circle is the summary of top-down approaches. The right circle is the summary of bottom-up approaches. Source: Zhang *et al.* [46]. Reproduced with permission of Elsevier.

2.3.1 Bottom-Up MS

Bottom-up MS involves the MS/MS of proteins at the peptide level. MS/MS of a peptide results in a ladder of ions with charges retained in the N- or C-terminus. Depending on where on the peptide backbone this fragmentation occurs, a series of N-terminally charge retained ions (*a*, *b*, or *c* ions) or C-terminally charged ion ladders (*x*, *y*, or *z* ions) are formed. Figure 2.13 shows the various ion series that can be produced after fragmentation of a peptide by MS/MS. The types of fragment ions observed in an MS/MS spectrum depend on many variables including the peptides' primary sequence, the amount of internal energy, how the energy was introduced to the ion for fragmentation, and the charge state. The recognized nomenclature for fragment ions was first proposed by Roepstorff and Fohlman [47] and was later modified by Johnson *et al.* [48].

Using CID, the most common type of fragmentation for peptide analysis by workhorse instruments such as the Orbitrap and Q-TOF families of mass spectrometers, backbone-based cleavage is the most useful for peptide sequence identification. In reality, samples that are subjected to bottom-up MS are typically separated online using reverse phase chromatography followed by MS/MS. For simple samples such as a single recombinant protein these analyses can be performed with rapid chromatography in the order of just a few minutes. For more complicated samples or indeed protein mixtures, at least an hour of chromatography is typically performed online with MS/MS analysis to ensure good sequence coverage is facilitated. For more high-throughput analysis of biotherapeutic samples there are several options. One can multiplex, for example, by employing a multiplexed LC system. These instruments are designed to have up to four separate chromatographic systems, which operate simultaneously, but in a somewhat overlapping manner. In addition, instruments have become sophisticated and with higher throughput with faster scanning capabilities. For example, hybrid instruments such as the Orbitrap

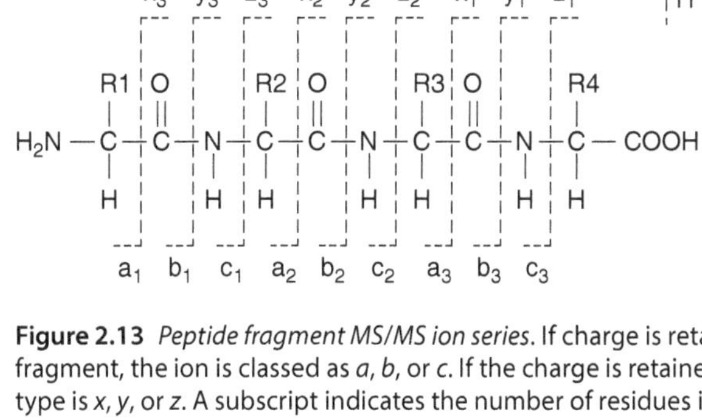

Figure 2.13 *Peptide fragment MS/MS ion series.* If charge is retained on the N-terminal fragment, the ion is classed as *a*, *b*, or *c*. If the charge is retained on the C-terminal, the ion type is *x*, *y*, or *z*. A subscript indicates the number of residues in the fragment.

series allow MS to happen in the high-resolution Orbitrap simultaneously with MS/MS in the ion trap part of the instrument—a process sometimes referred to as parallel acquisition. This produces less lag time in the analysis and overall increased sequence coverage of a protein.

The approach of using liquid chromatography–MS/MS analysis with CID for peptide mapping and sequencing was employed to compare the marketed monoclonal antibody trastuzumab and two biosimilar products, mAb A containing D359E and L361M variations at the Fc site, in addition to mAb B without variants. Hundred percent complete sequence coverage including disulfide linkages, glycosylations, deamidation, oxidation, dehydration, and C-terminal lysine clipping were identified using maps generated from multienzyme digestions. In addition to the targeted comparison for the relative populations of targeted modification forms, a nontargeted approach was used to globally compare ion intensities in tryptic maps. In this study by Chen *et al.*, a peptide containing the two variants in mAb A, D359E and L361M, was revealed. In contrast, no significant differences were observed when trastuzumab was self-compared or compared with mAb B [49]. Other groups have employed alternative fragmentation techniques for antibody and biotherapeutic moiety characterization. Zhang *et al.* showed the utility of ETD for the analysis of a monoclonal antibody including localization of PTMs and disulfide bonds. Here, limited digestion of fully denatured (reduced and alkylated) monoclonal antibodies was accomplished in seconds by flowing a sample in 8 M urea at a controlled flow rate through a microcolumn reactor containing the immobilized enzyme aspergillopepsin I. Post proteolysis the resultant peptidic mixture contained 3–9 kDa-sized peptides which were fractionated by capillary column liquid chromatography and analyzed online by both ETD and CID MS. This approach enabled identification of peptides that cover the complete sequence of a murine monoclonal antibody. With customized MS/MS and utilization of the proteo-informatic algorithm ProSightPC (Thermo Scientific), Zhang *et al.* characterized 95% amino acid residues of this antibody and identified numerous PTMs (oxidized methionine, pyroglutamylation, deamidation of Asn, and several forms of N-linked glycosylation) [50].

2.3.2 Proteoinformatic Analysis of Bottom-Up Proteomic Data Sets

Regardless of the techniques employed, interpretation of the results can be a daunting task to approach manually due to the sheer amounts of data produced. This is particularly true when analyzing MS/MS data from bottom-up shotgun sequencing experiments. Although it is possible to analyze such spectra *de novo* to generate likely sequence candidates, incomplete fragment ion series and varying data quality make such an approach difficult and computationally intensive. Instead, the reverse approach is often employed, relying on a database of theoretical spectra generated from a set of known protein sequences. This database-searching approach can be applied with more accuracy and speed to a data set when the search space can be well defined ahead of time.

One of the first of these database-searching tools to be developed was Sequest [51]. Taking as input a FASTA-formatted file of theoretical protein sequences, Sequest first produces an *in silico* digest of the protein sequences into peptides based upon the proteolytic enzyme used (e.g., trypsin). Each tandem mass spectrum is then examined one at a time. The list of putative peptide sequences is filtered to those peptides with overall masses equivalent to the MS/MS spectrum's precursor mass. Then, a peaklist of theoretical masses is generated from the sequence of each possible peptide and compared to the observed spectrum and the quality of the match is scored.

While most database-searching algorithms follow these general steps, they differ in their methods of scoring the peptide–spectral match (PSM). Sequest utilizes a dot product approach while other tools such as Mascot [40] are based upon probabilistic approaches. These early tools such as Sequest and Mascot still provide excellent performance and are widely used within the field, where they have been joined by a myriad of other tools in use today such as Tandem [52], PEAKS [53] and MSGF+ [54]. Although they differ in their implementation details, all the tools employ the same basic principle of generating theoretical spectra from a list of supplied peptide sequences and matching those theoretical spectra against actual spectra observed from the mass spectrometer. Each algorithm provides acceptable performance for most routine tasks for protein verification. Some tools, such as Mascot, also provide facilities for PMF analyses as well.

An important consideration is the quality of the PSM resulting from an analysis, which can vary widely. Although all algorithms report a score for the resulting PSM, the appropriate score cutoff can vary given the context of a particular data set. Instead of simple scoring thresholds, most bottom-up proteomics experiments today are analyzed in the context of a false discovery rate, which strikes a balance between sensitivity and specificity in the experiment. This is often accomplished through the use of a target decoy database search strategy [55], where the typical database of known peptide sequences (the target) is joined with a set of decoy sequences unlikely to occur naturally (e.g., reversed or randomized peptide sequences). The number of decoy hits assigned by the search algorithm then serves as a proxy for the number of false positives within the entire data set. Typically reported false discovery rate values are in the range of 1–5%, depending on the size of the data set and the stringency desired.

The assignment of PTMs to specific protein residues is another area where ambiguity is possible. Many algorithms, such as Ascore [56], attempt to address this problem by capturing the probability that a particular modification is correctly localized. Further work has extended this concept to the production of a false localization rate [57, 58], akin to a false discovery rate of peptide or protein identifications.

Nevertheless, although database-matching algorithms are very successful for the matching of many spectra to known sequences, they do have their drawbacks. Limitations in search space dictate a limit on the number of modifications that can be interrogated at once, and the requirement for *a priori* knowledge of the putative protein sequence means that such algorithms are ill-suited for the definition of novel proteins or protein variants. To tackle these challenges, specialized tools must be employed. Byonic [59] is one example of a tool for the interrogation of the novel modifications and point mutations, along with advanced processing modes provided by PEAKS and Mascot. These tools are frequently employed in combination with database search algorithms to identify the bulk of the unmodified spectra.

For large stretches of unknown sequence or where complete coverage is necessary, *de novo* algorithms must be used. These algorithms rely on large numbers of related spectra generated from different enzymatic digestion techniques to assemble the closely related spectra into "contigs" of novel sequence. From the combined application of these tools, one can completely sequence a novel protein or antibody. For additional informatics tools for the *de novo* sequencing or antibodies please refer to Chapter 7.

In experiments where complete sequence verification is not necessary and simple verification of protein termini is sufficient, more rapid techniques can be employed, such as MALDI In-Source Decay (MALDI-ISD) analysis of intact proteins. Through laser ablation and a proton-donating matrix, fragments of the protein termini are generated and analyzed using a TOF instrument. While such spectra can be interpreted manually, tools such as ISDetect [60] can provide objective scoring for terminal verification in a few minutes of total analysis time.

2.3.3 Top-Down MS

Top-down proteomics complements bottom-up methodologies for the characterization of proteins as it is capable of identifying and quantifying protein isoforms (proteoforms) through the fragmentation and analysis of intact proteins [61]. In top-down proteomics, after ionization proteins are fragmented typically using ECD or more commonly these days ETD, although any type of fragmentation such as CID can also be successfully applied. The main advantages of top-down analysis include the ability to characterize sequence variants, various combinations of PTMs, or proteolysis products. Off-line protein separation techniques such as gel elution-based fractionation or electrophoretic fractionation have been successfully employed as ways to separate proteins prior to top-down analysis. Top-down MS has been employed to characterize numerous recombinant proteins [62–64] of various levels of complexity and provides an additional mass spectrometric tool in the bioanalytical chemist's toolbox.

Middle-down analysis, which involves the generation of large polypeptides from an intact protein, and then performing fragmentation on these larger polypeptides, is a technique that, as the name implies, sits between bottom-up and top-down protein characterization. This method is also commonly employed as a tool for characterizing biomolecules and has been used to successfully validate therapeutic antibody sequences [65] and for comparative analyses of biosimilars [25, 66].

2.4 Conclusions and Perspectives

In this chapter we describe some of the various mass spectrometric tools available to researchers for the characterization of biotherapeutics. MS is a rapidly evolving field and new techniques as well as the limits of detection and resolution are continuously challenged. By coupling MS to other so-called hyphenated techniques (CE, HPLC, SEC, for example) more complex mixtures of therapeutic biomolecules can readily be analyzed, with heterogeneity due to proteolysis or PTM events being more easily characterized. As our biomolecules become more complex in design, analytical technologies will likewise adapt. The arrival of the new generation of TOFs and the EMR mass spectrometers, for example, have allowed the accurate analysis of bispecific antibody mispairs, PEGylated species, and other complex moieties that previously were unachievable to the standard proteomics laboratory. In parallel with the development of instrumentation, the software also needs to continuously evolve. This has been evident with the introduction of top-down proteomic analysis, whereby new software solutions have been developed to keep pace with the intact molecular weight (MW) and top-down ion series analysis. One thing that will remain constant in this field is that biotherapeutic design and mass spectrometric instrumentation evolution are both going to continue in parallel at an astonishing rate.

References

1 Carini, M., Regazzoni, L., & Aldini, G., 2011. Mass spectrometric strategies and their applications for molecular mass determination of recombinant therapeutic proteins. *Current Pharmaceutical Biotechnology*, 12(10), pp. 1548–1557.

2 Abdul, S. *et al.*, 2016. Natural heterogeneity of α2-antiplasmin: functional and clinical consequences. *Blood*, 127(5), pp. 538–545.

3 Bush, D.R. *et al.*, 2016. High resolution CZE-MS quantitative characterization of intact biopharmaceutical proteins: proteoforms of interferon-β1. *Analytical Chemistry*, 88(2), pp. 1138–1146.
4 Pimienta, G. *et al.*, 2007. Autophosphorylation properties of inactive and active JNK2. *Cell Cycle (Georgetown, Tex.)*, 6(14), pp. 1762–1771.
5 Schiel, J.E., Rogstad, S.M., & Boyne, M.T., 2015. Comparison of traditional 2-AB fluorescence LC-MS/MS and automated LC-MS for the comparative glycan analysis of monoclonal antibodies. *Journal of Pharmaceutical Sciences*, 104(8), pp. 2464–2472.
6 Kilian, T.-M. & Beck-Sickinger, A.G., 2015. Recombinant expression and characterization of biologically active protein delta homolog 1. *Protein Expression and Purification*, 110, pp. 72–78.
7 Hillenkamp, F. & Karas, M., 1990. Mass spectrometry of peptides and proteins by matrix-assisted ultraviolet laser desorption/ionization. *Methods in Enzymology*, 193, pp. 280–295.
8 Fenn, J.B. *et al.*, 1989. Electrospray ionization for mass spectrometry of large biomolecules. *Science (New York, NY)*, 246(4926), pp. 64–71.
9 Whitehouse, C.M. *et al.*, 1985. Electrospray interface for liquid chromatographs and mass spectrometers. *Analytical Chemistry*, 57(3), pp. 675–679.
10 Jaskolla, T.W. & Karas, M., 2011. Compelling evidence for Lucky Survivor and gas phase protonation: the unified MALDI analyte protonation mechanism. *Journal of the American Society for Mass Spectrometry*, 22(6), pp. 976–988.
11 Stutz, H., 2005. Advances in the analysis of proteins and peptides by capillary electrophoresis with matrix-assisted laser desorption/ionization and electrospray-mass spectrometry detection. *Electrophoresis*, 26(7–8), pp. 1254–1290.
12 Šesták, J., Moravcová, D., & Kahle, V., 2015. Instrument platforms for nano liquid chromatography. *Journal of Chromatography A*, 1421, pp. 2–17.
13 Dodonov, A.F., Chernushevich, I.V., & Laiko, V.V., 1993. Electrospray ionization on a reflecting time-of-flight mass spectrometer. In *Time-of-Flight Mass Spectrometry*. ACS Symposium Series. American Chemical Society, Washington, DC, pp. 108–123. Available at: http://dx.doi.org/10.1021/bk-1994-0549.ch007 [Accessed March 25, 2016].
14 Rose, R.J. *et al.*, 2012. High-sensitivity Orbitrap mass analysis of intact macromolecular assemblies. *Nature Methods*, 9(11), pp. 1084–1086.
15 Rosati, S. *et al.*, 2012. Exploring an orbitrap analyzer for the characterization of intact antibodies by native mass spectrometry. *Angewandte Chemie (International Ed. in English)*, 51(52), pp. 12992–12996.
16 Thompson, N.J., Rosati, S., & Heck, A.J.R., 2014. Performing native mass spectrometry analysis on therapeutic antibodies. *Methods (San Diego, Calif.)*, 65(1), pp. 11–17.

17 Rosati, S. *et al.*, 2013. In-depth qualitative and quantitative analysis of composite glycosylation profiles and other micro-heterogeneity on intact monoclonal antibodies by high-resolution native mass spectrometry using a modified Orbitrap. *mAbs*, 5(6), pp. 917–924.

18 De Jong, R.N. *et al.*, 2016. A novel platform for the potentiation of therapeutic antibodies based on antigen-dependent formation of IgG hexamers at the cell surface. *PLoS Biology*, 14(1), p. e1002344.

19 Heck, A.J.R., 2008. Native mass spectrometry: a bridge between interactomics and structural biology. *Nature Methods*, 5(11), pp. 927–933.

20 Cumeras, R. *et al.*, 2015a. Review on ion mobility spectrometry. Part 1: current instrumentation. *The Analyst*, 140(5), pp. 1376–1390.

21 Cumeras, R. *et al.*, 2015b. Review on ion mobility spectrometry. Part 2: hyphenated methods and effects of experimental parameters. *The Analyst*, 140(5), pp. 1391–1410.

22 Bagal, D. *et al.*, 2010. Resolving disulfide structural isoforms of IgG2 monoclonal antibodies by ion mobility mass spectrometry. *Analytical Chemistry*, 82(16), pp. 6751–6755.

23 Pacholarz, K.J. *et al.*, 2016. Molecular insights into the thermal stability of mAbs with variable-temperature ion-mobility mass spectrometry. *Chembiochem: A European Journal of Chemical Biology*, 17(1), pp. 46–51.

24 Huang, R.Y.-C. *et al.*, 2015. Utility of Ion Mobility Mass Spectrometry for Drug-to-Antibody Ratio Measurements in Antibody-Drug Conjugates. *Journal of the American Society for Mass Spectrometry*, 26(10), pp. 1791–1794.

25 Beck, A. *et al.*, 2015. Cutting-edge mass spectrometry characterization of originator, biosimilar and biobetter antibodies. *Journal of Mass Spectrometry: JMS*, 50(2), pp. 285–297.

26 Ferrige, A.G. *et al.*, 1991. Maximum entropy deconvolution in electrospray mass spectrometry. *Rapid Communications in Mass Spectrometry*, 5(8), pp. 374–377.

27 Horn, D.M., Zubarev, R.A., & McLafferty, F.W., 2000. Automated reduction and interpretation of high resolution electrospray mass spectra of large molecules. *Journal of the American Society for Mass Spectrometry*, 11(4), pp. 320–332.

28 Marty, M.T. *et al.*, 2015. Bayesian deconvolution of mass and ion mobility spectra: from binary interactions to polydisperse ensembles. *Analytical Chemistry*, 87(8), pp. 4370–4376.

29 Zhang, Z. & Marshall, A.G., 1998. A universal algorithm for fast and automated charge state deconvolution of electrospray mass-to-charge ratio spectra. *Journal of the American Society for Mass Spectrometry*, 9(3), pp. 225–233.

30 Stoll, D.R., 2010. Recent progress in online, comprehensive two-dimensional high-performance liquid chromatography for non-proteomic applications. *Analytical and Bioanalytical Chemistry*, 397(3), pp. 979–986.

31 Smith, R.D. *et al.*, 1990. Sensitivity considerations for large molecule detection by capillary electrophoresis-electrospray ionization mass spectrometry. *Journal of Chromatography*, 516(1), pp. 157–165.
32 Landers, J.P., 2007. *Handbook of Capillary and Microchip Electrophoresis and Associated Microtechniques*, 3rd Edition. CRC Press, Boca Raton.
33 Basu, A. *et al.*, 2006. Structure-function engineering of interferon-beta-1b for improving stability, solubility, potency, immunogenicity, and pharmacokinetic properties by site-selective mono-PEGylation. *Bioconjugate Chemistry*, 17(3), pp. 618–630.
34 Gahoual, R. *et al.*, 2016. Independent highly sensitive characterization of asparagine deamidation and aspartic acid isomerization by sheathless CZE-ESI-MS/MS. *Journal of Mass Spectrometry: JMS*, 51(2), pp. 150–158.
35 François, Y.-N. *et al.*, 2016. Characterization of cetuximab Fc/2 dimers by off-line CZE-MS. *Analytica Chimica Acta*, 908, pp. 168–176.
36 El Deeb, S. *et al.*, 2016. Recent advances in capillary electrophoretic migration techniques for pharmaceutical analysis (2013–2015). *Electrophoresis*, 37(12), pp. 1591–1608.
37 Bynum, M.A. *et al.*, 2009. Characterization of IgG N-glycans employing a microfluidic chip that integrates glycan cleavage, sample purification, LC separation, and MS detection. *Analytical Chemistry*, 81(21), pp. 8818–8825.
38 Henzel, W.J. *et al.*, 1993. Identifying proteins from two-dimensional gels by molecular mass searching of peptide fragments in protein sequence databases. *Proceedings of the National Academy of Sciences of the United States of America*, 90(11), pp. 5011–5015.
39 Chalkley, R.J. *et al.*, 2005. Comprehensive analysis of a multidimensional liquid chromatography mass spectrometry dataset acquired on a quadrupole selecting, quadrupole collision cell, time-of-flight mass spectrometer II. New developments in protein prospector allow for reliable and comprehensive automatic analysis of large datasets. *Molecular & Cellular Proteomics*, 4(8), pp. 1194–1204.
40 Perkins, D.N. *et al.*, 1999. Probability-based protein identification by searching sequence databases using mass spectrometry data. *Electrophoresis*, 20(18), pp. 3551–3567.
41 Gahoual, R. *et al.*, 2013. Rapid and multi-level characterization of trastuzumab using sheathless capillary electrophoresis-tandem mass spectrometry. *mAbs*, 5(3), pp. 479–490.
42 Wells, J.M. & McLuckey, S.A., 2005. Collision-induced dissociation (CID) of peptides and proteins. *Methods in Enzymology*, 402, pp. 148–185.
43 Brodbelt, J.S., 2014. Photodissociation mass spectrometry: new tools for characterization of biological molecules. *Chemical Society Reviews*, 43(8), pp. 2757–2783.
44 Syka, J.E.P. *et al.*, 2004. Peptide and protein sequence analysis by electron transfer dissociation mass spectrometry. *Proceedings of the National Academy of Sciences of the United States of America*, 101(26), pp. 9528–9533.

45 Zubarev, R.A., Kelleher, N.L., & McLafferty, F.W., 1998. Electron capture dissociation of multiply charged protein cations. A nonergodic process. *Journal of the American Chemical Society*, 120(13), pp. 3265–3266.

46 Zhang, H., Cui, W., & Gross, M.L., 2014. Mass spectrometry for the biophysical characterization of therapeutic monoclonal antibodies. *FEBS Letters*, 588(2), pp. 308–317.

47 Roepstorff, P. & Fohlman, J., 1984. Proposal for a common nomenclature for sequence ions in mass spectra of peptides. *Biomedical Mass Spectrometry*, 11(11), p. 601.

48 Johnson, R.S. *et al.*, 1987. Novel fragmentation process of peptides by collision-induced decomposition in a tandem mass spectrometer: differentiation of leucine and isoleucine. *Analytical Chemistry*, 59(21), pp. 2621–2625.

49 Chen, S.-L. *et al.*, 2013. A global comparability approach for biosimilar monoclonal antibodies using LC-tandem MS based proteomics. *Journal of Pharmaceutical and Biomedical Analysis*, 80, pp. 126–135.

50 Zhang, L. *et al.*, 2015. Analysis of monoclonal antibody sequence and post-translational modifications by time-controlled proteolysis and tandem mass spectrometry. *Molecular & Cellular Proteomics: MCP*, 15(4), pp. 1479–1488.

51 Eng, J.K., McCormack, A.L., & Yates, J.R., 1994. An approach to correlate tandem mass spectral data of peptides with amino acid sequences in a protein database. *Journal of the American Society for Mass Spectrometry*, 5(11), pp. 976–989.

52 Craig, R. & Beavis, R.C., 2004. TANDEM: matching proteins with tandem mass spectra. *Bioinformatics (Oxford, England)*, 20(9), pp. 1466–1467.

53 Ma, B. *et al.*, 2003. PEAKS: powerful software for peptide de novo sequencing by tandem mass spectrometry. *Rapid Communications in Mass Spectrometry: RCM*, 17(20), pp. 2337–2342.

54 Kim, S. & Pevzner, P.A., 2014. MS-GF+ makes progress towards a universal database search tool for proteomics. *Nature Communications*, 5, p. 5277.

55 Elias, J.E. & Gygi, S.P., 2007. Target-decoy search strategy for increased confidence in large-scale protein identifications by mass spectrometry. *Nature Methods*, 4(3), pp. 207–214.

56 Beausoleil, S.A. *et al.*, 2006. A probability-based approach for high-throughput protein phosphorylation analysis and site localization. *Nature Biotechnology*, 24(10), pp. 1285–1292.

57 Chalkley, R.J. & Clauser, K.R., 2012. Modification site localization scoring: strategies and performance. *Molecular & Cellular Proteomics: MCP*, 11(5), pp. 3–14.

58 Wiese, H. *et al.*, 2014. Comparison of alternative MS/MS and bioinformatics approaches for confident phosphorylation site localization. *Journal of Proteome Research*, 13(2), pp. 1128–1137.

59 Bern, M., Kil, Y.J., & Becker, C., 2012. Byonic: advanced peptide and protein identification software. *Current Protocols in Bioinformatics*, Chapter 13, p. Unit13.20.

60 Bakalarski, C.E., Gan, Y., Wertz, I., Lill, J.R., & Sandoval, W.N., 2016. Rapid, semi-automated protein terminal characterization using ISDetect. *Nature Biotechnology*, 34, 811–813.

61 Tran, J.C. *et al.*, 2011. Mapping intact protein isoforms in discovery mode using top-down proteomics. *Nature*, 480(7376), pp. 254–258.

62 Peng, Y. *et al.*, 2013. In-depth proteomic analysis of human tropomyosin by top-down mass spectrometry. *Journal of Muscle Research and Cell Motility*, 34(3–4), pp. 199–210.

63 Tsybin, Y.O. *et al.*, 2011. Structural analysis of intact monoclonal antibodies by electron transfer dissociation mass spectrometry. *Analytical Chemistry*, 83(23), pp. 8919–8927.

64 Zhang, Z. & Shah, B., 2007. Characterization of variable regions of monoclonal antibodies by top-down mass spectrometry. *Analytical Chemistry*, 79(15), pp. 5723–5729.

65 Resemann, A. *et al.*, 2016. Full validation of therapeutic antibody sequences by middle-up mass measurements and middle-down protein sequencing. *mAbs*, 8(2), pp. 318–330.

66 Wang, B. *et al.*, 2013. Structural comparison of two anti-CD20 monoclonal antibody drug products using middle-down mass spectrometry. *The Analyst*, 138(10), pp. 3058–3065.

3

Characterizing the Termini of Recombinant Proteins

Nestor Solis and Christopher M. Overall

Center for Blood Research, University of British Columbia, Vancouver, British Columbia, Canada

Abbreviations

1DE/SDS-PAGE	one-dimensional gel electrophoresis/sodium dodecyl sulfate polyacrylamide gel electrophoresis
2DE	two-dimensional gel electrophoresis
ATOMS	Amino terminal orientated mass spectrometry
BMP-1	bone morphogenetic protein-1
CCR-1	CC-chemokine receptor-1
CID	collision-induced dissociation
COFRADIC	combined fractional diagonal chromatography
DNFB	1-fluoro-2,4-dinitrobenzene
ECD	electron capture dissociation
ESI	electrospray ionization
ETD	electron transfer dissociation
FT-ICR	Fourier transform-ion cyclotron resonance
GELFrEE	gel-eluted liquid fraction entrapment electrophoresis
HCD	higher-energy collisional dissociation
IEF	isoelectric focusing
IPG	immobilized pH gradient
IRMPD	infrared multiphoton dissociation
iTRAQ	isobaric tags for relative and absolute quantification
LC-MS/MS	liquid chromatography–tandem mass spectrometry
MCP-3	monocyte chemoattractant protein 3 (also known as CCL7)
MMP	matrix metalloproteinase
NAT	N-acetyltransferase

Analytical Characterization of Biotherapeutics, First Edition. Edited by Jennie R. Lill and Wendy Sandoval.

PTH	phenylthiohydantoin
PVDF	polyvinylidene difluoride
QqQ	triple quadrupole
Q-TOF	quadrupole-time-of-flight
RP	reversed-phase
TEV	tobacco etch virus
TNBS	2,4,6-trinitrobenzenesulfonic acid
XIC	extracted ion chromatogram

3.1 Introduction

The central dogma of molecular biology stipulated that for each gene there was one protein that executed one function [1]. For many years this was the focus of most biological and biochemical research and with the advent of protein sequencing and DNA sequencing there was a paradigm shift more toward our current understanding: one gene can exhibit a *set* of functions [2]. What does this mean? Gene products, proteins, do not act in isolation. They are spatio-temporally expressed and have different interacting partners dependent on cell type, tissue, and time. Genes can therefore be expressed as different splice variants and can exist in differently modified isoforms known as proteoforms that encompass all PTMs of a protein [3, 4]. Consequently, many proteins are multifunctional and PTMs are strictly regulated in order to achieve the correct folding, localization, and functionality. Biological systems employ several control mechanisms that ensure that proteins are correctly translated and posttranslationally modified, which gives rise to the diversity of proteins from comparatively "simple" genomes. The latest drafts of the human proteome show that there are codes for approximately 20 066 genes (http://www.nextprot.org/) [5, 6]. However the protein diversity in humans has at *least* 100 000 different forms [3, 4] that increases to many more when PTMs are also considered.

There are ~>400 PTMs of proteins including the well-known and more familiar phosphorylation, glycosylation, acetylation (both co- and posttranslational [7]), and lipidation [8]. All of these chemical groups modify proteins in ways that affect their function, but all of these also have a defining feature: a particular chemical group has been added [8]. The sites at which these modifications occur is also dependent on surrounding sequences and enzyme specificity, but many of these modifications can be reversible, thus modulating protein function. There is one modification, however, that does not have the addition of a defined chemical group that enables selective purification. In fact, this PTM is not additive at all; rather, it is deletive: proteolytic processing generates new N- and/or C-termini, often with new functions depending on the position of the truncation.

Proteolytic processing is the hydrolysis of amide bonds between amino acid residues across the backbone of the protein or peptide. The enzymes responsible for catalyzing such reactions are proteases and exist in two broad functional classes: exoproteases and endoproteases [9]. Exoproteases remove terminal residues from a peptide either N-terminally (aminopeptidases) [10] or C-terminally (carboxypeptidases) [11]. In contrast, endoproteases cleave amide bonds in the middle of protein or peptide backbones with varying degrees of specificity. Such a modification does not generate a chemical group that allows for isolation and is irreversible. There are over 500 proteases in humans [12] that interact together as a protease web [13] in complex pathways to activate, deactivate, and modulate their substrates. Thus, proteolytic processing is a pervasive powerful driving modifying force in many proteins in many biological processes.

Proteolytic processing exists across all kingdoms of life as a means of modulating protein function. For example, one of the simplest (smallest genome) life forms, *Mycoplasma hyopneumoniae* (the etiological agent of porcine pneumonia [14] incurs tremendous costs for the porcine industry and does not have a successful vaccine [15]. One potential candidate for development of a novel vaccine is the cilium adhesin P97, a surface-bound proteolytic fragment of the gene Mhp183 from virulent strains of *M. hyopneumoniae* [16]. P97 is a cilium adhesin that binds heparin in pigs and is one of the main effectors of virulence. It contains two heparin-binding domains and binds heparin in a concentration-dependent manner. However, it has been recently shown that P97 is proteolytically processed yielding a final of 13 isoforms that are all fragments from one single gene. This is not only significant from a proteolytic perspective, but it also highlights the diversity generated by proteolytic cleavage particularly in a pathogenic organism that only codes 692 genes [17]. Moreover, each fragment has unique binding properties to different surface-exposed host factors that contribute to different aspects of pathogen binding thus enhancing its virulence [16].

A separate example is the striking change of activity of the chemokine monocyte chemoattractant protein 3 (MCP-3, CCL7). Chemokines are signaling molecules secreted in response to injury that recruit immune cells to mediate a number of processes such as inflammation and tissue resolution [18]. The careful spatio-temporal orchestration of these responses is critical for the effective recruitment of the effector cells to eliminate the cause of injury and dampen inflammation. CCL7 is recognized by CC-chemokine receptor-1 (CCR-1) on monocytes and upon binding triggers a calcium-mediated chemotactic response that can then initiate inflammation. However, matrix metalloproteinase 2 (MMP2) was shown to cleave CCL7 near the N-terminus clipping off the first four residues (in close proximity to the CCR-1 recognition motif) [19, 20]. This molecular event triggered a complete change of function of CCL7. Not only did it abolish the agonist response of CCL7, it elicited an antagonist response in a dose-dependent manner. Chemotactic responses and immune cell recruitment are mediated in part by proteolytic cleavage of specific chemokines and as such MMPs can enhance or abrogate inflammatory responses. This exemplifies the nuanced nature of some of these modifications, albeit with distinct profound biological effects and demands for technologies to directly identify *bona fide* termini of proteins and peptides [19]. Consequently, detection of the correct isoforms when generating recombinant proteins or isolating proteins from natural sources is needed to ensure the desired functionality of the molecule. Here we describe some of the current approaches for the characterization of N-termini of simple protein mixtures and their advantages and limitations. The focus is placed mainly on mass spectrometric approaches, as these have been the most recent technologies to directly address the issue of true starts of termini.

3.2 Gel Electrophoresis and Edman Sequencing

For many years gel electrophoresis has been the classical method of separating proteins and their isoforms due to the utility and ease of execution of sodium dodecyl sulfate polyacrylamide gel electrophoresis (SDS-PAGE), also known as 1D gels, and the two-dimensional variant (two-dimensional gel electrophoresis, 2DE) [21]. A single protein with multiple isoforms resolved by gel electrophoresis can appear as multiple bands of different molecular weights in a 1D gel, but more often migrates as one band masking the true makeup of the mixture. In contrast, on a 2D gel a protein will appear as different spots of different molecular weights and different charges; 2DE resolves proteins in the first dimension by isoelectric focusing (IEF) on an immobilized pH gradient (IPG) strip, thus separating by charge. In the second dimension the IPG strip is overlaid at the top of a slab of polyacrylamide gel and performing standard SDS-PAGE [22]. Although 2DE provides higher resolution and charge information on protein fragments, 1D gels require less sample loading and with multiple lanes for separate analyses favors its use for the characterization of protein termini.

After electrophoretic resolution on a 1D or 2D gel the proteins are visualized with sensitive dyes to distinguish low-intensity fragments following which they can be excised and extracted for further analysis [22]. Alternatively, and preferentially for sequencing, the resolved proteins are transferred onto a polyvinylisodene difluoride (PVDF) membrane and then subject to microsequencing. The most successful form of determining N-terminal sequences prior to the advent of biological mass spectrometry (MS) was Edman degradation. This biochemical method was first published by Pehr Edman in 1949 as a means of determining the amino acid sequence in peptides by a combination of protection and hydrolysis steps of individual amino acids from the

N-terminus of a peptide [23] and is still widely used as an effective method, yet there is uncertainty regarding its continuation as the vendor has indicated that they will halt production of the sequencers.

The chemical mechanism for Edman sequencing relies on reacting the N-terminal residue of a peptide with a protective chemical group, phenylisothiocyanate (PITC). PITC reacts selectively with the terminal primary amine of peptides in alkaline conditions to generate a PITC–peptide derivative. Following rearrangement, the peptide is split from the now modified N-terminal amino acid moiety as a thiozalinone group [24]. This moiety contains the side chain specific to the original N-terminal residue and is extracted in organic solvent to treat with acid and heat and generate a stable phenylthiohydantoin (PTH). As the PTH was isolated from the peptide, the peptide is now available for a new cycle of PITC treatment and hence provides the next amino acid in the sequence (Figure 3.1). In fact, utilization of any chemical group is possible provided that they follow this mechanistic scheme and modern protein sequencers utilize Sanger's reagent (1-fluoro-2,4-dinitrobenzene, DNFB) [25] or dansyl chloride [26] as an alternative to PITC. Current sequencers are automated and are capable of performing electrophoresis or chromatography to identify the specific PTH to an amino acid side chain and infer N-terminal sequences [24, 27].

There are drawbacks to this methodology that can limit its utility in profiling the true starts of proteins. Edman degradation is reliant on the reactivity of the N-terminal residue with PITC, namely, if it has a freely available primary amine. While a large number of proteins possess a freely available N-terminus, there is a substantially large number of proteins, mainly intracellular, and peptides that do not. Perhaps the most common modification of the true start of proteins is acetylation occurring both cotranslationally and posttranslationally that replaces a primary amine with an acetyl group yielding a mass increase of +42.010565 Da [7]. There are several enzymes that replace the N-terminal primary amine with a stable nonreactive acetyl moiety (the N-acetyltransferases, NATs [28]) and can modify up to 60% of all translated proteins in simple eukaryotic organisms alone [7]. Other estimates indicate that ~90% of mammalian intracellular proteins are acetylated. The next most common occurrence of N-terminal blocking is pyroglutamation of N-terminal glutamines. N-terminal glutamines can spontaneously cyclize under mildly acidic conditions to form a pyroglutamic acid residue, which is unreactive with PITC as it has a secondary amine and not a primary amine (as is the case with proline). In addition to spontaneous formation, glutamine to pyroglutamate conversion is also mediated enzymatically (for a review on N-terminal modifications refer to Marino and colleagues [29]).

The largest drawback of Edman degradation is the size of the sequence reads that can be generated. In practice, only the first 30 amino acids of a protein can be confidently determined due to the protection and deprotection reactions

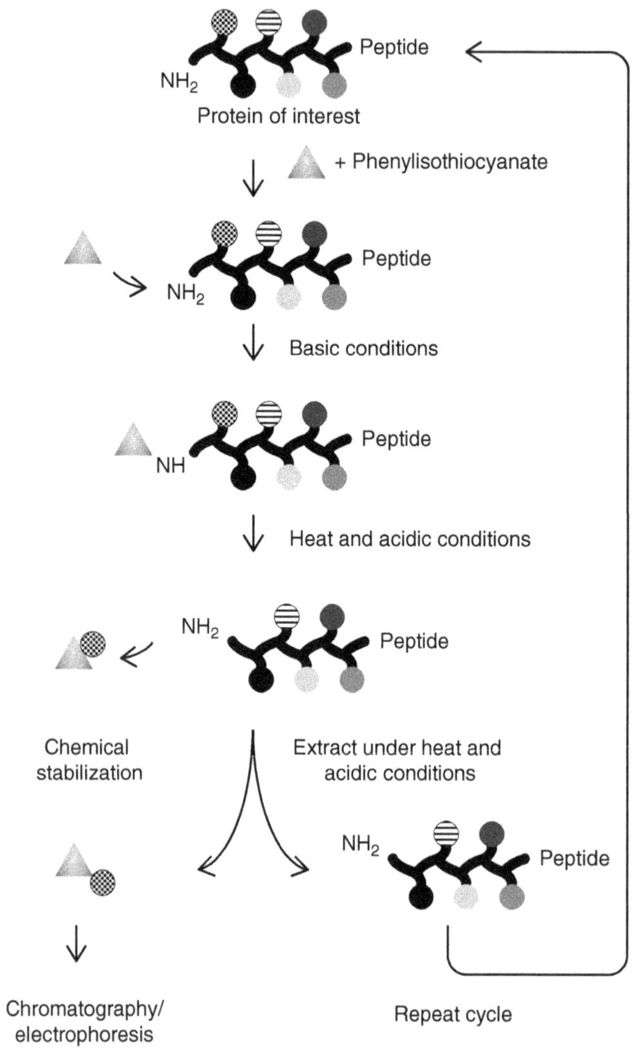

Figure 3.1 Edman sequencing. Proteins and peptides (shown with each separate amino acid residue as circle) with freely available N-termini can be reacted with phenylisothiocyanate (PITC) (marked as a triangle) under alkaline conditions. This generates a PITC–peptide derivative which after rearrangement and treatment with acid and heat results in the peptide (now one residue shorter) and a thiozalinone group (specific to the first residue of the original peptide). This group can be extracted in organic solvent and rearranged into a phenylthiohydantoin (PTH) group (triangle/circle) with acid and heat treatment. The residue-specific PTH group can be identified by chromatography or electrophoresis and the isolated peptide is available for the subsequent cycle of sequencing as indicated.

not always going to completion. As such, Edman sequencing has limitations in achieving complete protein sequences and is best suited for peptide sequences. Nonetheless, achieving the first 30 residues is mostly sufficient to identify the protein with genome sequence information and hence contributes to Edman sequencing remaining largely applicable today.

The current N-terminal sequencing strategy employed in Edman degradation is a modification of an older method of sequencing known as Bergmann sequencing [30]. In contrast to Edman sequencing, Bergmann degradation generated reads from the C-terminal end of the protein. The mechanism of action of C-terminal sequencing will not be discussed but is analogous to the Edman method, albeit with much lower efficiency. The reasons involved are due to the relatively low reactivity of carboxylic acids in comparison to primary amines and the need for activating the C-terminus for reaction with other reagents to generate individual amino acids which can be used for identification. Lower reaction efficiencies in this method made it prohibitive for large sequencing projects as only a few amino acids were generated for identification. Consequently, for many years the C-terminus of proteins has been considered intractable in practical terms. However, two recent mass spectrometric methods have tackled this problem, one a variant of terminal amino isotopic labeling of substrates (TAILS) known as C-TAILS [31] and, in a back-to-back paper, by utilizing chromatographic separation [32] that will be discussed in the following sections.

3.3 Mass Spectrometric Approaches for Characterizing True Starts of Proteins

Current proteomic techniques and methodologies are the result of the marriage of large-scale genomic sequencing projects and the development of ionization methods for biological molecules to keep them intact for mass spectrometric analysis [33–35]. There are currently two broad forms of performing biological MS analysis on proteins: top-down and bottom-up proteomics [36].

3.3.1 Top-Down Approaches

The prevalence for bottom-up approaches (discussed here) during the initial years of proteomic research was due mainly as a technological constraint. For many years the workhorse mass spectrometers were quadrupole-based instruments, that is, triple quadrupoles (QqQ) and quadrupole-time-of-flight (Q-TOF) instruments. Their operational mass-to-charge range is amenable for tryptic peptides; they had high sensitivity and low to moderate resolution and were considered to have fast scanning rates [37]. Most important was that they were hybrid instruments capable of performing tandem mass spectrometric (MS/MS) analyses to enable peptide identification from fragment ion spectra [37]. As described, having full genome sequences and computer algorithms to interrogate these databases, bottom-up approaches had widespread popularity. In contrast, top-down approaches focus on the identification of intact protein ions by MS [38]. Perhaps the largest consideration for characterizing N-terminal sections of proteins is the size. Currently the best top-down methodologies are only best suited to target proteins up to 40 kDa in size and the efficiency for identifying larger molecules drops dramatically [39]. For a full explanation on mass spectrometric instrumentation please refer to Chapter 2.

Identification of intact proteins by MS is substantially different from using peptide sequences to infer the original protein. Top-down approaches were traditionally, but not exclusively limited to identification of simple protein mixtures using Fourier transform–ion cyclotron resonance (FT-ICR) mass spectrometers of exquisite mass accuracy and extremely high resolution. One of the first top-down MS/MS analyses was in a QqQ instrument [40]. FT-ICR MSs were the instruments of choice for performing top-down analyses due to their ability to capture ions in a well-defined trajectory and perform Fourier transform analysis in order to produce mass spectra of sufficiently high resolution to resolve isotopic peaks in protein ion patterns [41]. The necessity for extremely high resolution derives from the fact that due to the many varied sizes and charges of proteins, they can ionize with a vast array of charge states. This is not as pronounced for peptides due to the fact that digestion of proteins with trypsin often generates fragments of similar size and thus they preferentially only have charge states +2, +3 or +4, which are ideal for collision-induced dissociation (CID) [42]. Furthermore, the mass-to-charge values of these ions are within the optimal operational range for most mass spectrometers and have simple isotopic patterns that require only moderate to high resolution. In addition to this, peptides typically only exist in one or two charged forms. Proteins on the other hand can ionize in electrospray ionization (ESI, [34]) with multiple protons. Thus, proteins can easily exist in charge states of +2 through to +30 and take many conformation states with a single protein existing in charge states +8, +9, +10, +11, and so on [38]. This means that a single protein generates a cloud of ions in different abundances and each individual charge state is in fact composed of a complex distribution of isotopic forms where the monoisotopic peak (the first peak in an isotopic series), which is required for correct mass assignation, can be difficult to assign (Figure 3.2). That is, data deconvolution of protein ions in top-down approaches is nontrivial and so determining the N-terminal or C-terminal sequences is not always unequivocal. Most importantly, FT-ICR instruments require large powerful superconducting magnets, which are hard to maintain to facilitate production of extremely stable ion trajectories in the mass spectrometer. Consequently, their cost is very high, they have high upkeep, and acquiring MS/MS spectra with hybrid instrumentation is difficult due to ion pathing trajectories [41].

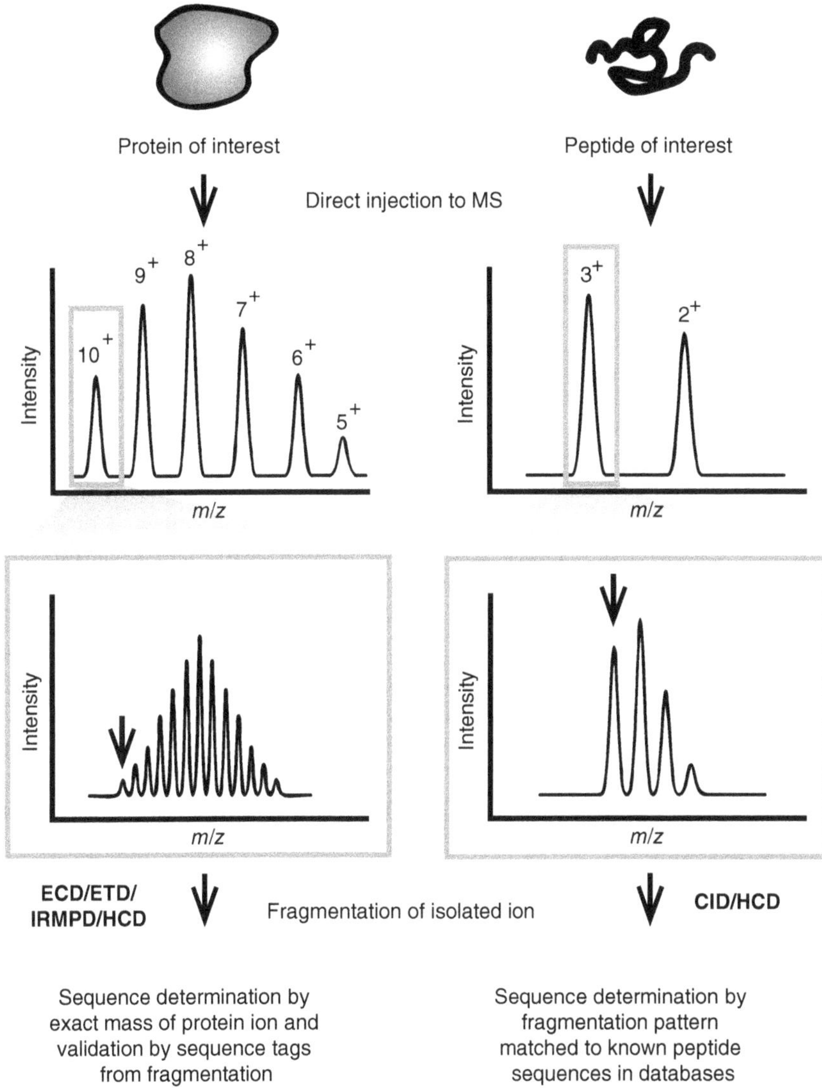

Figure 3.2 Top-down proteomics. Intact protein mass spectrometry (left hand side) is contrasted to peptide mass spectrometry (on the right hand side). Biological molecules are directly infused to the mass spectrometer (shown here by ESI). Protein ions often exist in multiple different charge states (seen here between +5 and +10), thus increasing mass spectra complexity, whereas peptides often exist in one or two charge states only (+2 and +3 shown here). In addition, when zoomed in it can be seen that the +3 ion shows a simple isotopic distribution where the monoisotopic peak (marked by an arrow) is easily determined for correct mass-to-charge calculations. Contrastingly, the isotopic pattern for the +10 ion is far more complex and the monoisotopic peak (marked by an arrow) is not as easily determined due to the requirements for higher-resolution instrumentation to differentiate across all peaks and high sensitivity to determine the monoisotopic peak with a confident signal-to-noise ratio. Each isolated precursor ion can be fragmented with a variety of methods. For top-down mass spectrometry, the exact mass of the ion is used (hence the critical need for correct monoisotopic peak determination) in combination with sequence information generated by fragmentation events, which can be used to interrogate databases for the N-terminal sequence of the protein.

Advancements in MS technology such as the Orbitrap mass spectrometer have now circumvented the issues associated with FT-ICR instrumentation and now allow frequent and high-throughput analyses of top-down approaches [43]. Orbitrap instruments operate on different principles than FT-ICR. Orbitraps do not maintain ions in trajectory by means of a large magnet, but by two opposing charged poles inside a small curved chamber where ions can orbit in defined trajectories. Hence, this allows for fast Fourier transform analyses that are required for top-down analyses, but alleviates the usage of high-powered magnets. Current advances in Orbitrap technologies are moving to bench top configurations making it accessible for most laboratories [44–46]. Not only have Orbitrap instruments allowed top-down analyses to move forward [39], but many instrument types are improving resolution and mass accuracy to allow analyses of intact proteins. One of the recent advances is the Exactive Plus EMR (extended mass range) instrument, which has enhanced resolution and a mass range larger or comparable to that of time-of-flight instruments [47]. Utilization of "native" MS is now an emerging technology to identify proteins without the usage of acidic buffers commonly employed in ESI as it does not generate highly protonated protein ion forms, thus minimizing resolution requirements as well as maintaining intact molecules (i.e., not disrupting or breaking during ionization) [48]. The power of native MS with EMR was recently demonstrated in the characterization of different highly glycosylated forms of a protein [49]. Baseline separation with high resolution and mass accuracy was achievable between differently glycosylated forms of proteins using native MS ionization methods and highlights the potential for it to identify the true starts and ends of recombinant proteins.

In addition to mass spectrometric instrumentation, technological advancements in computational power and algorithms that can deconvolute and identify proteins from mass spectra have propelled forward the field of top-down proteomics. Similar to peptide identification that utilizes MS/MS spectra to infer sequences, proteins are fragmented in a collision cell using a variety of methods such as CID [50], electron capture dissociation (ECD) [51], infrared multiphoton dissociation (IRMPD) [52], electron transfer dissociation (ETD) [53], or higher-energy collisional dissociation (HCD) [54].

Top-down methodologies allow for the identification of isoforms of proteins and can be used to profile the natural N-termini of proteins by direct measurement of the intact protein. Protein fragments are separated by charge utilizing methods such as IEF and mass by utilization of gel-eluted liquid fraction entrapment electrophoresis (GELFrEE) and can be subsequently loaded on to a liquid chromatography–mass spectrometry (LC-MS) system that utilizes high-resolution scanning [39]. Once separated, protein isoforms that require N-terminal characterization can be selected in full scan mode, have their exact (subparts-per-million accuracy) mass calculated, and be fragmented to generate MS/MS spectra. These spectra generate sequence tags to interrogate the database and the entire protein sequence to localize the cleavage site that would correspond to the exact mass calculated. The mass accuracy and resolution offered by top-down methodologies can allow not only the profiling of N-terminal sequences of proteins, but they can also identify PTMs and polymorphisms that can make database searching cumbersome. Software such as ProsightPTM can allow the identification of proteins from liquid chromatography–tandem mass spectrometry (LC-MS/MS) data as well as their PTMs [55].

Perhaps one of the most classic examples of the applicability of top-down MS for characterization of protein modifications was demonstrated by Zabrouskov and colleagues in 2006 [56]. In this study the authors detected the stepwise deamidation of ribonuclease A (13 689 Da) which is a modification characterized by an increase in only 1 Da. Ribonuclease A was found to be deamidated at five potential sites at different rates and this was confirmed by MS/MS (CID and ECD) utilizing a 6 T FT-ICR instrument. The two most competitive sites for deamidation were found to be Asn71 and Asn94, thus showing the ability of top-down MS to interrogate single amino acids and find subtle changes in protein sequence. Characterization of entire protein molecules by top-down MS using various types of fragmentation is now possible and allows not only the determination of true start sites of proteins but also the determination of modifications as well as amino acid substitutions [57].

Exemplifying the power of top-down MS for the analysis of N-terminal truncations of proteins was demonstrated in a recent report by Laitaoja and colleagues [58]. N-terminal truncations of the bovine seminal plasma protein PDC-109, a 109-mer polypeptide, were determined utilizing a 12 T FT-ICR and MS/MS. PDC-109 is a major glycoprotein of bovine seminal plasma and has been proposed to have important roles in fertilization as its molecular mechanisms involve cholesterol efflux on spermatozoa membranes and facilitate mobility [59]. Deconvolution of protein ions revealed different proteoforms: proteoform V1 an N-terminal truncation of the acidic region (residues 15–109) as well as a proteoform (V2) with two polymorphisms, a proteoform (V3) which appears to be a truncation of V1 with putative N-terminal sequence L(I)VTSTEPTQDG.

Matrix assisted laser desorption ionization in-source decay (MALDI-ISD, or ISD) has been demonstrated to be a quick and versatile technique for the determination of the protein terminal sequence, and in recent years has replaced the traditional Edman chemistries for N-terminal sequencing. ISD is a top-down technique wherein a purified protein or antibody is subjected to laser ablation and fragments predominantly at molecular termini. These fragment ions can be sequentially read and compared to a theoretical sequence, and actual terminal start position may be confirmed. Software has been developed which allows terminal sequencing in a high-throughput manner and has reduced sequence time from hours to minutes [60].

Top-down proteomics is still in its infancy and has continued growth as a field and although improving each year there are a few drawbacks to consider when attempting to characterize the N-termini of proteins. Membrane proteins can offer significant problems for these analyses as their solubility in aqueous buffers often employed for fractionation or solubilization for LC-MS/MS is cumbersome. As mentioned earlier the largest consideration for characterizing N-terminal sections of proteins is the size where the current limit is ~40 kDa [39]. In addition to this, ionization of proteins by ESI, native mass spectrometric methods, or other forms of ionization is not always suitable for top-down analyses with sufficient resolution, sensitivity, or baseline separation.

3.3.2 Current Caveats in Mass Spectrometric Identification of Protein Termini

Despite the current progress in top-down technologies for the characterization of intact proteins, the issues of size, PTMs, and technological resolution make the identification of N-terminal and C-terminal peptides unreliable. Bottom-up approaches have evolved to identify proteins on the basis of the peptides generated upon digestion with a site-specific enzyme (discussed later) but general proteomic approaches do not often identify N- and C-terminal peptides. Most bottom-up experiments focused on the large-scale identification of proteins rely on tryptic peptides as these have properties that are amenable for MS analysis and database search identification. Typically, tryptic peptides are termed "fully specific." That is, their N-terminus and C-terminus are a trypsin digestion site. Yet the vast majority of N-terminal and C-terminal peptides are only "semi-specific," where only one end of the peptide is the digestion site. Semi-specific peptides are often less amenable to MS analysis, less abundant, and mainly ignored by most researchers in large-scale analyses. Furthermore, positional information of peptides is lost due to digestion with enzymes such as trypsin, thus rendering the availability of the true N- or C-terminus often destroyed.

In order to retain positional information, the true starts of proteins are protected prior to digestion and techniques such as TAILS [61] and COFRADIC [62] have emerged as the dominant approaches to enrich for terminal peptides and characterize the true termini of proteins. However, these techniques are for large-scale experiments, unlike the characterization of single recombinant proteins. In principle, TAILS and COFRADIC can be performed on single proteins. However, this is not necessary due to the low complexity of single protein digests. Consequently, we describe ATOMS [63], which is a procedure that is fast, cheap, and effective at determining the true start of proteins on principles similar to those found in TAILS, here. In addition, we describe the use of a recently introduced protease we termed lysargiNase [64] (http://www.ibmb.csic.es/LysargiNase/) to characterize C-terminal peptides as a complement to

ATOMS. Due to the nature of the digestion with lysargiNase (discussed later), C-terminal peptides are now highly amenable to mass spectrometric analysis and its use can identify the true C-terminal end of proteins, which has been a technological problem plaguing recombinant protein characterization for many years.

3.3.3 Bottom-up Approaches for Identification of N- and C-Terminal Peptides

Bottom-up approaches refer to proteomic approaches where protein mixtures are digested with site-specific proteases to generate peptides. In doing so it increases the complexity of the sample and loses the protein context. That is, if there are homologs or protease cleavage events, those common peptides will no longer be discrete to which parent protein it belonged [37]. Nevertheless, the advantage of generating and analyzing peptides is that their physicochemical behavior is far superior to that of proteins, they are less prone to aggregation or precipitation, and they have readily predictable chromatographic and mass spectrometric properties [65, 66]. Thus, although the protein context is lost with bottom-up approaches, the benefits gained by the manipulation of peptides rather than proteins far outweigh this drawback. In light of the fact that identifying the true start of termini is absolutely dependent on the protein context, specialized approaches have been developed to exclusively target, identify, and quantify the natural N- and C-termini of proteins as well as the "neo-termini" generated after proteolysis of substrates in complex protein samples. Three main strategies have emerged to target N-terminal peptides from proteins in complex mixtures: TAILS [61], subtiligase-mediated enrichment [67], and the first approach, combined fractional diagonal chromatography (COFRADIC) [62]. One of the two key conceptually key steps in TAILS is the blocking of protein primary amines at the whole protein level. As the protein N-terminal α-amine is blocked, any peptide with a blocked α-amine must represent an N-terminus of a protein present in the sample and so can be distinguished from the unblocked α-amines of internal tryptic peptides. The second key step is to enrich for the N-terminal peptides by selective removal of the internal tryptic peptides to achieve N-terminal peptide enrichment, which greatly increases the depth of coverage.

In TAILS, amine blocking can be by reductive dimethylation [61] or use of iTRAQ [68] or tandem mass tags (TMT) [69]. Digestion with a site-specific enzyme (trypsin, GluC, or lysargiNase) generates internal peptides with free primary amines. These internal peptides with newly formed primary amines are removed by utilizing the reactivity of the amines by their coupling to a polyaldehyde derivatized hyperbranched polyglycerol water-soluble polymer under reductive conditions whereas blocked peptides, that is, the N-termini of proteins, remain unbound for identification by LC-MS/MS, usually by one or one up to four separate analyses only. The more complex COFRADIC works using protein acetylation of proteins, digestion to generate new primary amines, and resolving the complex peptide mixture by reversed-phase high performance liquid chromatography (RP-HPLC). Each fraction is then treated with 2,4,6-trinitrobenzenesulfonic acid (TNBS), which reacts with the newly formed primary amines making them extremely hydrophobic. Peptides are again separated by RP-HPLC. Peptides corresponding to the true start of proteins (blocked or acetylated peptides) elute at the same retention time as in the first step, but internal peptides now exhibit a strong hydrophobic shift after TNBS treatment and can be discarded. As can be gathered, COFRADIC is not so simple and requires considerable high performance liquid chromatography (HPLC) and chemical steps as well as a large number of MS analyses per proteome sample. A third technique which is useful is biotin/subtiligase enrichment of N-termini functions by labeling proteins with a TEV (tobacco etch virus) cleavable biotin tag exclusively at the N-terminal primary amine. Proteins are digested with trypsin and chromatographed through a streptavidin column, where N-terminally-labeled peptides (true N-termini) bind but internal peptides do not. Bound peptides are removed by TEV protease treatment and analyzed by LC-MS/MS. However, the approach suffers from sensitivity issues requiring up to 50 mg of sample and for quantitation must rely upon less reliable and MS-intensive label-free methods.

Both TAILS and COFRADIC are able to identify the true starts of proteins in complex mixtures even if they are naturally blocked (i.e., not have a freely available primary amine at the N-terminus) whereas biotin/subtiligase does not, which for cytosolic proteins can represent up to 90% of the proteins, and indeed is a common reason for failure of Edman sequencing projects. For protein termini determination of complex protein mixtures enrichment of the N-termini is required by separating these peptides from the internal tryptic peptides. However, we have also developed a simple variant of TAILS for more simple mixtures of proteins termed amino terminal orientated mass spectrometry (ATOMS) [70]. ATOMS is very well suited as a complement to Edman degradation because of its simplicity, speed, and hence general applicability in many labs even those not specializing in proteomics.

3.3.4 Amino Terminal Orientated Mass Spectrometry

Identification of true N-termini by Edman sequencing is hindered by the size of the fragments being sequenced as well as any modifications at the N-terminus that block the sequencing reaction. Peptides smaller than 4 kDa are difficult to visualize by SDS-PAGE, and present a technical challenge for excision and Edman degradation. As such, complementary methodologies are required to adequately profile these peptides and ATOMS provides a bottom-up mass spectrometric approach to tackle this.

ATOMS takes advantage of the ability to label entire proteins with combinations of isotopically labeled formaldehyde and sodium cyanoborohydride at their primary amines (thus "amino terminal oriented"). Unlike common quantitative shotgun proteomics approaches where tryptic peptides are labeled, in ATOMS labeling is performed *before* digestion and thus the exposed N-terminus is labeled, whether this be the natural N-terminus of the recombinant protein or purified protein or protease cleavage products. The highly specific, rapid, easy, and very affordable nature of dimethylation reactions coupled to proteases with strict sequence specificity, such as trypsin [71] and lysargiNase [64], allow the profiling of multiple start sites for a protein of interest in a single analysis. Combined with the ease and availability of tools to analyze, these data make N-terminal profiling by ATOMS effective, accurate, and fast. The workflow behind ATOMS is conceptually simple and easy to perform in any lab and the sample is then sent for MS analysis at a core facility, which does not require any specialized equipment other than what is normally used for proteomics experiments (Figure 3.3 and Box 3.1).

(a)

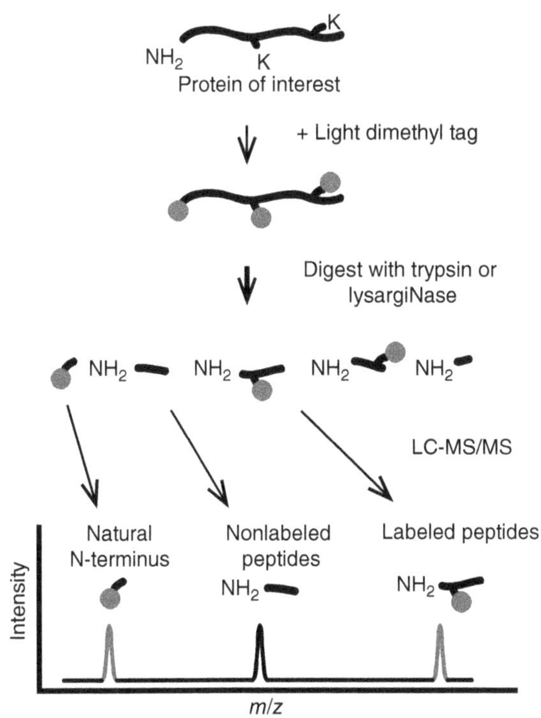

Figure 3.3 ATOMS. (a) ATOMS can be utilized to identify the true N-terminus of a recombinant protein by labeling the entire protein by reductive dimethylation.

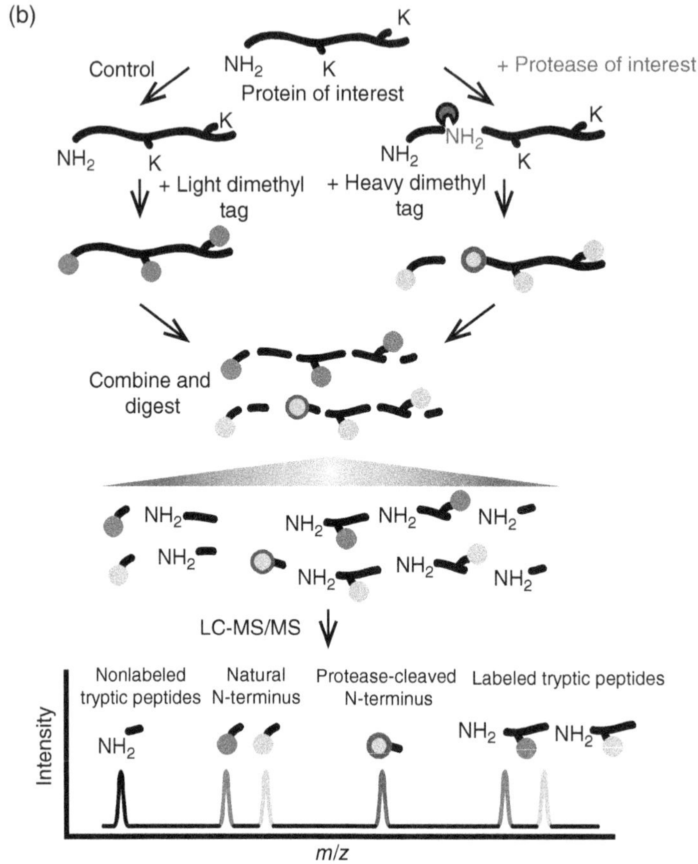

Figure 3.3 (Continued) Retaining positional information by means of the tag allows digestion with trypsin (and a separate aliquot with lysargiNase) to generate peptides that can be analyzed by LC-MS/MS. Peptides with a dimethylated N-terminus are indicative of the true start of the protein. (b) A protein N-terminus or a protease-cleaved recombinant protein substrate of interest can be analyzed by ATOMS as shown on the right hand side. Primary amines are blocked at the protein level with different combinations of isotopically labeled formaldehyde and cyanoborohydride (shown as circles). In this example the newly generated N-terminus from protease cleavage is indicated in dark gray. Each labeled protein sample can then be mixed and digested with trypsin (and a separate aliquot with lysargiNase) to generate peptides, which as highlighted are a combination of freely available primary amines or blocked by dimethylation. Note that natural N-termini are blocked at equal ratios of dimethyl tags whereas the newly formed N-terminus from protease cleavage is a singlet. Following LC-MS/MS analysis it can be seen that nonlabeled tryptic peptides cannot be quantified, the protease-cleaved N-terminal peptide exists in a heavy: light ratio $\gg 1$, while natural protein N-termini are dimethylated at their peptide N-termini in a ratio of 1 : 1. Internal tryptic peptides that have been labeled also are present in a ratio of 1 : 1. Database searching and quantification allows the identification of the protease-cleaved N-terminus as well as the natural N-termini. However, a simple rule is that any peptide with a labeled N-terminal primary amine represents the N-terminus or Neo-termini present in the sample.

Box 3.1 ATOMS protocol for identification of N-termini in a recombinant protein.

1) Dissolve protein of interest in 4 M GuCl (guanidine chloride), 100 mM HEPES pH 8.0.
2) Reduce and alkylate protein with reagents of choice (e.g., 10 mM dithiothreitol (DTT) with 15 mM iodoacetic acid (IAA)).
3) Perform reductive dimethylation. Add 20 mM formaldehyde and 40 mM sodium cyanoborohydride and incubate at 37°C for 16 h.
4) Quench by addition of 20 mM Tris–HCl and incubate for 1 h at 37°C.
5) Split sample in equal parts:
   a) Aliquot A: Add sequencing-grade trypsin (ratio 1 : 100 trypsin : substrate) and incubate for 16 h at 37°C.
   b) Aliquot B: Add $CaCl_2$ to a final 10 mM; then add lysargiNase (ratio 1 : 50 lysargiNase : substrate) and incubate for 16 h at 37°C.
6) Peptide mixtures from each aliquot are purified by offline C_{18} chromatography and analyzed by LC-MS/MS.
7) Bioinformatic analysis reveals peptides that are dimethylated at their N-termini (mass difference of +28.03130 Da) as the N-termini of the test protein. Additional information about the true C-terminus of the protein inferred from lysargiNase digests.
8) Determination of the C-terminus of the protein from lysargiNase peptide identifications:
   a) Higher priority is given to semi-specific peptides as they are often indicative of the true C-terminus of the recombinant protein
   b) Having a non-lysargiNase site at the C-terminus of the peptide is a high-probability candidate for the true C-terminus of the protein. Manual validation of the MS/MS spectrum to identify intense a- and b-ion fragment ladders with high mass accuracy confirms the sequence assignation.
   c) If the true C-terminus of the protein is arginine or lysine, then comparison with trypsin-generated peptides is required for confirmation and the usage of a third enzyme such as GluC.

Depending on the question being asked, either a recombinant protein can be sequenced by ATOMS to identify the protein's N-terminus or alternatively a protease cleavage site can be determined in a protein after treatment with a protease. If the aim is to identify a protease cleavage site then the candidate protein is analyzed with or without incubation with the test protease. An equal amount of candidate protein that is not digested is also kept for analysis to identify the true start of the protein and any cleavage products present in the protein sample before digestion with the test protease. Each sample is then denatured with guanidine hydrochloride and their cysteines reduced and alkylated.

Converting cysteines to their reduced thiol forms prevents unwanted crosslinking between proteins and peptides as well as facilitates LC-MS/MS analysis. A blocking agent against the free thiols from cysteines is employed (such as iodoacetamide) to prevent reformation of disulfide bonds. With proteins available in a semi-unfolded state each sample is then treated with formaldehyde and cyanoborohydride in different isotopic forms. Typically the control untreated protein is incubated with "light" formaldehyde (CH_2O) in combination with "light" sodium cyanoborohydride ($NaBH_3CN$) at neutral to basic pH [72]. Such a reaction targets almost exclusively primary amines (free N-termini and lysine side chains) in a rapid manner. Cyanoborohydride catalyzes the replacement of each hydrogen atom in the primary amine with a $-CH_3$ group, thus dimethylating the nitrogen yielding a final +28.03130 Da change in the mass of the original primary amine. Similarly, the treated protein sample is dimethylated with heavy formaldehyde ($^{13}CD_2O$) and cyanoborohydride resulting in a mass change of +34.06312 Da [72]. Due to the fact that each protein has been labeled with the same chemical group and only differs in their isotopic forms, there is no difference in retention time or ionization efficiencies between the peptides derived from the target protein and as such this allows for true relative quantification [72]. Moreover, dimethylation can be multiplexed easily with a multiplicity of 3 (light, medium, and heavy labeled samples) in a single experiment by the use of deuterated cyanoborohydride and formaldehyde forms. Not only do the dimethyl tags provide a means for quantification, but they also allow for the detection of isolated forms (singlets) as follows.

Dimethylated forms of each sample are combined and digested with a site-specific protease, most often trypsin and also the recently introduced lysargiNase [64]. Trypsin has high specificity and cleaves the C-terminal to arginines and lysines not preceded by prolines, and consequently always generates a basic residue at the C-terminus, which enhances ionization and identification by database searching [71, 73]. Dimethylated lysines are not recognized by trypsin and hence only C-terminal arginines are encountered, which indirectly generate longer peptides (compared to lysine- and arginine-cleaved forms) and thus also improve sequence coverage and identification by MS/MS. Digestion with a site-specific protease generates new primary amines at the N-termini of these peptides enabling their bioinformatic differentiation from the blocked termini in later steps. The peptide mixture can be purified by C18 chromatography and analyzed by LC-MS/MS. Like trypsin, lysargiNase also cleaves at arginines and lysines with high specificity and completion but, in contrast to trypsin, lysargiNase cleaves N-terminally to arginines and lysines. This specificity results in a positively charged N-terminal residue on peptides and allows lysargiNase-generated peptides to have enhanced ionization properties by having a basic residue at the N-terminus. Furthermore, these peptides have fragmentation properties that allow for confident identification by database searching as in the case with trypsin, albeit with different fragmentation patterns observed in MS/MS (discussed later).

Peptides elute according to their hydrophobicity in reversed-phase (RP) liquid chromatography, which is the most widely used platform in LC-MS/MS due to its ease in implementation and extremely high resolution. As each sample was combined after dimethylation, peptides will appear in doublets separated by 6 Da (or whichever isotopic labeling strategy is employed) and the peptides will elute at the same time at separate mass-to-charge values. This allows for the construction of extracted ion chromatograms (XICs) for each peptide ion and the areas under the curve of each of these provides the relative quantification between heavy and light labeled forms of a peptide. Each peptide ion (precursor ion) can then be isolated for CID, a process where precursor ions are collided with gas molecules to generate fragment ions required for identification of the peptide sequence [50]. Database search engines then extract precursor ion mass-to-charge information as well as its corresponding set of fragment ions to interrogate genome sequences to determine the unique peptide sequence that would simultaneously satisfy the requirements for both fragment and precursor ion mass-to-charge values [74–77]. Current modern mass spectrometers can fragment up to 50 precursor ions in a cycle of ~3 s, which for a single standard LC run of 90 min of separation can easily generate 90 000 MS/MS spectra for database searching. Hence, using a combination of site-specific enzymes to generate peptides (and correct databases), modern LC-MS/MS systems, and powerful computer algorithms, profiling *bona fide* N-terminal sequences of candidate proteins has unprecedented depth, accuracy, and reliability.

3.3.5 Determining the True Start of Proteins from ATOMS LC-MS/MS Data

ATOMS data will generate four or five classes of peptides identified by LC-MS/MS:

For trypsin digests:

a) Natural N-termini blocked by pyroglutamate or acetylation
b) Natural N-termini blocked by dimethylation in the ATOMS procedure
c) Nonlabeled internal tryptic peptides not containing lysine
d) Lysine-labeled internal tryptic peptides but with unlabeled N-terminal α-amine
e) If the protein was subjected to proteolytic digestion by a test protease, then the protease-generated neo-N and neo-C-terminal peptides will also be identified as blocked internal peptides.

For lysargiNase digests:

a) Natural C-terminal peptide inferred by sequence matching and position in relation to the known C-terminus of the protein to infer the actual C-terminus of the protein being characterized
b) Natural N-termini blocked naturally by pyroglutamate or acetylation (but in low abundance due to the absence of lysine or arginine in the lysargiNase digested N-terminus)

c) Natural N-termini blocked by dimethylation (but in low abundance due to the absence of lysine or arginine in the lysargiNase digested N-terminus)
d) Nonlabeled internal lysargiNase peptides not containing lysine
e) Nonlabeled N-terminal α-amine internal lysargiNase peptides that are N-terminal lysine ε-amine-labeled.
f) If the protein was subjected to proteolytic digestion by a test protease, then the protease-generated neo-N- and neo-C-terminal peptides will also be identified as internal peptides: the neo-C-terminal peptide with a N-terminal lysine or arginine; and sometimes the neo-N-terminal peptide displaying a neo-N-terminal lysine or arginine if the test protease cuts with that specificity (before the lysargiNase step). In this case the α-amine is labeled.

It is important to note the distinction in applications of ATOMS for the purpose of the experiment. That is, if it is being used to identify the true N-terminus of a recombinant protein or, alternatively, if protease cleavage sites are under investigation for a protease substrate couple. The reason for isotope coding (usage of light and heavy labels) is for detecting protease cleavage sites whereas for the characterization of the true N-termini in a single recombinant protein, utilization of unlabeled dimethylation reagents is all that is needed (Box 3.1). In the case of a single recombinant protein, identification of the dimethylated (or acetylated, cyclized, etc.) peptides at their N-terminus is indicative of the N-terminus of the protein.

Interpretation of LC-MS/MS data generated in ATOMS is straightforward and easily determined by the following scheme [63, 70]:

N-terminal peptides:

- Peptides with N-terminal acetylation or pyroglutamation that *do not* contain a lysine are the *naturally blocked N-terminus of the protein* and manifest as a singlet ion. That is, no heavy or light isotopic variants of the peptide. These species are those in cases (a) in the trypsin digests and (b) in the lysargiNase digests.
- Peptides with N-terminal acetylation or pyroglutamation that *do* contain a lysine are also the *naturally blocked N-terminus of the protein* and these manifest as a doublet with H : L ratio = 1 : 1. These species are those in cases (a) in the trypsin digests and (b) in the lysargiNase digests.
- Peptides that contain a dimethylated N-terminus observed in both light and heavy isotopic forms are the *naturally unblocked N-terminus of the protein* and manifest as a doublet with H : L ratio = 1 : 1. Note the presence of lysine is not required. These species are those in cases (b) in the trypsin digests and (c) in lysargiNase digests.
- Peptides that contain a dimethylated N-terminus observed *only in the heavy isotopic form* is the *test protease-generated N-terminus of the protein* and manifests as a doublet with H : L ratio ≫ 1. Note that the presence of lysine is not required. These species are those in cases (e) from trypsin digests and (f) in lysargiNase digests.

C-terminal peptides:

- Peptides that have a lysargiNase N-terminal cleavage site. Inference of the true C-terminus is based on fragmentation pattern validation at high mass accuracy and validation with digestion with other site-specific enzymes such as GluC. These species are those in case (a) from lysargiNase digests.

Internal peptides:

- Peptides that have a *freely available N-terminal primary amine and no lysine* are *internal peptides* that do not correspond to the N-terminus of the protein. They manifest as a singlet ion and these species are those in cases (c) from trypsin digests and (d) in lysargiNase digests.
- Peptides that have a *freely available N-terminal primary amine and do contain lysine* are *internal peptides* that do not correspond to the N-terminus of the protein. They manifest as a doublet with H : L ratio = 1 : 1 and these species are those in cases (d) from trypsin digests and (e) in lysargiNase digests.

Due to the ability to identify N-termini from target proteins following dimethylation, ATOMS allows cross-validation of these peptides by utilization of different proteases. Indeed, this was proven in the original publication of ATOMS where the candidate proteins to be tested were fibronectin and laminin as models and new termini generated by matrix metalloproteinases 2 and 8 (MMP2 and MMP8) were explored. Utilizing ATOMS the effectiveness of the method was validated by employing a site-specific enzyme to generate new termini. Having validated the method the authors identified that both candidate proteins had 14 cleavage sites (novel termini) by ATOMS and 9 by Edman degradation with only 3 termini common to both. The lack of overlap was due to deficiencies in Edman degradation and the length of peptides generated by trypsin after the GluC digest, which were shorter than normal and hence had difficulties in MS identification. But this showed the complementary nature of Edman degradation and ATOMS. Since then ATOMS has been successfully utilized to profile the true N-termini generated by cleavage using other proteases, for example, the metalloproteinase bone morphogenetic protein-1 (BMP-1) in the context of extracellular matrix remodeling [78]. Here the new starts of betaglycans in rats and humans were examined using a modified ATOMS approach (isobaric tags for relative and absolute quantification–ATOMS; iTRAQ-ATOMS) and it was determined that there were five novel N-termini in humans (positions 19, 21, 364, 379, and 553) and four termini in rats (positions 24, 366, 554, and 556). As described earlier, the authors also performed Edman degradation and found excellent agreement with ATOMS, albeit iTRAQ-ATOMS being superior at identifying number of sequences and length of reads.

Characterization of C-terminal peptides is a challenge at the proteomic level utilizing trypsin as the protease for digestion in any bottom-up approach. Identification of tryptic peptides is often facilitated due to the high intensity of y-ions [79] generated during MS/MS due to the presence of the basic arginine or lysine at the C-terminus. However, unless the C-terminus of a protein ends with an arginine or lysine that enhances ionization and y-ion formation trypsin cleavage removes the N-terminal most positive residue of the C-terminus, usually rendering the peptide opaque to conventional mass spectrometric identification due to the combination of poor fragmentation patterns resulting in poor identification. Moreover, most bottom-up approaches do not consider N-terminal or C-terminal peptides as typically researchers perform fully tryptic searches as opposed to semi-tryptic for higher confidence identifications of proteins. How can the C-termini of recombinant proteins then be explored? As discussed earlier, we have recently introduced lysargiNase, a highly site-specific protease with cleavage specificity inverse to trypsin [64]. LysargiNase cleaves N-terminally to arginines and lysines and in doing so provides a basic N-terminal residue to C-terminal peptides (Figure 3.4). Thus, C-terminal peptides now have enhanced ionization, intense a- and b-ion series, making MS/MS identification significantly more reliable, which in combination with semi-cleavage specificity allows for the profiling of C-termini [64]. LysargiNase provides complementary coverage to trypsin and in combination with ATOMS offers a powerful avenue for the characterization of C-termini of recombinant proteins. However, without C-terminal labeling in ATOMS to specifically identify the terminus, the C-terminus must be inferred, but with high probability, from the parallel lyasargiNase digests.

3.4 Conclusions

Identifying the N-terminus of proteins is an important biological characterization. The vast majority of proteins are proteolytically processed in some form and these processing events can alter the function of the protein. As such, knowing the true start of the protein as opposed to simply identifying the internal peptides that are often used for identification in proteomics experiments is of great importance. Most importantly, whereas most PTMs have a chemical group that can be utilized as bait for isolation by a variety of enrichment methods such as strong cation exchange [80] or titanium dioxide [81] for phosphopeptides, antibodies against acetyl-peptides [82], hydrophilic interaction chromatography for glycopeptides [83], or disulfide exchange columns against redox-modified peptides [84], protease-cleaved proteins do not. This makes their characterization intrinsically more challenging and requires specialized approaches and workflows for correct sequence assignation.

Current tools for characterization of N-termini are biochemical-based or MS-based and are complementary with each other dependent on the candidate protein under investigation. The most successful biochemical-based method is Edman N-terminal sequencing due to its automation, specificity,

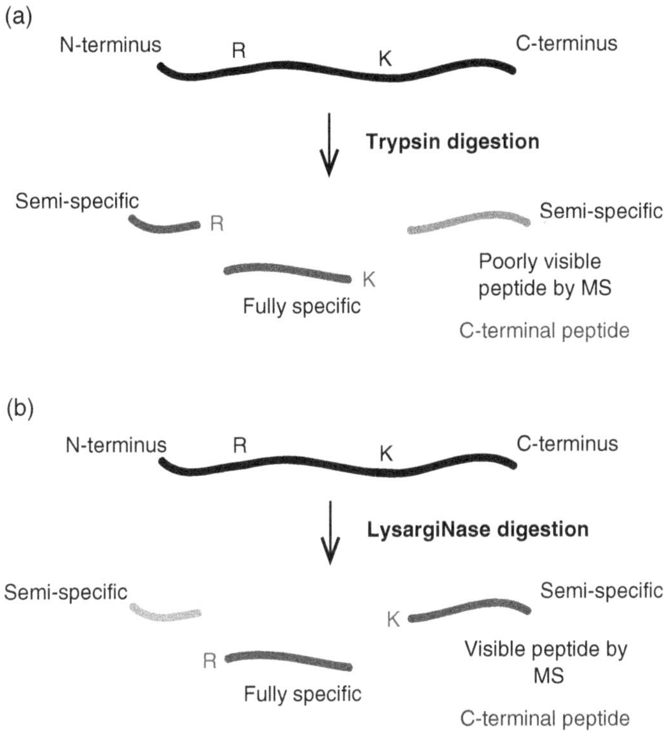

Figure 3.4 LysargiNase digestion to profile true C-termini of recombinant proteins. (a) Digestion of a recombinant protein with trypsin yields fully specific internal peptides (protease cleavage site at the N-terminus and C-terminus of the peptide) and semi-specific terminal peptides (either the N-terminus or the C-terminus have the protease cleavage site). Most proteins do not contain a positively charged C-terminal residue, thus making tryptic-generated C-terminal peptides unfavorable for mass spectrometric analysis. (b) Digestion of the same recombinant protein with lysargiNase generates N-terminal basic residues that are amenable for mass spectrometric analysis. In this case the C-terminus of the protein has a peptide that can be observable by MS analysis and provides the true end of the protein.

and efficiency. However, if fragments are too large and/or blocked at their N-terminus then this method is ineffective. MS-based tools have emerged with the advent of biological molecule ionization, whole genome sequencing, and computational search engines. Bottom-up techniques such as ATOMS and top-down methodologies can bypass the limitations encountered by Edman sequencing and provide additional orthogonal information to characterize N-termini from target proteins. Top-down methodologies excel at obtaining high-resolution, high-mass accuracy spectra in characterizing intact proteins, thus providing a full true picture of the modifications of the target protein.

However, protein size and solubility are often constraints for these technologies as well as more intense data analysis to deconvolute mass spectra. In contrast, ATOMS excels at determining N-terminal sequences from peptides generated by site-specific digestion (by trypsin) of the target protein and can be performed on any MS platform with cheap, specific, and fast chemical reactions even in laboratories that do not specialize in proteomics. It also bypasses the limitation of blocked N-termini as in Edman degradation and the size/solubility issues in top-down MS, but lacks the ability to view all isoforms of the target protein as in top-down methodologies. In complement to ATOMS, digestion with lysargiNase allows the identification of true C-termini in recombinant proteins as well as complex mixtures with high confidence and reliability. The simplicity and coverage provided by ATOMS and lysargiNase digestion approaches make it the ideal method of choice for most applications.

Hence, with each methodology having its own unique strengths and limitations there are now orthogonal and effective options for determining *bona fide* initiation sites of proteins by technologies that are improving at an extremely rapid rate, thus making these characterizations more routine and effective than ever before.

References

1 Crick F (1970) Central dogma of molecular biology. *Nature* 227(5258):561–563.
2 Stearns FW (2010) One hundred years of pleiotropy: a retrospective. *Genetics* 186(3):767–773.
3 Wilhelm M, *et al.* (2014) Mass-spectrometry-based draft of the human proteome. *Nature* 509(7502):582–587.
4 Kim MS, *et al.* (2014) A draft map of the human proteome. *Nature* 509(7502):575–581.
5 Ezkurdia I, Vázquez J, Valencia A, & Tress M (2014) Analyzing the first drafts of the human proteome. *J Proteome Res* 13(8):3854–3855.
6 Ezkurdia I, *et al.* (2014) Multiple evidence strands suggest that there may be as few as 19,000 human protein-coding genes. *Hum Mol Genet* 23(22):5866–5878.
7 Arnesen T, *et al.* (2009) Proteomics analyses reveal the evolutionary conservation and divergence of N-terminal acetyltransferases from yeast and humans. *Proc Natl Acad Sci U S A* 106(20):8157–8162.
8 Mann M & Jensen ON (2003) Proteomic analysis of post-translational modifications. *Nat Biotechnol* 21(3):255–261.
9 Barrett AJ & McDonald JK (1986) Nomenclature: protease, proteinase and peptidase. *Biochem J* 237(3):935.
10 Taylor A (1993) Aminopeptidases: structure and function. *FASEB J* 7(2):290–298.

11 Skidgel RA & Erdös EG (1998) Cellular carboxypeptidases. *Immunol Rev* 161:129–141.
12 Overall CM & Blobel CP (2007) In search of partners: linking extracellular proteases to substrates. *Nat Rev Mol Cell Biol* 8(3):245–257.
13 Fortelny N, *et al.* (2014) Network analyses reveal pervasive functional regulation between proteases in the human protease web. *PLoS Biol* 12(5):e1001869.
14 Baskerville A & Wright CL (1973) Ultrastructural changes in experimental enzootic pneumonia of pigs. *Res Vet Sci* 14(2):155–160.
15 Villarreal I, *et al.* (2012) Effect of challenge of pigs previously immunised with inactivated vaccines containing homologous and heterologous *Mycoplasma hyopneumoniae* strains. *BMC Vet Res* 8:2.
16 Raymond BB, *et al.* (2015) Proteolytic processing of the cilium adhesin MHJ_0194 (P123J) in *Mycoplasma hyopneumoniae* generates a functionally diverse array of cleavage fragments that bind multiple host molecules. *Cell Microbiol* 17(3):425–444.
17 Minion FC, *et al.* (2004) The genome sequence of *Mycoplasma hyopneumoniae* strain 232, the agent of swine mycoplasmosis. *J Bacteriol* 186(21):7123–7133.
18 Rossi D & Zlotnik A (2000) The biology of chemokines and their receptors. *Annu Rev Immunol* 18:217–242.
19 McQuibban GA, *et al.* (2002) Matrix metalloproteinase processing of monocyte chemoattractant proteins generates CC chemokine receptor antagonists with anti-inflammatory properties in vivo. *Blood* 100(4):1160–1167.
20 McQuibban GA, *et al.* (2000) Inflammation dampened by gelatinase A cleavage of monocyte chemoattractant protein-3. *Science* 289(5482):1202–1206.
21 O'Farrell PH (1975) High resolution two-dimensional electrophoresis of proteins. *J Biol Chem* 250(10):4007–4021.
22 Görg A, Weiss W, & Dunn MJ (2004) Current two-dimensional electrophoresis technology for proteomics. *Proteomics* 4(12):3665–3685.
23 Edman P (1949) A method for the determination of amino acid sequence in peptides. *Arch Biochem* 22(3):475.
24 Edman P & Begg G (1967) A protein sequenator. *Eur J Biochem* 1(1):80–91.
25 Sanger F (1945) The free amino groups of insulin. *Biochem J* 39(5):507–515.
26 Hartley BS (1970) Strategy and tactics in protein chemistry. *Biochem J* 119(5):805–822.
27 Edman P (1970) Sequence determination. *Mol Biol Biochem Biophys* 8:211–255.
28 Polevoda B, Arnesen T, & Sherman F (2009) A synopsis of eukaryotic Nalpha-terminal acetyltransferases: nomenclature, subunits and substrates. *BMC Proc* 3 Suppl 6:S2.

29 Marino G, Eckhard U, & Overall CM (2015) Protein termini and their modifications revealed by positional proteomics. *ACS Chem Biol* 10(8):1754–1764.

30 Bergmann M (1934) Synthesis and degradation of proteins in the laboratory and in metabolism. *Science* 79(2055):439–445.

31 Schilling O, Barré O, Huesgen PF, & Overall CM (2010) Proteome-wide analysis of protein carboxy termini: C terminomics. *Nat Methods* 7(7):508–511.

32 Van Damme P, *et al.* (2010) Complementary positional proteomics for screening substrates of endo- and exoproteases. *Nat Methods* 7(7):512–515.

33 Karas M & Hillenkamp F (1988) Laser desorption ionization of proteins with molecular masses exceeding 10,000 daltons. *Anal Chem* 60(20):2299–2301.

34 Fenn JB, Mann M, Meng CK, Wong SF, & Whitehouse CM (1989) Electrospray ionization for mass spectrometry of large biomolecules. *Science* 246(4926):64–71.

35 Patterson SD & Aebersold RH (2003) Proteomics: the first decade and beyond. *Nat Genet* 33(Suppl):311–323.

36 Chait BT (2006) Chemistry. Mass spectrometry: bottom-up or top-down? *Science* 314(5796):65–66.

37 Aebersold R & Mann M (2003) Mass spectrometry-based proteomics. *Nature* 422(6928):198–207.

38 Kelleher N (2004) Top-down proteomics. *Anal Chem* 76(11):196A–203A.

39 Tran JC, *et al.* (2011) Mapping intact protein isoforms in discovery mode using top-down proteomics. *Nature* 480(7376):254–258.

40 Loo JA, Edmonds CG, & Smith RD (1990) Primary sequence information from intact proteins by electrospray ionization tandem mass spectrometry. *Science* 248(4952):201–204.

41 Marshall AG, Hendrickson CL, & Jackson GS (1998) Fourier transform ion cyclotron resonance mass spectrometry: a primer. *Mass Spectrom Rev* 17(1):1–35.

42 Good DM, Wirtala M, McAlister GC, & Coon JJ (2007) Performance characteristics of electron transfer dissociation mass spectrometry. *Mol Cell Proteomics* 6(11):1942–1951.

43 Makarov A (2000) Electrostatic axially harmonic orbital trapping: a high-performance technique of mass analysis. *Anal Chem* 72(6):1156–1162.

44 Michalski A, *et al.* (2011) Mass spectrometry-based proteomics using Q Exactive, a high-performance benchtop quadrupole Orbitrap mass spectrometer. *Mol Cell Proteomics* 10(9):M111.011015.

45 Hebert AS, *et al.* (2014) The one hour yeast proteome. *Mol Cell Proteomics* 13(1):339–347.

46 Senko MW, *et al.* (2013) Novel parallelized quadrupole/linear ion trap/Orbitrap tribrid mass spectrometer improving proteome coverage and peptide identification rates. *Anal Chem* 85(24):11710–11714.

47 Rose RJ, Damoc E, Denisov E, Makarov A, & Heck AJ (2012) High-sensitivity Orbitrap mass analysis of intact macromolecular assemblies. *Nat Methods* 9(11):1084–1086.

48 Heck AJ (2008) Native mass spectrometry: a bridge between interactomics and structural biology. *Nat Methods* 5(11):927–933.

49 Maple HJ, *et al.* (2014) Application of the Exactive Plus EMR for automated protein-ligand screening by non-covalent mass spectrometry. *Rapid Commun Mass Spectrom* 28(13):1561–1568.

50 Wells JM & McLuckey SA (2005) Collision-induced dissociation (CID) of peptides and proteins. *Methods Enzymol* 402:148–185.

51 Zubarev RA (2004) Electron-capture dissociation tandem mass spectrometry. *Curr Opin Biotechnol* 15(1):12–16.

52 Little DP, Speir JP, Senko MW, O'Connor PB, & McLafferty FW (1994) Infrared multiphoton dissociation of large multiply charged ions for biomolecule sequencing. *Anal Chem* 66(18):2809–2815.

53 Syka JE, Coon JJ, Schroeder MJ, Shabanowitz J, & Hunt DF (2004) Peptide and protein sequence analysis by electron transfer dissociation mass spectrometry. *Proc Natl Acad Sci U S A* 101(26):9528–9533.

54 Olsen JV, *et al.* (2007) Higher-energy C-trap dissociation for peptide modification analysis. *Nat Methods* 4(9):709–712.

55 LeDuc RD, *et al.* (2004) ProSight PTM: an integrated environment for protein identification and characterization by top-down mass spectrometry. *Nucleic Acids Res* 32(Web Server issue):W340–W345.

56 Zabrouskov V, *et al.* (2006) Stepwise deamidation of ribonuclease A at five sites determined by top down mass spectrometry. *Biochemistry* 45(3):987–992.

57 Shaw JB, *et al.* (2013) Complete protein characterization using top-down mass spectrometry and ultraviolet photodissociation. *J Am Chem Soc* 135(34):12646–12651.

58 Laitaoja M, Sankhala RS, Swamy MJ, & Jänis J (2012) Top-down mass spectrometry reveals new sequence variants of the major bovine seminal plasma protein PDC-109. *J Mass Spectrom* 47(7):853–859.

59 Thérien I, Moreau R, & Manjunath P (1998) Major proteins of bovine seminal plasma and high-density lipoprotein induce cholesterol efflux from epididymal sperm. *Biol Reprod* 59(4):768–776.

60 Bakalarski C, Gan Y, Wertz I, Lill J, Sandoval W. (2016) Rapid, semi-automated protein terminal characterization using ISDetect. *Nat Biotechnol* 34(8):811–813.

61 Kleifeld O, *et al.* (2010) Isotopic labeling of terminal amines in complex samples identifies protein N-termini and protease cleavage products. *Nat Biotechnol* 28(3):281–288.

62 Gevaert K, *et al.* (2002) Chromatographic isolation of methionine-containing peptides for gel-free proteome analysis: identification of more than 800 Escherichia coli proteins. *Mol Cell Proteomics* 1(11):896–903.

63 Doucet A & Overall CM (2011) Broad coverage identification of multiple proteolytic cleavage site sequences in complex high molecular weight proteins using quantitative proteomics as a complement to edman sequencing. *Mol Cell Proteomics* 10(5):M110.003533.
64 Huesgen PF, *et al.* (2015) LysargiNase mirrors trypsin for protein C-terminal and methylation-site identification. *Nat Methods* 12(1):55–58.
65 Washburn MP, Wolters D, & Yates JR (2001) Large-scale analysis of the yeast proteome by multidimensional protein identification technology. *Nat Biotechnol* 19(3):242–247.
66 Wolters DA, Washburn MP, & Yates JR (2001) An automated multidimensional protein identification technology for shotgun proteomics. *Anal Chem* 73(23):5683–5690.
67 Mahrus S, *et al.* (2008) Global sequencing of proteolytic cleavage sites in apoptosis by specific labeling of protein N termini. *Cell* 134(5):866–876.
68 Prudova A, *et al.* (2014) TAILS N-terminomics of human platelets reveals pervasive metalloproteinase-dependent proteolytic processing in storage. *Blood* 124(26):e49–60.
69 Klein T, *et al.* (2015) The paracaspase MALT1 cleaves HOIL1 reducing linear ubiquitination by LUBAC to dampen lymphocyte NF-κB signalling. *Nat Commun* 6:8777.
70 Doucet A & Overall CM (2011) Amino-Terminal Oriented Mass Spectrometry of Substrates (ATOMS) N-terminal sequencing of proteins and proteolytic cleavage sites by quantitative mass spectrometry. *Methods Enzymol* 501:275–293.
71 Olsen JV, Ong SE, & Mann M (2004) Trypsin cleaves exclusively C-terminal to arginine and lysine residues. *Mol Cell Proteomics* 3(6):608–614.
72 Boersema PJ, Raijmakers R, Lemeer S, Mohammed S, & Heck AJ (2009) Multiplex peptide stable isotope dimethyl labeling for quantitative proteomics. *Nat Protoc* 4(4):484–494.
73 Rodriguez J, Gupta N, Smith RD, & Pevzner PA (2008) Does trypsin cut before proline? *J Proteome Res* 7(1):300–305.
74 Perkins DN, Pappin DJ, Creasy DM, & Cottrell JS (1999) Probability-based protein identification by searching sequence databases using mass spectrometry data. *Electrophoresis* 20(18):3551–3567.
75 Craig R & Beavis RC (2003) A method for reducing the time required to match protein sequences with tandem mass spectra. *Rapid Commun Mass Spectrom* 17(20):2310–2316.
76 Cox J, *et al.* (2011) Andromeda: a peptide search engine integrated into the MaxQuant environment. *J Proteome Res* 10(4):1794–1805.
77 Eng JK, McCormack AL, & Yates JR (1994) An approach to correlate tandem mass spectral data of peptides with amino acid sequences in a protein database. *J Am Soc Mass Spectrom* 5(11):976–989.

78 Delolme F, *et al.* (2015) Proteolytic control of TGF-β co-receptor activity by BMP-1/tolloid-like proteases revealed by quantitative iTRAQ proteomics. *Cell Mol Life Sci* 72(5):1009–1027.

79 Steen H & Mann M (2004) The ABC's (and XYZ's) of peptide sequencing. *Nat Rev Mol Cell Biol* 5(9):699–711.

80 Villén J & Gygi SP (2008) The SCX/IMAC enrichment approach for global phosphorylation analysis by mass spectrometry. *Nat Protoc* 3(10):1630–1638.

81 Larsen MR, Thingholm TE, Jensen ON, Roepstorff P, & Jørgensen TJ (2005) Highly selective enrichment of phosphorylated peptides from peptide mixtures using titanium dioxide microcolumns. *Mol Cell Proteomics* 4(7):873–886.

82 Choudhary C, *et al.* (2009) Lysine acetylation targets protein complexes and co-regulates major cellular functions. *Science* 325(5942):834–840.

83 Mysling S, Palmisano G, Højrup P, & Thaysen-Andersen M (2010) Utilizing ion-pairing hydrophilic interaction chromatography solid phase extraction for efficient glycopeptide enrichment in glycoproteomics. *Anal Chem* 82(13):5598–5609.

84 Paulech J, *et al.* (2013) Large-scale capture of peptides containing reversibly oxidized cysteines by thiol-disulfide exchange applied to the myocardial redox proteome. *Anal Chem* 85(7):3774–3780.

4

Assessing Activity and Conformation of Recombinant Proteins

Diego Ellerman[1], *Till Maurer*[1], *and Justin M. Scheer*[1,2]

[1] *Department of Protein Chemistry and Structural Biology, Genentech Inc., South San Francisco, CA, USA*
[2] *Antibody Engineering Department, Boerhinger Ingelheim, Ridgefield, Connecticut*

Abbreviations

Ab	antibody
BSA	bovine serum albumin
CD	circular dichroism
CMC	carboxymethylcellulose
CSF-1	colony stimulating factor
DMSO	dimethyl sulfoxide
DOSY	diffusion-ordered spectroscopy
DSC	differential scanning calorimetry
DSF	differential scanning fluorimetry
EGFR	epithelial growth factor receptor
EPD	empirical phase diagram
FcRn	neonatal Fc receptor
HDX	hydrogen–deuterium exchange
HOS	higher order structure
HPMC	hydroxylpropyl methyl cellulose
HSA	human serum albumin
HSQC	heteronuclear single quantum correlation
IGF1R	insulin-like growth factor receptor 1
ITC	isothermal titration calorimetry
K_a	association constant
mAb	monoclonal antibody
MHC	major histocompatibility complex

Analytical Characterization of Biotherapeutics, First Edition. Edited by Jennie R. Lill and Wendy Sandoval.

NMR	nuclear magnetic resonance
PTM	posttranslational modification
rhVEGF	recombinant human vascular endothelial growth factor
SRCD	synchrotron radiation circular dichroism
TCR	T-cell receptor
T_m	melting temperature
TROSY	transversed relaxation optimized spectroscopy
UV	ultraviolet

4.1 Introduction

The higher order structure (HOS) of a protein includes the secondary, tertiary, and quaternary structures of a protein that are required for its function. There is a diversity of biophysical methods available for the characterization of a protein HOS, each of them with their associated benefits and limitations. An important distinction between the different methods is whether they are sensitive to small, local conformation changes or to large, global folding modifications (Figure 4.1). In general, the techniques with the lowest level of resolution are simpler and faster to run and therefore they are amenable to high-throughput processing. On the other hand, techniques providing more information require more time for acquiring and interpreting the data and thus are less compatible with automated, high-throughput processing. Besides the ease of use, resolution, and analysis time, there is also a difference in how widely applicable the techniques are: assays with a higher degree of resolution such as nuclear magnetic resonance (NMR) or X-ray crystallography are limited in regard to the subset of proteins that can be analyzed. For example, X-ray crystallography can only be used on well-ordered proteins and NMR has been typically used on proteins smaller than 25 kDa.

Different stages in the development of a biotherapeutic usually require varying levels of in-depth structural characterization and breadth of samples or conditions tested, so different techniques find applications in different settings and in some cases they provide complementary information. A comprehensive review of most of the techniques available for the analysis of the HOS of biotherapeutics is beyond the scope of this chapter, but interested readers are referred to other sources [1, 2]. In this chapter we describe the fundamental principles and applications in the biotherapeutic field including four techniques representing different stages along the range of resolution and complexity described in Figure 4.1. We have chosen circular dichroism (CD), differential scanning calorimetry (DSC), hydrogen–deuterium exchange–mass spectrometry (HDX-MS) and NMR as examples because they are arguably the techniques most widely used within that range of the resolution spectrum.

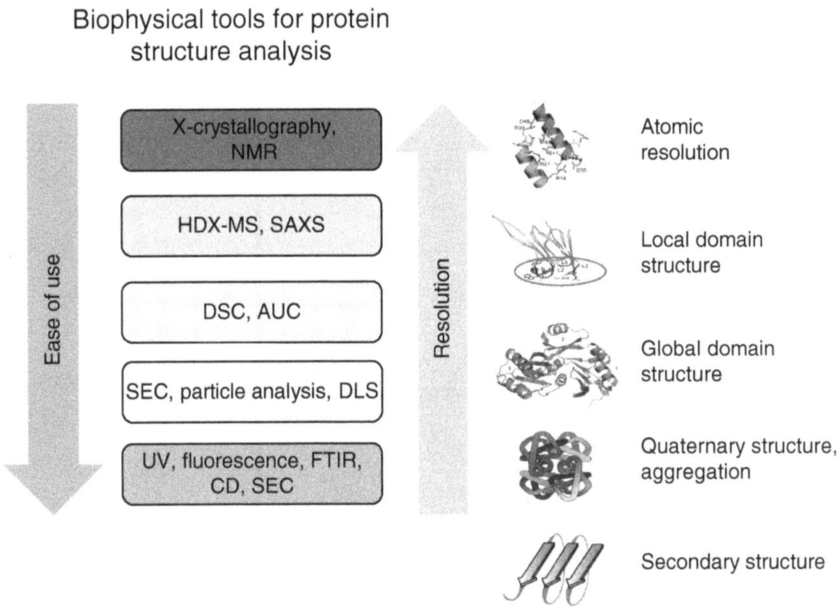

Figure 4.1 Hierarchy of biophysical methods for the characterization of higher order structure of proteins. The methods are organized in four tiers according to the degree of complexity in the experimental setup and level of resolution they provide. AUC, analytical ultracentrifugation; CD, circular dichroism; DLS, dynamic light scattering; DSC, differential scanning calorimetry; FTIR, Fourier transform infrared; HDX-MS, hydrogen–deuterium exchange–mass spectrometry; NMR, nuclear magnetic resonance; SAXS, small angle X-ray scattering; SEC, size-exclusion chromatography; UV, ultraviolet.

4.2 Circular Dichroism

CD is an excellent tool for characterizing the secondary structure content of a protein. CD occurs when molecules absorb left or right circularly polarized light at different levels [3]. Differential absorption is observed for chiral molecules or molecules that are placed in a chiral environment. In proteins, two chemical groups are active in absorbing circularly polarized light: the amide group and aromatic residues. The amide groups involved in different secondary structures show different maxima of wavelength absorption in the far ultraviolet (UV) spectra (below 250 nm) (Figure 4.2). Aromatic residues absorb circularly polarized light in the near-UV (250–300 nm) region depending on conditions of the environment and therefore analysis of CD in this region of the spectrum is used to inform on the tertiary structure of proteins. The use of CD for characterizing the tertiary structure is not as widespread as for secondary structure analysis but some researchers have evaluated the use of tryptophan CD spectroscopy for that purpose [5, 6].

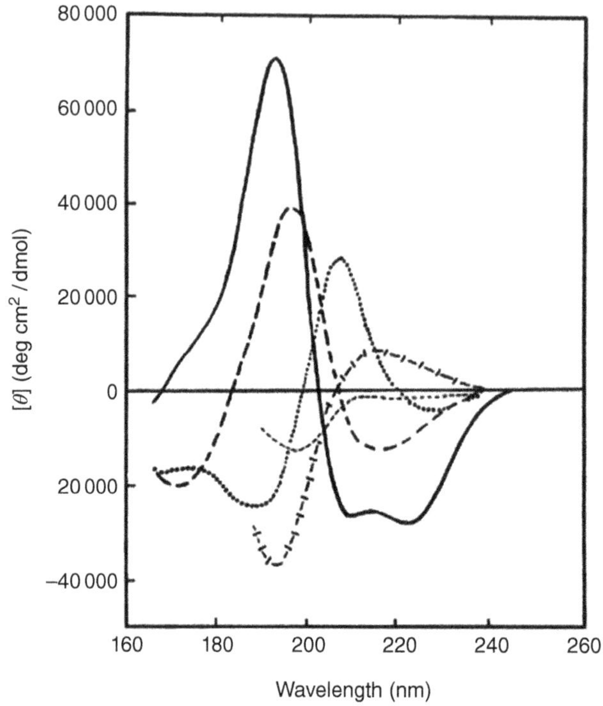

Figure 4.2 Far UV CD spectra associated with various types of secondary structure. Solid line, α-helix; long dashed line, antiparallel β-sheet; dotted line, type I β-turn; cross dashed line, extended 3_1 helix or poly (Pro) II helix; short dashed line, irregular structure. Source: Kelly *et al.* [4]. Reproduced with permission of Elsevier.

Even though the CD spectra can be deconvoluted to quantify the content of α-helixes, β-sheets, turns, and random coils, the results for proteins containing a high percentage of β-sheets are not always reliable. Improvements in the methods for secondary structure prediction have been recently introduced to make the projections more accurate, and as a consequence they may spread this predictive use of CD [7]. However, so far CD is more often used to assess changes in the secondary structure of a protein in different conditions, for example, in different formulations or as a result of different processes. Although CD provides only low-resolution structural information it is very sensitive to changes in conformation and it requires relatively small amounts of material (~200 µl of a 0.1–0.15 mg/ml solution). One limitation of this technique is the incompatibility with buffers and additives that absorb light in the UV range. The majority of simple buffer components are compatible with CD measurements below 200 nm. However, high concentrations of chloride, nitrate, some solvents (dimethyl sulfoxide (DMSO), dioxane), some biological buffer (HEPES, PIPES, MES), and reducing agents should be avoided [8].

4.2.1 Applications of CD

CD has been used in the pharmaceutical industry for many decades, for example, for the characterization of the stability of insulin and for the study of the relationship between structure and activity [9, 10]. More recently, CD is used in combination with other low-resolution biophysical tools to provide complementary information on different aspects of the protein structure.

4.2.1.1 Thermal Stability Analysis

Even though the preferred method for studying the thermal stability of proteins is DSC (described later), CD can also be used. In such experiments the secondary study content of a protein is recorded at increasing temperatures, either by analyzing the entire near-UV spectrum or by choosing a specific wavelength indicative of a particular structure. One limitation of this application is that analysis of multidomain proteins is difficult. Determination of the melting temperatures by CD has been used, for example, for studying prion proteins from different species with different misfolding susceptibilities [11] or for comparing the stability of Epoetin alfa from different manufacturers [12].

4.2.1.2 Characterization of the Effect of PEGylation

PEGylation is a strategy used to extend the half-life of proteins in circulation as a consequence of increasing the hydrodynamic radius (Rh) and consequently reducing the rate of glomerular filtration. One important requirement for PEGylated proteins for therapeutic use is that after the chemical modification the protein must retain its natural conformation and activity. Interferon 2β is a protein that has been PEGylated to extend its serum half-life and CD analyses in both near-UV and far-UV were used to confirm the lack of significant modifications in the structure [13, 14]. Similarly, recombinant human factor VIIa PEGylated [15] and human growth hormone [16] showed no major modifications in its secondary and tertiary structure when analyzed by CD.

4.2.1.3 Formulation and Stability Studies

CD can be used for studying the effect of different formulation excipients on the secondary structure of proteins. One example is the characterization of recombinant human vascular endothelial growth factor (rhVEGF) formulated in different gel compositions. The changes in secondary structure in the presence of different polymers was studied by Ji *et al.* [17], and the analysis revealed that formulation in hydroxypropyl methylcellulose (HPMC) produced fewer changes than formulation in carboxymethylcellulose (CMC). Similarly, insulin has also been formulated in a polymer gel for potential oral administration. In this case it was also important to study whether the process of formulation insulin in a cyclodextrin polymer affected the structure of the protein and the examination of the CD spectra showed the absence of significant differences [18].

In the field of Ab therapeutics, CD is sometimes used in combination with other orthogonal assays amenable to high-throughput processing for formulation purposes. This approach is referred to as empirical phase diagram (EPD). The involvement of different biophysical techniques providing information on different aspects of the protein structure (secondary structure, tertiary structure, aggregation propensity) allows for a more comprehensive characterization of the protein structure. This methodology has been used for finding stabilizing excipients for IgG_1 [19] and IgG_2 [20] Abs.

CD is also often used for analyzing changes in the secondary structure in studies aimed at characterizing the stability of modified proteins. Examples include the study of the effect of deglycosylation of an IgG1 [21], the impact of reducing the intramolecular disulfide bonds in the C_H3 domain of an Ab [22], the stability of individual recombinant domains of an IgG [23], or the conformation of aggregates formed under stressing conditions of Ab–drug conjugates with different levels of payloads [24].

4.2.1.4 Analysis of Biosimilars

Once a successful biotherapeutic comes off patent, companies other than the one owning the intellectual property for the original process are allowed to pursuit the production and commercialization of the drug. A distinctive aspect of biosimilars is that they are produced using manufacturing methods different to the original product, which may lead to subtle or significant differences in potency and safety. It is important then to compare the drug produced by the variant method to the original one in terms of structure and activity. Because it is a relatively simple technique, CD is useful for initial comparative analysis of the secondary and tertiary structure, sometimes in conjunction with other techniques. This approach was recently used by Levy *et al.* [25] in studying granulocyte colony stimulating factor from different manufacturers.

4.2.2 Technical Improvements

Synchrotron radiation circular dichroism (SRCD) is an emergent technique that expands the range of wavelengths where CD can be measured effectively. SRCD permits analyzing CD in the 160–190 nm range [26], which could be important for detecting subtle changes useful for drug binding studies of high-throughput analysis. For example, a method combining SCRD and magnetic circular dichroism (MCD) has been proposed for studying the photo-stability of biopharmaceutical products like Abs [27].

4.3 DSC and Isothermal Titration Calorimetry

These techniques measure the thermodynamic driving forces of biochemical processes such as association, dissociation, or unfolding. Although both techniques characterize the energetics of a reaction, they differ in the experimental design and main applications. DSC measures changes in the heat capacity (ΔCp) as a function of temperature and changes in enthalpy (ΔH) can be derived from this information. Isothermal titration calorimetry (ITC) measures the heat of a reaction at constant temperature; the heat exchanged can then be used to derive Gibbs free energy (ΔG), entropy (ΔS), enthalpy (ΔH), the association constant (K_a), and stoichiometry. Readers interested in more in-detail literature on thermodynamics, experimental design, and data analysis of calorimetry experiments are directed elsewhere [28–30].

In the biotherapeutic field DSC is commonly used for studying the thermal stability of proteins and ITC for characterizing protein–protein or protein–small molecule interactions. In a typical analysis of thermal stability by DSC, the protein sample is heated along with a reference cell and the differences in heat transfer are recorded. As the protein is heated and starts unfolding, a peak in the heat capacity of the protein sample is detected, as thermal unfolding is an endothermic process. The temperature corresponding to the apex of the peak is referred to as the melting temperature or T_m, and it is a property of the protein domain as a whole. The integration of Cp over temperature yields the enthalpy of the transition (ΔH_m) and is correlated to the content of ordered secondary structure of a protein [31], being the net value of endothermic contributions such as the disruption of hydrogen bonds, and exothermic contributions like the breakup of hydrophobic interactions. The sharpness of the transition peak can be measured as the width at half the maximum height, and it is indicative of the cooperativity of the reaction. If denaturation occurs within a narrow range of temperature the process is considered highly cooperative. Multidomain proteins may have more than one T_m corresponding to the different domains.

Some advantages of both ITC and DSC are that they are label-free techniques, they are performed in solution, they do not present molecular weight limitations, and they do not require extensive assay development. Conversely, in comparison to DSF (see later), DSC requires more sample and it takes longer to run. One disadvantage of ITC is that since the concentration of the titrant and the sample are used for calculating the stoichiometry, having a fraction of inactive material of either molecule will impact on the accuracy of the reported ratios. Typically ITC can accurately characterize interactions with affinities between the nM and mM range, but lower affinities can also be measured by changing the experimental design [32].

In a typical ITC experiment small aliquots of a reactant are introduced successively into a larger volume of protein sample at constant temperature and the heat exchange is recorded. After each injection of reactant, the fraction of unbound protein diminishes and the heat exchanged is reduced consequently. The heat exchanged is often plotted as a function of the concentration of the ligand introduced and knowing the concentration of the sample protein it is possible to derive K_a and the stoichiometry of the interaction. In addition, the information obtained on the enthalpy and entropy changes may shed light on the mechanistic involved in the molecular interaction.

4.3.1 Use of DSC and ITC in Therapeutics Discovery

During the research stage DSC is often used for analyzing the effect of introducing mutations in the candidate Ab on the general packing and thermal stability of a therapeutic candidate protein. Ab engineering is a good example of a field that uses DSC copiously for analyzing the effect of introducing mutations in the candidate Ab. The rise of bispecific Abs as a new type of therapeutic molecule has prompted the design of novel Ab formats that facilitate their production (for more details refer to Chapter 10). The new bispecific Ab formats usually emerge after several rounds of engineering the protein sequence. During this process it is important to verify that the introduced mutations do not significantly affect the thermal stability of the molecule as this may have consequences on the colloidal stability and half-life. A number of recent publications illustrate the use of DSC in the engineering of bispecific Abs [33–37]. Figure 4.3 shows the DSC analysis of different engineered constructs developed by Choi *et al.* [34] with the goal of directing the formation of heterodimeric IgG molecules and disfavoring the formation of homodimers. The analysis shows that while the T_m for the C_H2 domain in all constructs is fairly similar, the engineered proteins have a C_H3

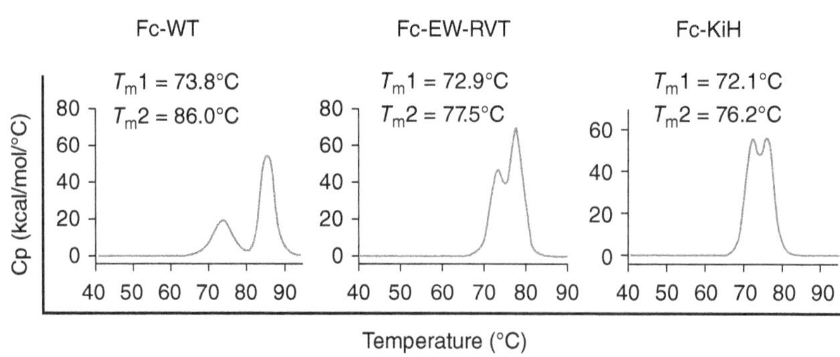

Figure 4.3 Thermal stability analysis of WT (left panel) and engineered Fc fragments (center and right panels). The lower T_m corresponds to the melting of the C_H2 domain and the higher T_m corresponds to the C_H3 domain. Source: Choi *et al.* [34]. Reproduced with permission of American Association for Cancer Research.

domain with reduced T_m, suggesting that the gain in heterodimeric selectivity occurred at the expense of slightly reducing the stability of the protein.

DSC is used not only in the field of Ab engineering but also for other biotherapeutics like protein-based vaccines. Two different *Neisseria* species cause serious human diseases, namely, meningitis and gonorrhea. Currently there is a vaccine effective for some strains of *Neisseria meningitides* but there is no vaccine available against *Neisseria gonorrhoeae.* In order to increase the coverage of the immune response researchers are trying to generate recombinant proteins containing antigens for both species. The hybrid protein, however, must retain a conformation similar to the native proteins in order to elicit a full range of cross-reacting Abs. Together with other biophysical assays, DSC was used to characterize the hybrid antigens, showing that the epitope grafting process did not impact on the thermal stability [38]. In another example, Crespillo *et al.* [39] used thermal stability analysis by DSC in their biophysical characterization of different HIV vaccines candidates based on mimetics of the gp41 protein. A first generation of vaccine constructs showed propensity to aggregate at the physiological pH, so a second generation was designed. DSC analysis of the constructs from the second generation showed an improved stability, and additional ITC experiments characterizing the interaction with a known ligand confirmed a native-like structure.

ITC is a valuable technique for the characterization of protein complexes and the study of cooperative effects [40]. One protein complex from snake venoms with potential anticoagulant application was analyzed by ITC. Blood coagulation is a complex biological process that starts after the exposure of subendothelial tissue factor (TF) to plasma factor VII (FVII) and leads to fibrin formation. This initial step is crucial for blood coagulation and considered an ideal target for the treatment of thromboembolic disorders. Hemextin AB complex is a heterotetrameric protein complex isolated from snake venom that potently and specifically inhibits the TF-FVII complex. Using ITC Banarjee and collaborators [41] characterized the formation of the hemextin Ab complex. Fitting of the ITC data showed a 1 : 1 stoichiometry, confirming previous data on the complex composition. The analysis also showed a favorable ΔH change and an unfavorable ΔS change, suggesting that the complex formation is enthalpically driven and resulting in a less flexible or less disordered structure. The study of the Ka in the presence of different salt or glycerol concentrations indicated the importance of ionic and hydrophobic interactions for the formation of the complex and led to a complex formation model. ITC has also been used to gain insights on the mechanisms of immunosuppression by the Epstein–Barr protein BARF1. A series of titration experiments, along with other biochemical assays, showed that BARF1 binds to the colony stimulating factor (CSF-1) locking it in a conformation unable to interact with its cognate receptor [42]. Other examples of the use of ITC to characterize the interactions of proteins with therapeutic potential include binding of a nanobody to α-synuclein [43], the binding of an Ab against the oncological target ROBO1 [44], the interaction of the alternative binding scaffold "repebody" with IL-6 [45, 46], and the interaction of small artificial binding proteins Affitins to their targets [47]. Characterization of protein complexes formed by more than two different units is also possible, but obviously more complex, requiring multiple titration experiments. In one publication ITC was used to characterize a trivalent Ab with specificity toward the epithelial growth factor receptor (EGFR), insulin-like growth factor receptor1 (IGF1R), and Her3. By sequentially titrating the trivalent Ab with the three targets and fitting the data the authors confirmed the 1 : 1 stoichiometry and therefore the simultaneous binding of all three targets [48].

4.3.2 Protein Conjugation

As mentioned earlier, PEGylation is often used to improve the pharmacokinetic and solubility properties of certain therapeutic proteins. In addition to CD, DSC is also used for characterizing the effect of PEGylation on the structure and stability of proteins. A recent study investigated the effect of PEGylation on two small therapeutic proteins, an Ab fragment and Adnectin, an engineered human fibronectin domain III. The characterization by DSC showed that PEGylation slightly increased the thermal stability of both proteins [49] and therefore provided support for the use of PEGylation in this type of molecules.

4.3.3 Formulation and Stability

At later stages of biotherapeutic development both DSC and ITC are used frequently for optimizing the composition of formulation buffers. Some applications require the development of alternative formulations to adapt the protein to oral, nasal, or transdermal administration to improve the delivery to the intended tissue or organ. Among others, these formulations include polymer or liposome encapsulation to form nanoparticles. The process of making the nanoparticle may compromise the structural integrity of the protein; therefore it is important to scrutinize the effects of the production process on the conformation of the protein. DSC along with other analytical assays is used for this purpose. For example, the effect of including polyethylene glycol (PEG) as a protective excipient in polylactide-*co*-glycolide particles containing the model protein bovine serum albumin (BSA) was studied. The results showed that BSA released from the particles formulated in the presence of PEG had an increased thermal stability [50].

DSC is often used to study the stabilizing or destabilizing effect of different buffer systems, or the inclusion of excipients or preservatives. One published example of the use of DSC in formulation showed the effect of different buffers and excipients on the T_m of interleukin-1 receptor [51]. In addition, the authors also evaluated the DSC stability predictions by studying the percentages of aggregation in the different conditions after a 60-day stress period, finding a good correlation between the T_m found by DSC and the levels of aggregation.

Although the general mechanisms involved in the protein stabilization effect by different excipients is known [52], getting a better understanding of the excipient–protein interactions that lead to increased stability may contribute to enabling a more rational formulation optimization approach. Several studies using ITC and DSC for this purpose have been reviewed elsewhere [53]. In one example Garidel *et al.* studied the stabilization of human serum albumin (HSA) and human IgGs by polysorbate 20 and 80. While DSC analysis showed a stabilizing effect on both HSA and the IgGs, ITC experiments showed binding of one to three molecules of detergents per albumin but no interactions where detected between the detergents and the IgGs [54]. The authors concluded that the colloidal stabilization observed for the IgGs was due to a different mechanism.

A technique associated with DSC is differential scanning fluorimetry (DSF). DSF also provides information about the temperature of the protein unfolding transition, but rather than measuring heat exchange it detects the fluorescence of the environmentally sensitive dye SYPRO Orange. The dye becomes fluorescent when it associates with hydrophobic patches exposed as the protein denatures. Although this technique does not characterize the thermodynamic parameters of the unfolding reaction, it is very sensitive and amenable for high-throughput processing. A recent example of the use of this assay screened different formulation conditions for a therapeutic Ab [55]. First, the authors compared the melting temperatures for three Abs obtained by DSC or DSF (Figure 4.4) observing that although the T_m obtained by DSF were approximately 3° lower than the T_m calculated from DSC experiments, the good correlation between values from the two techniques allowed using DSF for making comparisons between different formulations. The subsequent DSF analysis showed the stabilizing effect of sucrose and polysorbate 80 on one Ab and the destabilizing effect of low pH or increasing salt concentrations. As with any other assay, DSF also has its own limitations. DSF is not a good method of choice to analyze proteins with high melting temperatures [56] or for multidomain proteins, as domains with higher T_m may not be clearly detected. Since the dye is fluorescent when associated to hydrophobic patches, the use of detergents in the solution can create high background readings.

4.3.4 Analysis of Biosimilars

Given the proliferation of biosimilars, calorimetry can also be used for comparing the folding of biotherapeutics produced by different processes. Lang and Cole [57] have recently reported the use of DSC for the characterization of different commercial products of human albumin (HSA). Their study revealed

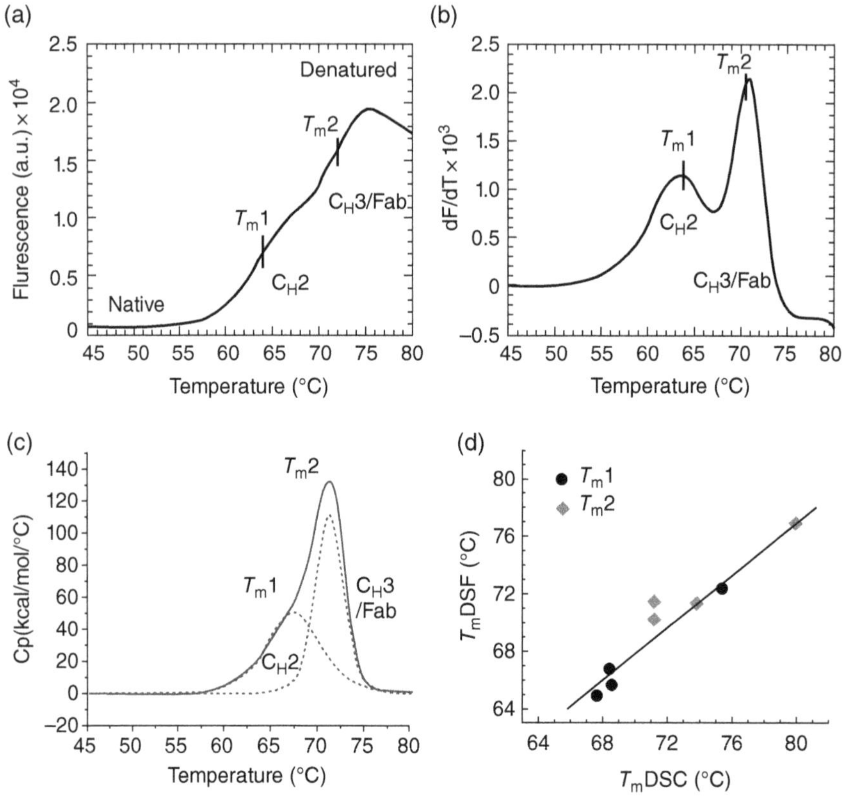

Figure 4.4 Analysis of thermal stability of antibodies by DSF compared to DSC. (a) Thermogram of an antibody obtained by DSF. The different T_m correspond to the inflection points of the curves, labeled as T_m1 and T_m2. (b) The different transition temperatures are identified more easily as the peaks of the first derivative function of the thermogram. (c) DSC analysis of the same antibody. (d) Correlation between the T_m obtained by DSC or DSF for three antibodies in different formulations. Source: Niedziela-Majka *et al.* [55]. Reproduced with permission of Sage Publication.

differences in the DSC profile of HSA from different sources. Upon further research it was found that those differences were due to variability in the fatty acid associated to HSA. This example, along with other studies of Ab biosimilars [58, 59], show that DSC is a valuable assay to compare biophysical properties of biosimilars.

DSC and ITC are powerful techniques that are used in all stages of biotherapeutics development. Continuous improvements in DSC equipment and technology are allowing for faster temperatures ramps; increased sensitivity and instruments more amenable to high-throughput processing will likely help to extend their use in the industry further.

4.4 Hydrogen–Deuterium Exchange–Mass Spectrometry

The basis of this methodology is the differential solvent exposure of the amino acids in the properly folded polypeptide. Hydrogens in the amide backbone are capable of spontaneous exchange with deuterium when the protein is diluted in a deuterated buffer. Exchange of amide protons for deuterium occurs in peptide regions that are solvent accessible and not involved in stable secondary structure. At different time points after incubating the protein in deuterated buffer the protein is enzymatically digested and the peptides analyzed by mass spectrometry. Because of the mass difference between hydrogen and deuterium the exchange causes a shift in the mass of the peptides that can be detected and the relative abundance quantified. The combined improvement in mass spectrometry equipment and the development of more sophisticated and automated analyses have paved the way for an increased use of this technique in recent years. A detailed recount of the expansion of the use of HDX in recent years can be found in the paper by Prirrone *et al.* [60]. Some of the applications of HDX include the mapping of epitopes or protein–protein interactions in general and the study of folding mechanisms and allosteric structural effects. A scheme of the workflow for some of these applications is shown in Figure 4.5.

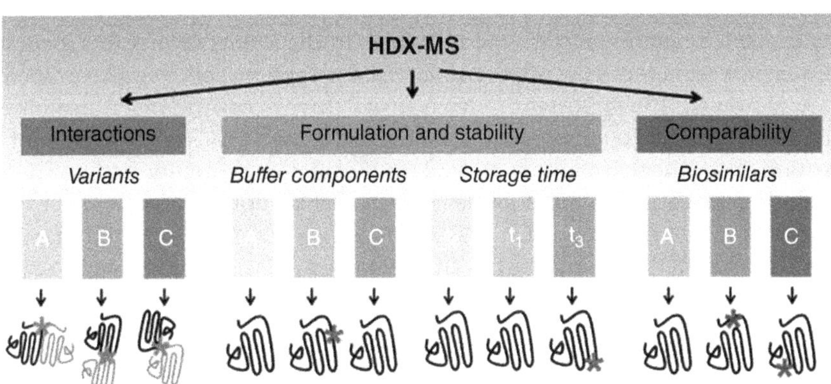

Figure 4.5 Applications of HDX-MS in biopharmaceutical drug development. Left panel: Mapping of protein–drug and protein–protein interaction sites and related conformational changes. Middle panel: Monitoring the conformational response of protein pharmaceutical to different formulations. Right panel: Comparing the structure of therapeutics produced by different processes. Source: Leurs *et al.* [61]. Reproduced with permission of Elsevier.

4.4.1 Applications of HDX

4.4.1.1 Ligand-induced Conformational Changes and Mapping Interaction Sites

The interaction of two proteins change the solvent accessibility of the interacting surfaces and this change can be detected by HDX. It is worth noting that the conformational changes detected by HDX may reflect sites directly involved in binding the other molecule or they may represent sites indirectly affected in an allosteric mode. Therefore, HDX is used to map sites on a protein interacting with another protein [62, 63], with a small molecule [64] or sites allosterically affected by an interaction [65–67]. In addition, it can also be used for studying the dynamics of protein conformational changes [68].

In the field of Ab therapeutics, HDX is emerging as a powerful technique for mapping epitopes [69–73]. In one study, different epitope mapping methodologies were used to characterize the interaction between the monoclonal antibody (mAb) 12C1 and the factor H binding protein (fHbp) from *N. meningitides* with the aim of developing vaccines against the pathogenic bacteria. HDX experiments identified a series of fHbp peptides that showed decreased deuteration in the presence of the Ab (Figure 4.6). As the authors subsequently obtained the structure of the Ab bound to fHbp by X-ray crystallography a direct comparison with the HDX results was possible. Interestingly, the results showed a very tight correlation between the sites identified by HDX and X-ray crystallography [74].

4.4.1.2 Applications in Protein Engineering

As discussed before for DSC, modifications in the sequence of a therapeutic protein are sometimes introduced to increase its potency or drug-like properties. However, it is important to analyze whether or not the mutations have unintended effects on the protein structure. Although DSC can provide information about the thermal stability and the global domain conformation, HDX is able to inform about more subtle changes that do not necessarily affect the global domain structure. HDX was used to gain insights into the different glycosylation patterns observed in an engineered Ab with a single amino acid substitution in the C_H3 domain (Y407). Since the mutation is located in a different protein domain to the one containing the glycosylation site (C_H2 domain), the cause of the modified glycosylation was not obvious. The HDX analysis showed allosteric conformational perturbations in the structure within the C_H2 domain that may explain the differential glycosylation observed in the engineered Ab [75]. HDX has also been used to identify conformational changes observed in a mAb after the removal of glycosylation [76]. Two regions of the IgG1 (residues 236–253 and 292–308) were found to have altered exchange properties upon deglycosylation. These results are consistent with

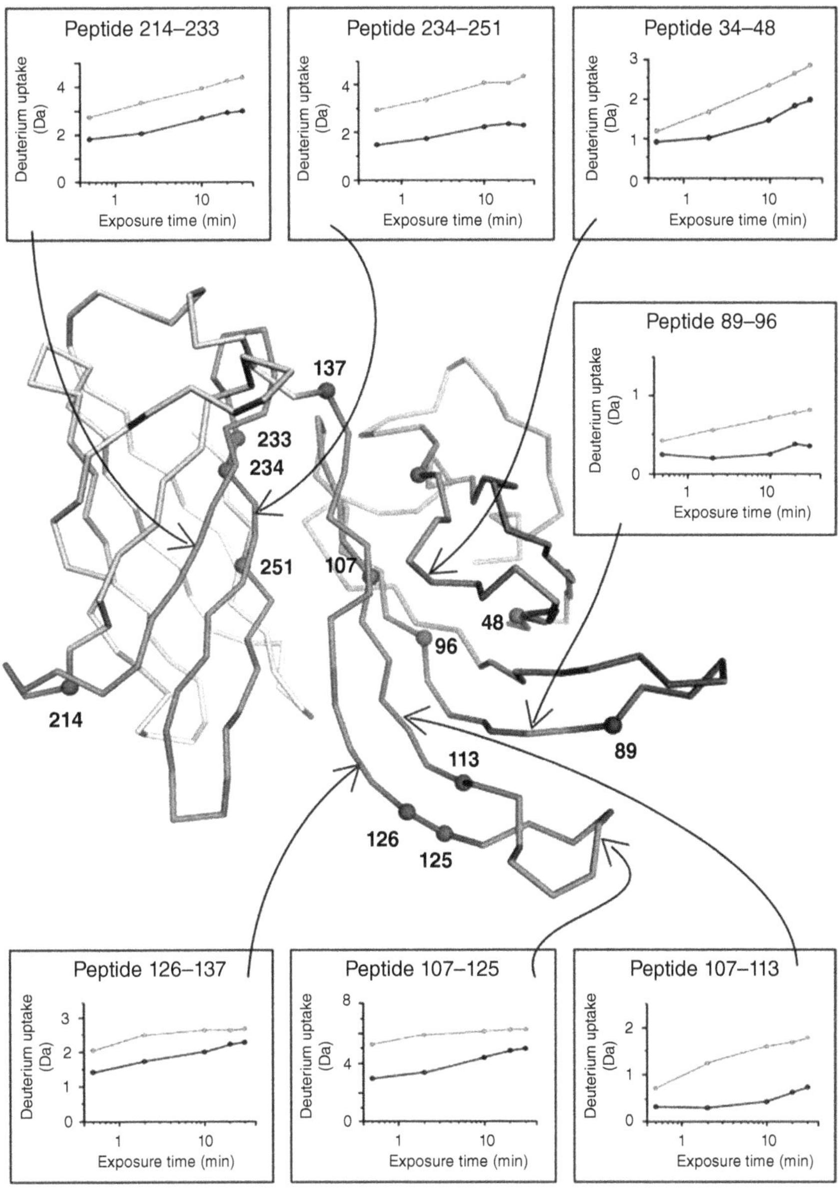

Figure 4.6 Epitope mapping by HDX. The inset show six different overlapping peptides that show reduced HDX over time in the presence of mAb 12C1. The segments of the protein corresponding to the protected peptides are highlighted in red. Source: Malito *et al.* [74]. Reproduced with permission of PNAS. (*See insert for color representation of the figure.*)

the impaired interaction of aglycosylated IgG1 with Fc receptors and recent results identifying residues on those regions as important elements for the interaction with the high-affinity FcRI [77].

One important requirement for most biotherapeutics is that they display an acceptable half-life in circulation as this impacts the dosing schedule. Mutations are sometimes introduced in therapeutic Abs to alter the half-life. An engineered Ab variant containing three mutations (M252Y/S254T/T256E, or YTE) in the second constant domain of the heavy chain domain (C_H2) has been shown to increase its affinity for the neonatal Fc receptor (FcRn) and thus increase the serum half-life [78]. Although the mutant Ab did not show any significant differences to the wild type (WT) in its crystal structure, it had a relatively lower physical stability than the WT [79]. Using HDX Majumdar *et al.* [80] observed the occurrence of subtle or no changes in peptides covering the mutations. However, they did observe increased flexibility in other regions of the C_H2 domain of the mutant Ab and even in other domains of the heavy chain (C_H1, V_H) as well as in the variable domain of the light chain (V_L). The authors noted that increased flexibility or perturbations in a specific region within the C_H2 domain (aa 244–255) has been correlated in many previous studies with decreased thermal stability and increased aggregations propensity [76, 81–83], suggesting perturbations in that area brought about by the YTE mutations, which caused the reduced stability observed in the engineered Ab.

Also related to protein engineering, HDX can serve screening the right constructs for crystallography. As chances of protein crystallization may be reduced in the presence of protein regions with high flexibility or lack of defined structure, HDX may serve as a tool to identify those areas and guide the design of constructs with better probability of producing crystals. An example of this approach is shown in efforts at crystallizing proteins from *Thermotoga maritima* [84, 85] where the use of HDX contributed toward overcoming unsuccessful production of protein crystals with previous constructs.

4.4.1.3 Comparability and Biosimilar Studies

During the different phases of production and commercialization it is critical to ensure that a therapeutic protein retains the desired active conformation. Similarly, the development of a biosimilar requires showing that the proposed new drug has structural properties similar to the original therapeutic, as mentioned before. HDX is emerging as a characterization technique that can provide more detailed information than other methods usually used for this purpose. Examples in the literature include the analysis of interferon 1β [86], the structural characterization of granulocyte colony stimulating factor after PEGylation [87], and a therapeutic IgG [76]. Two detailed studies used HDX to study whether different frequent posttranslational modifications (PTMs) in Abs have an effect on the conformation of the molecule. They found that asparagine deamidation and aspartic acid isomerization had little impact on the structure as revealed by the HDX pattern, whereas methionine oxidation could have global structure effects depending on the position of the oxidized methionine [88, 89]. Because slight changes in the expression and purification conditions can easily impact the extent and nature of PTMs, this type of studies using HD/X or related techniques [90] may become more common in the analysis of biosimilars.

4.4.1.4 Formulation and Aggregation Analysis

Long-term stability is an important aspect for biopharmaceutical products because storage conditions may affect the quality and potency of the drug over time. Thus, different formulation conditions need to be tested to find the ones that keep the function and structure integrity of the product over time. DSC is a useful technique to characterize thermal stability of proteins in different buffers. However, only differences are evident as alterations affect a given domain globally. HDX can be useful in screening formulation compositions, as it is a very sensitive assay that can provide more localized information. In one example, different Hoffmeister salts were tested for their ability to stabilize or destabilize the conformation of a mAb. Thyocyanate was found to have the strongest destabilizing effect, observed as increased deuteration of regions within the C_H2 domain, when compared to formulations containing chloride [81].

In addition to contributing toward finding the best formulation conditions, HDX experiments may help understand different protein aggregation pathways, which ultimately contribute in finding ways of preventing them. Ab aggregation may occur through different mechanisms depending on properties inherent to the protein as well as on environmental factors. Characterization of Ab aggregates induced by either thermal stress or cycles of freeze/thaw was studied using commercial bevacizumab as a model. The study found that aggregates generated by different methods differed in their native-like structure as shown by the HDX pattern [91].

The development of biotherapeutics intended for subcutaneous administration requires high protein concentration formulations, which could lead to aggregation or viscosity problems due to protein–protein interactions. Mapping the sites of protein–protein interactions occurring at high protein concentration could be useful in assisting engineering efforts to eradicate or mitigate aggregation or high viscosity problems. The change in deuterium incorporation in a mAb subjected to high concentrations was studied by Arora *et al.* [92]. They found that the main sites involved in reversible self-association (RSA) in a mAb mapped in the complementary determinant regions (CDRs).

4.4.2 Technical Improvements and Challenges

A persistent limitation of this technique is that some sites involved in the formation of complexes involve side chains of the amino acid residues which are not detected by HDX, and that not all interactions necessarily affect the amide backbone structure. On the other hand, one advantage of this method is that except for proline, it labels amide groups of all amino acid residues. Another limitation of this technique is that obtaining resolution at the single amino acid level is also a challenge. Current methodology allows characterizing the HDX at the peptide level, giving a coarse representation of the structure of the protein. However, ways to increase the resolution for protein conformation based on HDX are being developed by different labs. One approach to gain resolution involves using proteases with varying specificities such that different sets of peptides are generated. By analyzing the overlapping peptides it is possible to have a more precise characterization of the protected or exposed amino acid residues. The challenge of this approach is to generate a set of peptides such that every residue can be found as the overhanging single difference with another peptide. For the positions where this is not possible Englander *et al.* [93] developed a method to extract additional information from the m/z envelope with the help of an informatics program. The evaluation of the single residue resolution accuracy showed results were comparable to results obtained by 2D NMR.

An alternative approach to HDX is hydroxyl radical footprinting. An advantage over HDX is that the covalent modifications introduced by reaction with hydroxyl radicals are stable and not prone to loss during sample preparation, as is the case in HDX. Although hydroxyl radical can react with 14 out of the 20 amino acid side chains, unlike HDX exchange, the rates of modification are different for the different amino acid residues, with the rates varying up to 1000 fold. Another difference with the HDX exchange is that hydroxyl footprinting does not provide information about the backbone structure.

4.5 Nuclear Magnetic Resonance

NMR allows the determination of the structure of biological macromolecules in aqueous solution. It is possible to determine the structure of proteins in a native environment and observe and quantify protein dynamics as well as have the ability to see real-time effects of interactions or varying conditions. The basis of NMR is the measurement of energy state transitions by radiofrequency radiation of the nuclei of atoms with unpaired nuclear spins (i.e., ^{1}H, ^{13}C, ^{15}N). When placed in an external static magnetic field the Zeeman effect [94] induces energetic states that lead to characteristic absorbed frequencies in the MHz range. These are detected as “chemical shifts.” The frequency that a nucleus absorbs is a function of the strength of the external magnetic field and the characteristic gyromagnetic ratio of a given nucleus. It is highly influenced by the surrounding electronic environment. The term “chemical shift” is derived from the shielding effect of the electrons and provides a wealth of information about the nature of the electron cloud shielding the nucleus and thus on covalent and noncovalent interactions [95]. In addition to influencing the specific absorbed frequencies, adjacent atoms may produce multiplets in the detected signal due to the through-bond scalar coupling. The resulting signal patterns are used to inform about the presence of neighboring atoms [96]. Nuclei can also transfer energy through a through-space overall mechanism, an effect known as the nuclear Overhauser effect (NOE) [97]. As the intensity of this effect is dependent on the sixth power of the distance between nuclei, its magnitude can be used to determine interatomic distances and thus calculate structural models. Through the scalar and dipolar coupling a multitude of experiments can be realized that can detect different types of interactions and give rise to a wealth of NMR experiments and resulting data. The process of solving the 3D structure of a protein by NMR, for example, involves two main steps: first, the identification of the chemical shift of the relevant nuclei using through-bond coupled NMR experiments. In the second step, through-space (NOE-based) experiments allow the determination of interatomic distance "restraints." Additional information can be gained by using the angle dependence of scalar couplings to obtain dihedral angles as well as hydrogen bonding patterns through, for example, hydrogen/deuterium exchange experiments. Last but not the least, information on the orientation of a molecule, for example, a protein domain relative to another, can be gained by determination of residual dipolar couplings in orienting media [98]. All these data are fed into distance geometry and simulated annealing software packages that calculate possible structural solutions to a given set of "restraints." In spite of all the additional information that can be collected, structure calculations are increasingly based only on chemical shift information [99]. With a set of restraints and the linear sequence of the protein in question, the structure calculation software will produce sets of spatial conformations that best fit the experimental and empirical data, producing an NMR bundle (Figure 4.7). Whereas X-ray crystallography produces a single "image" of a rigid molecular structure, a structure based on NMR information yields several possible solutions to

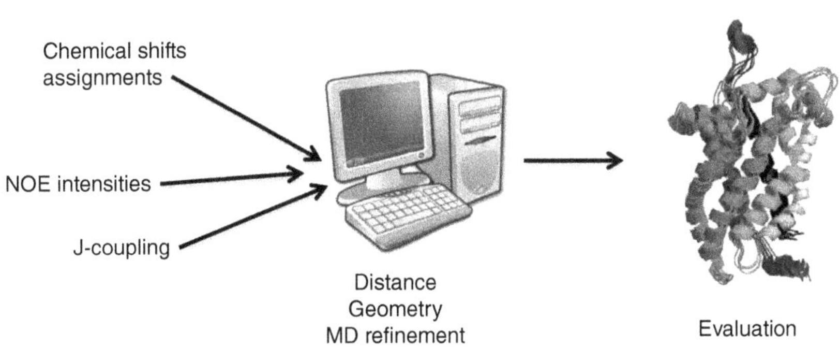

Figure 4.7 Process for determining the 3D structure in a NMR experiment. Experimental parameters are fed into a computer and run through a variety of optimizing programs to generate an ensemble of structures that satisfy the experimental constrains.

a given set of restraints. Thus, structures obtained by NMR, termed the structural ensemble, are usually represented as overlaid traces where the areas where there is poor alignment between the individual structures indicate regions of high flexibility.

NMR has been used to determine the structure of small molecules for several decades. However, its application to proteins has been challenging because of their large size. As the number of atoms increases so does the number of signals that overlap and make the interpretation difficult or not possible. A partial solution to this problem was the introduction of multidimensional NMR spectroscopy using other nuclei such as ^{13}C or ^{15}N. The introduction of 2D NMR spectroscopy partly resolved the problem of overlap by separating signals along two frequency axes. Similarly, 3D NMR introduces a third dimension allowing the analysis of increasingly large macromolecules.

In spite of multidimensional NMR, the use of NMR on proteins has historically been limited to proteins under 30 kDa. Above 30 kDa, the increase in line width that accompanies the decrease in overall tumbling due to the increased molecular weight makes full-fledged *de novo* structure determinations very laborious. TROSY (Transversed Relaxation Optimized Spectroscopy) helps mitigating these effects, allowing the study of larger proteins [100, 101]. This technical improvement together with novel data collection techniques and increased sensitivity can extend the size limit for protein structure determination up to approximately 80 kDa [102].

Experiments measuring the resonance of ^{13}C and ^{15}N usually require the isotopic labeling of proteins by expressing the recombinant proteins in media that contain the appropriately labeled precursors. This is because the natural abundance of ^{13}C and ^{15}N is so low (1.1 and 0.4%, respectively) that it greatly reduces the sensitivity of NMR detection.

4.5.1 Applications of NMR

NMR is a versatile technique used in the discovery phases of developing biotherapeutics to solve molecular structure at the atomic level. NMR is particularly useful in situations where proteins are resistant to crystallization, for instance, for the characterization of flexible polypeptides or protein complexes with very low affinity, as well as for studying dynamic processes. Because of its sensitivity and versatility, NMR has been widely used for studying target–ligand interactions. The wide range of protein concentrations that can be used in NMR experiments allow for detection of very weak binding interactions (in the mM range) [103]. In addition, indirect allosteric effects can also be observed and NMR titration experiments can be used to determine binding constants.

4.5.1.1 Flexible Proteins

Given the structural plasticity of poorly ordered proteins or proteins involved in dynamic processes, X-ray crystallography is not a viable option. It is not surprising then, that small, flexible proteins such as chemokines, growth factors or hormones, antimicrobial peptides, immunity proteins, and viral proteins are specially represented in the list of pharmaceutical proteins whose structures have been solved by NMR. A series of recent technical improvements have significantly contributed to advances in the study of intrinsically disordered proteins [104]. An intense area of growth within the intrinsically disordered proteins has been the study of proteins involved in amyloidopathies. The formation of fibrils by a number of structurally disordered proteins is implicated is several pathogenic conditions and subject of structural studies by NMR (see Section 4.5.1.4).

4.5.1.2 Mapping Protein–Protein Interactions

There are different NMR experiments that can be performed to study the interaction between two proteins in solution [105]; however, the most widely used is chemical shift perturbation mapping. It basically consists of a 2D heteronuclear single quantum correlation (HSQC) fingerprinting of one protein isotopically labeled in the absence or presence of the second protein or ligand. Because of the changing environment, those atoms that are isotopically labeled and in proximity to the second protein change their chemical shift, which can be identified comparing the spectra in the bound or unbound states. A change in the chemical shift represents a change in the atomic environment; this could be brought about by a direct contact between the two proteins or, alternatively, due to allosteric structural rearrangements or subtle changes in hydrogen bonding patterns. Additional, dedicated NMR experiments can be performed to distinguish between the two possibilities [106]. As in the case of other NMR applications, the size of the antigen is important because the number of signals for proteins over 300 amino acid residues becomes difficult to resolve and the intensity of the NMR signal decreases with increasing antigen size. Different partial or segmental labeling strategies can be used to solve this problem [107].

Some important protein–protein interactions in biology are weak, for example, cell surface receptors CD42 and CD58 have a k_d of 90 μM and therefore are not readily amenable to X-ray crystallography. One example of a weak interaction mapped by NMR is that of the T-cell receptor (TCR) with the peptide major histocompatibility complex (MHC). The low affinity between the MHC and the associated peptide (k_d >1 μM) poses significant hurdles for obtaining crystals. Moreover, answering questions regarding the docking orientation between TCRs with different variable regions to the same MHC–peptide complex may not require ultra-high resolution results. Thus, Varani *et al.* used in-solution NMR experiments to "paint" the surface of the MHC–peptide footprinted by the TCR [108]. Comparison of the NMR footprint with crystallographic data for the same complex showed a very close correspondence between the two methods [106].

4.5.1.3 Epitope Mapping

Although initially most epitopes mapped by NMR involved peptide antigens, examples of full-length antigens are also available: synuclein epitope [43], dengue virus epitope [109], epitope on rotavirus protein VP8*, epitope of factor H binding protein of Neisseria [110]. For further practical considerations on this topic readers are referred to Chapter 5 of this book and elsewhere [106].

4.5.1.4 Protein Dynamics

Proteins are dynamic molecules that experience different degrees of fluctuations in their conformation. One important difference between X-ray crystallography and NMR is that while crystallization provides information about the single structure of the formed crystal, in-solution NMR can capture the existence of different conformational subpopulations and the transitions between them. NMR is particularly useful for studying protein folding as it can characterize unstructured proteins at atomic resolution with minimal interference. Information about the kinetics of the process of folding/unfolding, relative populations of different folding states, and the structure of folding intermediates can be obtained.

One specific area within protein dynamics is protein folding. The misfolding of proteins into a toxic conformation is proposed as the cause of some neurodegeneration diseases like Alzheimer's and Parkinson's. For example, the formation of plaques of insoluble α-synuclein has been implicated in Parkinson's and amyloid in Alzheimer's. The unfolded states of several proteins of pharmaceutical interest have been described: alpha- and beta-synuclein [111–113]. Also, the dynamic process of forming supramolecular structures like protofibrils and fibrils has also been studied by NMR. Fawzi *et al.* used solution NMR to study the interaction between monomeric amyloid and heterogeneous protofibrils. They observed that the dynamics at the C-terminal hydrophobic region of the protein governs the rate of protofibril formation [114]. The interaction between Amyloid and HSP60 is required for translocation to the mitochondria and plays a role in Alzheimer's disease. The study of the association between Amyloid- and a GroEL, a homolog to HSP60, showed that binding involves two predominantly hydrophobic sites in the protein, [115], which may have implications in developing therapeutics for Alzheimer's disease.

4.5.1.5 Protein Conjugates and Complexes

Protein–polymer conjugates are being used in different therapeutic areas to increase half-life, reduce immunogenicity, and increase stability [116]. While this approach creates new molecular entities with improved properties, it also creates challenges for their thorough physicochemical characterization. Among many other techniques, NMR can be used for routine characterization of the polymers, but more specialized techniques like diffusion-ordered spectroscopy (DOSY) are used for the characterization of their size and morphology.

DOSY is based on a pulse-field gradient spin-echo NMR where the signal decay from the different components provides information about the diffusion coefficients, which in turn can be used to calculate Rh. DOSY has been used to characterize the different components in a dextrin–trypsin conjugation mixture in terms of composition and size [117]. Similarly, it has been used for the characterization of PEGylated proteins [118]. Polymer therapeutics and nanomedicine are areas of rapid expansion where DOSY NMR can become a powerful analytical technique.

4.5.1.6 Posttranslational Modifications

Some PTMs greatly influence the activity and stability of therapeutic proteins [119] and may also affect their immunogenic potential.

Glycosylation of Abs plays an important role in the interaction with Fc receptors that mediate effector functions. In addition, terminal addition of sialic acid may affect their half-life in circulation. Thus, better understanding of the conformational changes of the glycosidic moiety in IgGs may help design better therapeutics. In a study using spin relaxation experiments, it was shown that contrary to the view that carbohydrates are buried in a cavity of the IgG structure, the two terminal branches are considerably flexible [120]. This type of analyses may also help understand the interaction between carbohydrates in the IgGs and their receptors.

4.5.1.7 Biosimilars

A few hurdles have prevented the use of NMR in the production phase of the biopharmaceutical industry: the need for isotopically labeling the protein, the limitation to small (<25 kDa) proteins, and the fact that NMR is not a high-throughput method. As some of these limitations are being overcome, early examples of using NMR for process control and characterization are becoming available; we discuss some of them here.

Several low-resolution techniques have been proposed for the analysis of the structure of biosimilars. One limitation to the use of NMR for the characterization of biotherapeutics is that they cannot be isotopically labeled as the process would differ from the one used in manufacturing. However, the implementation of cryogenic probe technology and increase in spectrometer field strength has made it possible to analyze proteins at natural isotopes abundance and low concentrations. Although other biophysical techniques with lower resolution have been studied more intensely for their application to the analysis of the structure of biopharmaceuticals, some pioneer examples exist. A 2D NMR technique using isotopes at natural abundance has been used to characterize the structure of the recombinant human granulocyte macrophage stimulating factor (rhGM-CSF) [121] produced in two different hosts, *Escherichia coli* and *Streptomyces lividans*. The same method has also been applied to the study of the effects of pH and formulation on the structure of interferon α2 [122] and rhGM-CSF [123]. This method, however, is limited to small therapeutic proteins. An alternative 2D NMR procedure has been described recently for fingerprinting larger proteins such as a full-length IgG. The method is 2D ^{13}C–^{1}H NMR methyl fingerprinting. Because methyl groups are well distributed throughout the molecule and are sensitive to changes in their environment they are good reporters of protein conformation. This approach was used to compare the spectra of insulin from several commercial sources, including Actrapid® from Novo Nordisk and Umuline® and Humalog® from Eli Lilly [124]. While the fingerprints for Umuline and Actrapid were almost identical, the profiles for Umuline and Humalog showed many significant differences consistent with the fact that there is an inversion in two consecutive residues in their sequences (Figure 4.8). This example highlights the sensitivity of NMR to detect subtle changes in the structure and therefore the potential for its use for quality control purposes. The same method was recently used in combination with a nonuniform sampling approach and rapid acquisition method to obtain a 2D methyl fingerprint of an Ab [125]. The data collection strategy significantly reduces the time for this step to 30 min, which can help in implementing the assay to a process control setting. Analysis of the full-length Ab spectra showed that ~61% of the expected signals could be resolved while the remainder peaks showed some degree of overlap. To increase the resolution the authors also analyzed the profile of the Fab and Fc fragments after papain digestion. The combined Fab and Fc fractions produced a similar profile to the intact Ab, suggesting that the fragments could provide a substitute to the full-length Ab.

With the same goal of using NMR to characterize the HOS of biotherapeutics, other groups are pursuing alternative strategies. For example, Poppe *et al.* [126] proposed using a 1D ^{1}H NMR method that exploits differences between Abs and excipients in the diffusion coefficient to yield a profile with high resolution [126]. The authors also showed data supporting a better selectivity of this method over 2D NMR.

NMR is a very versatile and powerful technique than can be used for an increasing number of applications in the biopharmaceutical industry. Although its use in process development is not yet very broad, the increase in sensitivity and the development of novel data collection and analysis methods make its use more accessible and probably will soon be more widespread.

4.6 Concluding Remarks

Conformation is a critical feature of therapeutic protein that needs to be controlled throughout all stages of research and development. Different methods for assessing protein folding are available offering different benefits that may fit various specific needs, such as low sample requirement, high throughput, or single amino acid resolution. Accordingly, the use of the different methods

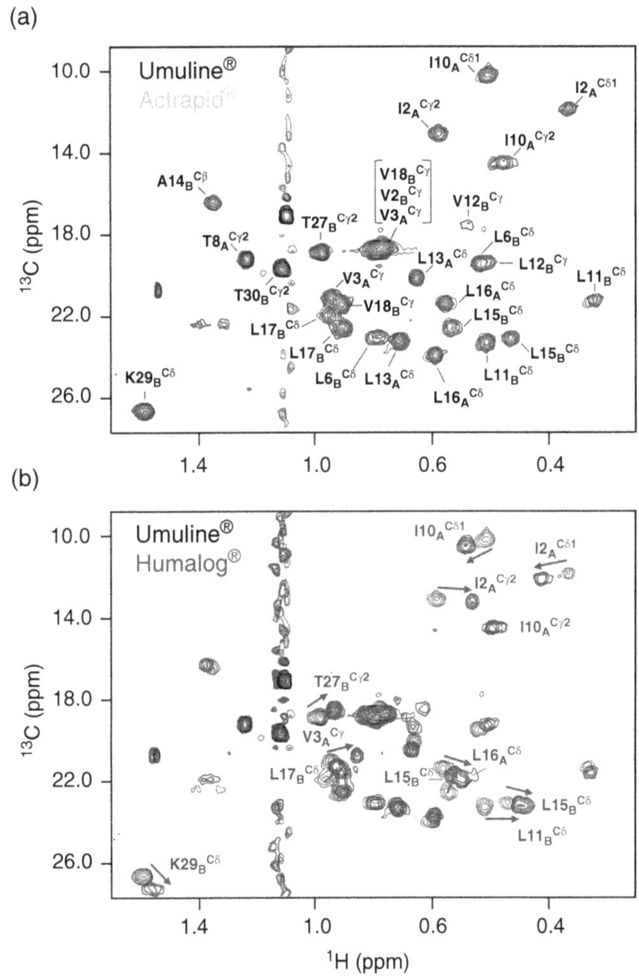

Figure 4.8 2D ^{1}H–^{13}C methyl SOFAST-HMQC NMR spectra for three different commercial insulin products. Top panel: overlay of Umuline® and Actrapid® showing very similar profile. Bottom panel: Overlay of Umuline and Humalog® spectra with several chemical shift changes indicated with arrows. Source: Quinternet *et al.* [124]. Reproduced with permission of Elsevier.

varies depending on the stage of the project and on the settings. In the discovery stages there is more flexibility and variability in regard to what assays are used more frequently because of a wider diversity of question that needs to be answered. Is my recombinant protein folded like the native? Does a given mutation introduce colloidal instability? What are the residues critical for a given interaction? Usually defining precisely the scientific query helps narrowing down the assay options to those providing the level of resolution required. Whenever it is possible the use of orthogonal methods is advised since they provide information on the same aspects of a given molecule from different physical perspectives. Once the development phase begins, it is common to sort through large numbers of clones or conditions; in these cases assays compatible with high-throughput processing are preferred. Once the number of conditions or clones is reduced, additional complementary assays are performed to validate the original findings.

In summary, throughout the entire process of developing a biotherapeutic different situations require characterizing different aspects of the protein HOS. Understanding the structural details of molecules as complex as large proteins often requires the combination of different biophysical approaches, but it is key to produce safe therapeutic proteins. Thus it is paramount to have access to a diverse protein characterization toolbox. The rapid technical improvements are allowing some of the techniques to be used in a higher-throughput way and also overcoming some of the limitations previously limiting their broader use. These changes will likely facilitate the implementation of more assays for assessing protein conformation during the development of biotherapeutics, allowing a deeper, more detailed structural characterization.

References

1 Narhi, L. O. (Ed.) (2013) *Biophysics for therapeutic protein development*, Vol. 4, Springer, New York.
2 Houde, D. B. S. (Ed.) (2015) *Biophysical characterization of proteins in developing biopharmaceuticals*, Elsevier, Amsterdam.
3 Fasman, G.D. (Ed.) (1996) *Circular dichroism and the conformational analysis of biomolecules*, Springer Science, New York.
4 Kelly, S. M., Jess, T. J., and Price, N. C. (2005) How to study proteins by circular dichroism. *Biochimica et biophysica acta*, 1751 (2), 119–139.
5 Nagatomo, S., Nagai, M., Ogura, T., and Kitagawa, T. (2013) Near-UV circular dichroism and UV resonance Raman spectra of tryptophan residues as a structural marker of proteins. *The journal of physical chemistry. B*, 117 (32), 9343–9353.
6 Gasymov, O. K., Abduragimov, A. R., and Glasgow, B. J. (2014) Probing tertiary structure of proteins using single Trp mutations with circular dichroism at low temperature. *The journal of physical chemistry. B*, 118 (4), 986–995.
7 Micsonai, A., Wien, F., Kernya, L., Lee, Y. H., Goto, Y., Refregiers, M., and Kardos, J. (2015) Accurate secondary structure prediction and fold recognition for circular dichroism spectroscopy. *Proceedings of the National Academy of Sciences of the United States of America*, 112 (24), E3095–E3103.

8 Martin, S. R., and Schilstra, M. J. (2008) Circular dichroism and its application to the study of biomolecules. *Methods in cell biology*, 84, 263–293.
9 Melberg, S. G., and Johnson, W. C., Jr. (1990) Changes in secondary structure follow the dissociation of human insulin hexamers: a circular dichroism study. *Proteins*, 8 (3), 280–286.
10 Yong, Z., Yingjie, D., Ming, L., Craig, D. Q., and Zhengqiang, L. (2009) A spectroscopic investigation into the interaction between bile salts and insulin in alkaline aqueous solution. *Journal of colloid and interface science*, 337 (2), 322–331.
11 Nystrom, S., and Hammarstrom, P. (2015) Generic amyloidogenicity of mammalian prion proteins from species susceptible and resistant to prions. *Scientific reports*, 5, 10101.
12 Deechongkit, S., Aoki, K. H., Park, S. S., and Kerwin, B. A. (2006) Biophysical comparability of the same protein from different manufacturers: a case study using Epoetin alfa from Epogen and Eprex. *Journal of pharmaceutical sciences*, 95 (9), 1931–1943.
13 Grace, M., Youngster, S., Gitlin, G., Sydor, W., Xie, L., Westreich, L., Jacobs, S., Brassard, D., Bausch, J., and Bordens, R. (2001) Structural and biologic characterization of pegylated recombinant IFN-alpha2b. *Journal of interferon & cytokine research: the official journal of the International Society for Interferon and Cytokine Research*, 21 (12), 1103–1115.
14 Wang, Y. S., Youngster, S., Grace, M., Bausch, J., Bordens, R., and Wyss, D. F. (2002) Structural and biological characterization of pegylated recombinant interferon alpha-2b and its therapeutic implications. *Advanced drug delivery reviews*, 54 (4), 547–570.
15 Plesner, B., Westh, P., and Nielsen, A. D. (2011) Biophysical characterisation of GlycoPEGylated recombinant human factor VIIa. *International journal of pharmaceutics*, 406 (1–2), 62–68.
16 da Silva Freitas, D., Mero, A., and Pasut, G. (2013) Chemical and enzymatic site specific PEGylation of hGH. *Bioconjugate chemistry*, 24 (3), 456–463.
17 Ji, J. A., Liu, J., Shire, S. J., Kamerzell, T. J., Hong, S., Billeci, K., Shen, Y., and Wang, Y. J. (2010) Characteristics of rhVEGF release from topical hydrogel formulations. *Pharmaceutical research*, 27 (4), 644–654.
18 Zhang, N., Li, J., Jiang, W., Ren, C., Li, J., Xin, J., and Li, K. (2010) Effective protection and controlled release of insulin by cationic beta-cyclodextrin polymers from alginate/chitosan nanoparticles. *International journal of pharmaceutics*, 393 (1–2), 212–218.
19 Bhambhani, A., Kissmann, J. M., Joshi, S. B., Volkin, D. B., Kashi, R. S., and Middaugh, C. R. (2012) Formulation design and high-throughput excipient selection based on structural integrity and conformational stability of dilute and highly concentrated IgG1 monoclonal antibody solutions. *Journal of pharmaceutical sciences*, 101 (3), 1120–1135.

20 Cheng, W., Joshi, S. B., He, F., Brems, D. N., He, B., Kerwin, B. A., Volkin, D. B., and Middaugh, C. R. (2012) Comparison of high-throughput biophysical methods to identify stabilizing excipients for a model IgG2 monoclonal antibody: conformational stability and kinetic aggregation measurements. *Journal of pharmaceutical sciences*, 101 (5), 1701–1720.

21 Alsenaidy, M. A., Kim, J. H., Majumdar, R., Weis, D. D., Joshi, S. B., Tolbert, T. J., Middaugh, C. R., and Volkin, D. B. (2013) High-throughput biophysical analysis and data visualization of conformational stability of an IgG1 monoclonal antibody after deglycosylation. *Journal of pharmaceutical sciences*, 102 (11), 3942–3956.

22 Sakurai, K., Nakahata, R., Lee, Y. H., Kardos, J., Ikegami, T., and Goto, Y. (2015) Effects of a reduced disulfide bond on aggregation properties of the human IgG1 CH3 domain. *Biochimica et biophysica acta*, 1854 (10), 1526–1535.

23 Yageta, S., Lauer, T. M., Trout, B. L., and Honda, S. (2015) Conformational and colloidal stabilities of isolated constant domains of human immunoglobulin G and their impact on antibody aggregation under acidic conditions. *Molecular pharmaceutics*, 12 (5), 1443–1455.

24 Beckley, N. S., Lazzareschi, K. P., Chih, H. W., Sharma, V. K., and Flores, H. L. (2013) Investigation into temperature-induced aggregation of an antibody drug conjugate. *Bioconjugate chemistry*, 24 (10), 1674–1683.

25 Levy, M. J., Gucinski, A. C., Sommers, C. D., Ghasriani, H., Wang, B., Keire, D. A., and Boyne, M. T., 2nd. (2014) Analytical techniques and bioactivity assays to compare the structure and function of filgrastim (granulocyte-colony stimulating factor) therapeutics from different manufacturers. *Analytical and bioanalytical chemistry*, 406 (26), 6559–6567.

26 Wallace, B. A., and Janes, R. W. (2001) Synchrotron radiation circular dichroism spectroscopy of proteins: secondary structure, fold recognition and structural genomics. *Current opinion in chemical biology*, 5 (5), 567–571.

27 Longo, E., Hussain, R., and Siligardi, G. (2015) Application of circular dichroism and magnetic circular dichroism for assessing biopharmaceuticals formulations photo-stability and small ligands binding properties. *International journal of pharmaceutics*, 480 (1–2), 84–91.

28 Jelesarov, I., and Bosshard, H. R. (1999) Isothermal titration calorimetry and differential scanning calorimetry as complementary tools to investigate the energetics of biomolecular recognition. *Journal of molecular recognition: JMR*, 12 (1), 3–18.

29 Freyer, M. W., and Lewis, E. A. (2008) Isothermal titration calorimetry: experimental design, data analysis, and probing macromolecule/ligand binding and kinetic interactions. *Methods in cell biology*, 84 (1), 79–113.

30 Privalov, P. L. (2012) *Microcalorimetry of macromolecules: the physical basis of biological structures*, John Wiley & Sons, Inc., Hoboken.

31 Bruylants, G., Wouters, J., and Michaux, C. (2005) Differential scanning calorimetry in life science: thermodynamics, stability, molecular recognition and application in drug design. *Current medicinal chemistry*, 12 (17), 2011–2020.
32 Velazquez-Campoy, A., and Freire, E. (2006) Isothermal titration calorimetry to determine association constants for high-affinity ligands. *Nature protocols*, 1 (1), 186–191.
33 Gunasekaran, K., Pentony, M., Shen, M., Garrett, L., Forte, C., Woodward, A., Ng, S. B., Born, T., Retter, M., Manchulenko, K., Sweet, H., Foltz, I. N., Wittekind, M., and Yan, W. (2010) Enhancing antibody Fc heterodimer formation through electrostatic steering effects: applications to bispecific molecules and monovalent IgG. *The Journal of biological chemistry*, 285 (25), 19637–19646.
34 Choi, H. J., Kim, Y. J., Lee, S., and Kim, Y. S. (2013) A heterodimeric Fc-based bispecific antibody simultaneously targeting VEGFR-2 and Met exhibits potent antitumor activity. *Molecular cancer therapeutics*, 12 (12), 2748–2759.
35 Mazor, Y., Oganesyan, V., Yang, C., Hansen, A., Wang, J., Liu, H., Sachsenmeier, K., Carlson, M., Gadre, D. V., Borrok, M. J., Yu, X. Q., Dall'Acqua, W., Wu, H., and Chowdhury, P. S. (2015) Improving target cell specificity using a novel monovalent bispecific IgG design. *mAbs*, 7 (2), 377–389.
36 Lewis, S. M., Wu, X., Pustilnik, A., Sereno, A., Huang, F., Rick, H. L., Guntas, G., Leaver-Fay, A., Smith, E. M., Ho, C., Hansen-Estruch, C., Chamberlain, A. K., Truhlar, S. M., Conner, E. M., Atwell, S., Kuhlman, B., and Demarest, S. J. (2014) Generation of bispecific IgG antibodies by structure-based design of an orthogonal Fab interface. *Nature biotechnology*, 32 (2), 191–198.
37 Muda, M., Gross, A. W., Dawson, J. P., He, C., Kurosawa, E., Schweickhardt, R., Dugas, M., Soloviev, M., Bernhardt, A., Fischer, D., Wesolowski, J. S., Kelton, C., Neuteboom, B., and Hock, B. (2011) Therapeutic assessment of SEED: a new engineered antibody platform designed to generate mono- and bispecific antibodies. *Protein engineering, design & selection: PEDS*, 24 (5), 447–454.
38 Rippa, V., Santini, L., Lo Surdo, P., Cantini, F., Veggi, D., Gentile, M. A., Grassi, E., Iannello, G., Brunelli, B., Ferlicca, F., Palmieri, E., Pallaoro, M., Arico, B., Banci, L., Pizza, M., and Scarselli, M. (2015) Molecular engineering of Ghfp, the gonococcal orthologue of *Neisseria meningitidis* factor H binding protein. *Clinical and vaccine immunology: CVI*, 22 (7), 769–777.
39 Crespillo, S., Camara-Artigas, A., Casares, S., Morel, B., Cobos, E. S., Mateo, P. L., Mouz, N., Martin, C. E., Roger, M. G., El Habib, R., Su, B., Moog, C., and Conejero-Lara, F. (2014) Single-chain protein mimetics of the N-terminal heptad-repeat region of gp41 with potential as anti-HIV-1 drugs. *Proceedings of the National Academy of Sciences of the United States of America*, 111 (51), 18207–18212.

40 Brown, A. (2009) Analysis of cooperativity by isothermal titration calorimetry. *International journal of molecular sciences*, 10 (8), 3457–3477.

41 Banerjee, Y., Lakshminarayanan, R., Vivekanandan, S., Anand, G. S., Valiyaveettil, S., and Kini, R. M. (2007) Biophysical characterization of anticoagulant hemextin AB complex from the venom of snake *Hemachatus haemachatus*. *Biophysical journal*, 93 (11), 3963–3976.

42 Elegheert, J., Bracke, N., Pouliot, P., Gutsche, I., Shkumatov, A. V., Tarbouriech, N., Verstraete, K., Bekaert, A., Burmeister, W. P., Svergun, D. I., Lambrecht, B. N., Vergauwen, B., and Savvides, S. N. (2012) Allosteric competitive inactivation of hematopoietic CSF-1 signaling by the viral decoy receptor BARF1. *Nature structural & molecular biology*, 19 (9), 938–947.

43 Guilliams, T., El-Turk, F., Buell, A. K., O'Day, E. M., Aprile, F. A., Esbjorner, E. K., Vendruscolo, M., Cremades, N., Pardon, E., Wyns, L., Welland, M. E., Steyaert, J., Christodoulou, J., Dobson, C. M., and De Genst, E. (2013) Nanobodies raised against monomeric alpha-synuclein distinguish between fibrils at different maturation stages. *Journal of molecular biology*, 425 (14), 2397–2411.

44 Nakayama, T., Mizohata, E., Yamashita, T., Nagatoishi, S., Nakakido, M., Iwanari, H., Mochizuki, Y., Kado, Y., Yokota, Y., Satoh, R., Tsumoto, K., Fujitani, H., Kodama, T., Hamakubo, T., and Inoue, T. (2015) Structural features of interfacial tyrosine residue in ROBO1 fibronectin domain-antibody complex: crystallographic, thermodynamic, and molecular dynamic analyses. *Protein science: a publication of the Protein Society*, 24 (3), 328–340.

45 Lee, S. C., Park, K., Han, J., Lee, J. J., Kim, H. J., Hong, S., Heu, W., Kim, Y. J., Ha, J. S., Lee, S. G., Cheong, H. K., Jeon, Y. H., Kim, D., and Kim, H. S. (2012) Design of a binding scaffold based on variable lymphocyte receptors of jawless vertebrates by module engineering. *Proceedings of the national academy of sciences of the United States of America*, 109 (9), 3299–3304.

46 Lee, J. J., Kim, H. J., Yang, C. S., Kyeong, H. H., Choi, J. M., Hwang, D. E., Yuk, J. M., Park, K., Kim, Y. J., Lee, S. G., Kim, D., Jo, E. K., Cheong, H. K., and Kim, H. S. (2014) A high-affinity protein binder that blocks the IL-6/STAT3 signaling pathway effectively suppresses non-small cell lung cancer. *Molecular therapy: the journal of the American Society of Gene Therapy*, 22 (7), 1254–1265.

47 Correa, A., Pacheco, S., Mechaly, A. E., Obal, G., Behar, G., Mouratou, B., Oppezzo, P., Alzari, P. M., and Pecorari, F. (2014) Potent and specific inhibition of glycosidases by small artificial binding proteins (affitins). *PLoS one*, 9 (5), e97438.

48 Castoldi, R., Jucknischke, U., Pradel, L. P., Arnold, E., Klein, C., Scheiblich, S., Niederfellner, G., and Sustmann, C. (2012) Molecular characterization of novel trispecific ErbB-cMet-IGF1R antibodies and their antigen-binding properties. *Protein engineering, design & selection: PEDS*, 25 (10), 551–559.

49 Palm, T., Esfandiary, R., and Gandhi, R. (2011) The effect of PEGylation on the stability of small therapeutic proteins. *Pharmaceutical development and technology*, 16 (5), 441–448.

50 Rawat, S., Kohli, N., Suri, C. R., and Sahoo, D. K. (2012) Molecular mechanism of improved structural integrity of protein in polymer based microsphere delivery system. *Molecular pharmaceutics*, 9 (9), 2403–2414.

51 Remmele, R. L., Jr., Nightlinger, N. S., Srinivasan, S., and Gombotz, W. R. (1998) Interleukin-1 receptor (IL-1R) liquid formulation development using differential scanning calorimetry. *Pharmaceutical research*, 15 (2), 200–208.

52 Kamerzell, T. J., Esfandiary, R., Joshi, S. B., Middaugh, C. R., and Volkin, D. B. (2011) Protein-excipient interactions: mechanisms and biophysical characterization applied to protein formulation development. *Advanced drug delivery reviews*, 63 (13), 1118–1159.

53 Khan, T. A., Mahler, H. C., and Kishore, R. S. (2015) Key interactions of surfactants in therapeutic protein formulations: a review. *European journal of pharmaceutics and biopharmaceutics: official journal of Arbeitsgemeinschaft fur Pharmazeutische Verfahrenstechnik e.V.*, 97 (Pt A), 60–67.

54 Garidel, P., Hoffmann, C., and Blume, A. (2009) A thermodynamic analysis of the binding interaction between polysorbate 20 and 80 with human serum albumins and immunoglobulins: a contribution to understand colloidal protein stabilisation. *Biophysical chemistry*, 143 (1–2), 70–78.

55 Niedziela-Majka, A., Kan, E., Weissburg, P., Mehra, U., Sellers, S., and Sakowicz, R. (2015) High-throughput screening of formulations to optimize the thermal stability of a therapeutic monoclonal antibody. *Journal of biomolecular screening*, 20 (4), 552–559.

56 Schaefer, J. V., and Pluckthun, A. (2012) Transfer of engineered biophysical properties between different antibody formats and expression systems. *Protein engineering, design & selection: PEDS*, 25 (10), 485–506.

57 Lang, B. E., and Cole, K. D. (2015) Unfolding properties of recombinant human serum albumin products are due to bioprocessing steps. *Biotechnology progress*, 31 (1), 62–69.

58 Visser, J., Feuerstein, I., Stangler, T., Schmiederer, T., Fritsch, C., and Schiestl, M. (2013) Physicochemical and functional comparability between the proposed biosimilar rituximab GP2013 and originator rituximab. *BioDrugs: clinical immunotherapeutics, biopharmaceuticals and gene therapy*, 27 (5), 495–507.

59 Lopez-Morales, C. A., Miranda-Hernandez, M. P., Juarez-Bayardo, L. C., Ramirez-Ibanez, N. D., Romero-Diaz, A. J., Pina-Lara, N., Campos-Garcia, V. R., Perez, N. O., Flores-Ortiz, L. F., and Medina-Rivero, E. (2015) Physicochemical and biological characterization of a biosimilar trastuzumab. *BioMed research international*, 2015,427235.

60 Pirrone, G. F., Iacob, R. E., and Engen, J. R. (2015) Applications of hydrogen/deuterium exchange MS from 2012 to 2014. *Analytical chemistry*, 87 (1), 99–118.

61 Leurs, U., Mistarz, U. H., and Rand, K. D. (2015) Getting to the core of protein pharmaceuticals—comprehensive structure analysis by mass spectrometry. *European journal of pharmaceutics and biopharmaceutics: official journal of Arbeitsgemeinschaft fur Pharmazeutische Verfahrenstechnik e.V.*, 93, 95–109.

62 Guttman, M., Garcia, N. K., Cupo, A., Matsui, T., Julien, J. P., Sanders, R. W., Wilson, I. A., Moore, J. P., and Lee, K. K. (2014) CD4-induced activation in a soluble HIV-1 Env trimer. *Structure*, 22 (7), 974–984.
63 Wei, H., Mo, J., Tao, L., Russell, R. J., Tymiak, A. A., Chen, G., Iacob, R. E., and Engen, J. R. (2014) Hydrogen/deuterium exchange mass spectrometry for probing higher order structure of protein therapeutics: methodology and applications. *Drug discovery today*, 19 (1), 95–102.
64 Hernychova, L., Man, P., Verma, C., Nicholson, J., Sharma, C. A., Ruckova, E., Teo, J. Y., Ball, K., Vojtesek, B., and Hupp, T. R. (2013) Identification of a second Nutlin-3 responsive interaction site in the N-terminal domain of MDM2 using hydrogen/deuterium exchange mass spectrometry. *Proteomics*, 13 (16), 2512–2525.
65 Resetca, D., Haftchenary, S., Gunning, P. T., and Wilson, D. J. (2014) Changes in signal transducer and activator of transcription 3 (STAT3) dynamics induced by complexation with pharmacological inhibitors of Src homology 2 (SH2) domain dimerization. *The Journal of biological chemistry*, 289 (47), 32538–32547.
66 Landgraf, R. R., Goswami, D., Rajamohan, F., Harris, M. S., Calabrese, M. F., Hoth, L. R., Magyar, R., Pascal, B. D., Chalmers, M. J., Busby, S. A., Kurumbail, R. G., and Griffin, P. R. (2013) Activation of AMP-activated protein kinase revealed by hydrogen/deuterium exchange mass spectrometry. *Structure*, 21 (11), 1942–1953.
67 Karageorgos, I., Wales, T. E., Janero, D. R., Zvonok, N., Vemuri, V. K., Engen, J. R., and Makriyannis, A. (2013) Active-site inhibitors modulate the dynamic properties of human monoacylglycerol lipase: a hydrogen exchange mass spectrometry study. *Biochemistry*, 52 (29), 5016–5026.
68 Shan, Y., Arkhipov, A., Kim, E. T., Pan, A. C., and Shaw, D. E. (2013) Transitions to catalytically inactive conformations in EGFR kinase. *Proceedings of the national academy of sciences of the United States of America*, 110 (18), 7270–7275.
69 Pandit, D., Tuske, S. J., Coales, S. J., E, S. Y., Liu, A., Lee, J. E., Morrow, J. A., Nemeth, J. F., and Hamuro, Y. (2012) Mapping of discontinuous conformational epitopes by amide hydrogen/deuterium exchange mass spectrometry and computational docking. *Journal of molecular recognition: JMR*, 25 (3), 114–124.
70 Lu, J., Witcher, D. R., White, M. A., Wang, X., Huang, L., Rathnachalam, R., Beals, J. M., and Kuhstoss, S. (2005) IL-1beta epitope mapping using site-directed mutagenesis and hydrogen-deuterium exchange mass spectrometry analysis. *Biochemistry*, 44 (33), 11106–11114.
71 Willison, L. N., Zhang, Q., Su, M., Teuber, S. S., Sathe, S. K., and Roux, K. H. (2013) Conformational epitope mapping of Pru du 6, a major allergen from almond nut. *Molecular immunology*, 55 (3–4), 253–263.
72 Zhang, Q., Willison, L. N., Tripathi, P., Sathe, S. K., Roux, K. H., Emmett, M. R., Blakney, G. T., Zhang, H. M., and Marshall, A. G. (2011) Epitope mapping of

a 95 kDa antigen in complex with antibody by solution-phase amide backbone hydrogen/deuterium exchange monitored by Fourier transform ion cyclotron resonance mass spectrometry. *Analytical chemistry*, 83 (18), 7129–7136.

73 Hao, G., Wesolowski, J. S., Jiang, X., Lauder, S., and Sood, V. D. (2015) Epitope characterization of an anti-PD-L1 antibody using orthogonal approaches. *Journal of molecular recognition: JMR*, 28 (4), 269–276.

74 Malito, E., Faleri, A., Lo Surdo, P., Veggi, D., Maruggi, G., Grassi, E., Cartocci, E., Bertoldi, I., Genovese, A., Santini, L., Romagnoli, G., Borgogni, E., Brier, S., Lo Passo, C., Domina, M., Castellino, F., Felici, F., van der Veen, S., Johnson, S., Lea, S. M., Tang, C. M., Pizza, M., Savino, S., Norais, N., Rappuoli, R., Bottomley, M. J., and Masignani, V. (2013) Defining a protective epitope on factor H binding protein, a key meningococcal virulence factor and vaccine antigen. *Proceedings of the national academy of sciences of the United States of America*, 110 (9), 3304–3309.

75 Rose, R. J., van Berkel, P. H., van den Bremer, E. T., Labrijn, A. F., Vink, T., Schuurman, J., Heck, A. J., and Parren, P. W. (2013) Mutation of Y407 in the CH3 domain dramatically alters glycosylation and structure of human IgG. *mAbs*, 5 (2), 219–228.

76 Houde, D., Arndt, J., Domeier, W., Berkowitz, S., and Engen, J. R. (2009) Characterization of IgG1 conformation and conformational dynamics by hydrogen/deuterium exchange mass spectrometry. *Analytical chemistry*, 81 (7), 2644–2651.

77 Kiyoshi, M., Caaveiro, J. M., Kawai, T., Tashiro, S., Ide, T., Asaoka, Y., Hatayama, K., and Tsumoto, K. (2015) Structural basis for binding of human IgG1 to its high-affinity human receptor FcgammaRI. *Nature communications*, 6, 6866.

78 Dall'Acqua, W. F., Kiener, P. A., Wu, H. (2006) Properties of human IgG1s engineered for enhanced binding to the neonatal Fc receptor (FcRn). *Journal of Biological Chemistry*, Aug 18, 281 (33), 23514–23524.

79 Tavakoli-Keshe, R., Phillips, J. J., Turner, R., and Bracewell, D. G. (2014) Understanding the relationship between biotherapeutic protein stability and solid-liquid interfacial shear in constant region mutants of IgG1 and IgG4. *Journal of pharmaceutical sciences*, 103 (2), 437–444.

80 Majumdar, R., Esfandiary, R., Bishop, S. M., Samra, H. S., Middaugh, C. R., Volkin, D. B., and Weis, D. D. (2015) Correlations between changes in conformational dynamics and physical stability in a mutant IgG1 mAb engineered for extended serum half-life. *mAbs*, 7 (1), 84–95.

81 Majumdar, R., Manikwar, P., Hickey, J. M., Samra, H. S., Sathish, H. A., Bishop, S. M., Middaugh, C. R., Volkin, D. B., and Weis, D. D. (2013) Effects of salts from the Hofmeister series on the conformational stability, aggregation propensity, and local flexibility of an IgG1 monoclonal antibody. *Biochemistry*, 52 (19), 3376–3389.

82 Manikwar, P., Majumdar, R., Hickey, J. M., Thakkar, S. V., Samra, H. S., Sathish, H. A., Bishop, S. M., Middaugh, C. R., Weis, D. D., and Volkin, D. B. (2013) Correlating excipient effects on conformational and storage stability of an IgG1 monoclonal antibody with local dynamics as measured by hydrogen/deuterium-exchange mass spectrometry. *Journal of pharmaceutical sciences*, 102 (7), 2136–2151.
83 Pan, L. Y., Salas-Solano, O., and Valliere-Douglass, J. F. (2014) Conformation and dynamics of interchain cysteine-linked antibody-drug conjugates as revealed by hydrogen/deuterium exchange mass spectrometry. *Analytical chemistry*, 86 (5), 2657–2664.
84 Spraggon, G., Pantazatos, D., Klock, H. E., Wilson, I. A., Woods, V. L., Jr., and Lesley, S. A. (2004) On the use of DXMS to produce more crystallizable proteins: structures of the *T. maritima* proteins TM0160 and TM1171. *Protein science: a publication of the Protein Society*, 13 (12), 3187–3199.
85 Pantazatos, D., Kim, J. S., Klock, H. E., Stevens, R. C., Wilson, I. A., Lesley, S. A., and Woods, V. L., Jr. (2004) Rapid refinement of crystallographic protein construct definition employing enhanced hydrogen/deuterium exchange MS. *Proceedings of the national academy of sciences of the United States of America*, 101 (3), 751–756.
86 Houde, D., Berkowitz, S. A., and Engen, J. R. (2011) The utility of hydrogen/deuterium exchange mass spectrometry in biopharmaceutical comparability studies. *Journal of pharmaceutical sciences*, 100 (6), 2071–2086.
87 Wei, H., Ahn, J., Yu, Y. Q., Tymiak, A., Engen, J. R., and Chen, G. (2012) Using hydrogen/deuterium exchange mass spectrometry to study conformational changes in granulocyte colony stimulating factor upon PEGylation. *Journal of the American Society for Mass Spectrometry*, 23 (3), 498–504.
88 Houde, D., Peng, Y., Berkowitz, S. A., and Engen, J. R. (2010) Post-translational modifications differentially affect IgG1 conformation and receptor binding. *Molecular & cellular proteomics: MCP*, 9 (8), 1716–1728.
89 Zhang, A., Hu, P., MacGregor, P., Xue, Y., Fan, H., Suchecki, P., Olszewski, L., and Liu, A. (2014) Understanding the conformational impact of chemical modifications on monoclonal antibodies with diverse sequence variation using hydrogen/deuterium exchange mass spectrometry and structural modeling. *Analytical chemistry*, 86 (7), 3468–3475.
90 Zhang, H., Cui, W., and Gross, M. L. (2014) Mass spectrometry for the biophysical characterization of therapeutic monoclonal antibodies. *FEBS letters*, 588 (2), 308–317.
91 Zhang, A., Singh, S. K., Shirts, M. R., Kumar, S., and Fernandez, E. J. (2012) Distinct aggregation mechanisms of monoclonal antibody under thermal and freeze-thaw stresses revealed by hydrogen exchange. *Pharmaceutical research*, 29 (1), 236–250.
92 Arora, J., Hickey, J. M., Majumdar, R., Esfandiary, R., Bishop, S. M., Samra, H. S., Middaugh, C. R., Weis, D. D., and Volkin, D. B. (2015) Hydrogen exchange

mass spectrometry reveals protein interfaces and distant dynamic coupling effects during the reversible self-association of an IgG1 monoclonal antibody. *mAbs*, 7 (3), 525–539.
93 Kan, Z. Y., Walters, B. T., Mayne, L., and Englander, S. W. (2013) Protein hydrogen exchange at residue resolution by proteolytic fragmentation mass spectrometry analysis. *Proceedings of the national academy of sciences of the United States of America*, 110 (41), 16438–16443.
94 Zeeman, P. (1897) The effect of magnetisation on the nature of light emitted by a substance. *Nature*, 55,347.
95 Wishart, D. S., and Case, D. A. (2001) Use of chemical shifts in macromolecular structure determination. *Methods in enzymology*, 338, 3–34.
96 Case, D. A. (2000) Interpretation of chemical shifts and coupling constants in macromolecules. *Current opinion in structural biology*, 10 (2), 197–203.
97 Overhauser, A. W. (1953) Polarization of nuclei in metals. *Physical reviews*, 92 (2), 411–415.
98 Brunner, E. (2001) Residual dipolar couplings in protein NMR. *Concepts in magnetic resonance*, 13 (4), 238–259.
99 Shen, Y., Lange, O., Delaglio, F., Rossi, P., Aramini, J. M., Liu, G., Eletsky, A., Wu, Y., Singarapu, K. K., Lemak, A., Ignatchenko, A., Arrowsmith, C. H., Szyperski, T., Montelione, G. T., Baker, D., and Bax, A. (2008) Consistent blind protein structure generation from NMR chemical shift data. *Proceedings of the national academy of sciences of the United States of America*, 105 (12), 4685–4690.
100 Pervushin, K., Riek, R., Wider, G., and Wuthrich, K. (1997) Attenuated T2 relaxation by mutual cancellation of dipole-dipole coupling and chemical shift anisotropy indicates an avenue to NMR structures of very large biological macromolecules in solution. *Proceedings of the national academy of sciences of the United States of America*, 94 (23), 12366–12371.
101 Tugarinov, V., and Kay, L. E. (2004) An isotope labeling strategy for methyl TROSY spectroscopy. *Journal of biomolecular NMR*, 28 (2), 165–172.
102 Frueh, D. P., Goodrich, A. C., Mishra, S. H., and Nichols, S. R. (2013) NMR methods for structural studies of large monomeric and multimeric proteins. *Current opinion in structural biology*, 23 (5), 734–739.
103 Vinogradova, O., and Qin, J. (2012) NMR as a unique tool in assessment and complex determination of weak protein-protein interactions. *Topics in current chemistry*, 326, 35–45.
104 Konrat, R. (2014) NMR contributions to structural dynamics studies of intrinsically disordered proteins. *Journal of magnetic resonance*, 241, 74–85.
105 Zuiderweg, E. R. (2002) Mapping protein-protein interactions in solution by NMR spectroscopy. *Biochemistry*, 41 (1), 1–7.
106 Bardelli, M., Livoti, E., Simonelli, L., Pedotti, M., Moraes, A., Valente, A. P., and Varani, L. (2015) Epitope mapping by solution NMR spectroscopy. *Journal of molecular recognition: JMR*, 28 (6), 393–400.

107 Michel, E., Skrisovska, L., Wuthrich, K., and Allain, F. H. (2013) Amino acid-selective segmental isotope labeling of multidomain proteins for structural biology. *Chembiochem: a European journal of chemical biology*, 14 (4), 457–466.
108 Varani, L., Bankovich, A. J., Liu, C. W., Colf, L. A., Jones, L. L., Kranz, D. M., Puglisi, J. D., and Garcia, K. C. (2007) Solution mapping of T cell receptor docking footprints on peptide-MHC. *Proceedings of the national academy of sciences of the United States of America*, 104 (32), 13080–13085.
109 Simonelli, L., Pedotti, M., Beltramello, M., Livoti, E., Calzolai, L., Sallusto, F., Lanzavecchia, A., and Varani, L. (2013) Rational engineering of a human anti-dengue antibody through experimentally validated computational docking. *PLoS one*, 8 (2), e55561.
110 Scarselli, M., Cantini, F., Santini, L., Veggi, D., Dragonetti, S., Donati, C., Savino, S., Giuliani, M. M., Comanducci, M., Di Marcello, F., Romagnoli, G., Pizza, M., Banci, L., and Rappuoli, R. (2009) Epitope mapping of a bactericidal monoclonal antibody against the factor H binding protein of *Neisseria meningitidis. Journal of molecular biology*, 386 (1), 97–108.
111 Binolfi, A., Rasia, R. M., Bertoncini, C. W., Ceolin, M., Zweckstetter, M., Griesinger, C., Jovin, T. M., and Fernandez, C. O. (2006) Interaction of alpha-synuclein with divalent metal ions reveals key differences: a link between structure, binding specificity and fibrillation enhancement. *Journal of the American Chemical Society*, 128 (30), 9893–9901.
112 Bertoncini, C. W., Rasia, R. M., Lamberto, G. R., Binolfi, A., Zweckstetter, M., Griesinger, C., and Fernandez, C. O. (2007) Structural characterization of the intrinsically unfolded protein beta-synuclein, a natural negative regulator of alpha-synuclein aggregation. *Journal of molecular biology*, 372 (3), 708–722.
113 Orcellet, M. L., and Fernandez, C. O. (2011) Structures behind the amyloid aggregation of alpha-synuclein: an NMR based approach. *Current protein & peptide science*, 12 (3), 188–204.
114 Fawzi, N. L., Ying, J., Ghirlando, R., Torchia, D. A., and Clore, G. M. (2011) Atomic-resolution dynamics on the surface of amyloid-beta protofibrils probed by solution NMR. *Nature*, 480 (7376), 268–272.
115 Libich, D. S., Fawzi, N. L., Ying, J., and Clore, G. M. (2013) Probing the transient dark state of substrate binding to GroEL by relaxation-based solution NMR. *Proceedings of the national academy of sciences of the United States of America*, 110 (28), 11361–11366.
116 Vicent, M. J., Dieudonne, L., Carbajo, R. J., and Pineda-Lucena, A. (2008) Polymer conjugates as therapeutics: future trends, challenges and opportunities. *Expert opinion on drug delivery*, 5 (5), 593–614.
117 Duncan, R., Gilbert, H. R., Carbajo, R. J., and Vicent, M. J. (2008) Polymer masked-unmasked protein therapy. 1. Bioresponsive dextrin-trypsin and -melanocyte stimulating hormone conjugates designed for alpha-amylase activation. *Biomacromolecules*, 9 (4), 1146–1154.

118 Frank Biedermann, U. R., Zayed, J. M., Scherman, O. A. (2010) A supramolecular route for reversible protein-polymer conjugations. *Chemical science*, 2, 279–286.

119 Zhong, X., and Wright, J. F. (2013) Biological insights into therapeutic protein modifications throughout trafficking and their biopharmaceutical applications. *International journal of cell biology*, 2013, 273086.

120 Barb, A. W., and Prestegard, J. H. (2011) NMR analysis demonstrates immunoglobulin G N-glycans are accessible and dynamic. *Nature chemical biology*, 7 (3), 147–153.

121 Aubin, Y., Gingras, G., and Sauve, S. (2008) Assessment of the three-dimensional structure of recombinant protein therapeutics by NMR fingerprinting: demonstration on recombinant human granulocyte macrophage-colony stimulation factor. *Analytical chemistry*, 80 (7), 2623–2627.

122 Panjwani, N., Hodgson, D. J., Sauve, S., and Aubin, Y. (2010) Assessment of the effects of pH, formulation and deformulation on the conformation of interferon alpha-2 by NMR. *Journal of pharmaceutical sciences*, 99 (8), 3334–3342.

123 Aubin, Y., Hodgson, D. J., Thach, W. B., Gingras, G., and Sauve, S. (2015) Monitoring effects of excipients, formulation parameters and mutations on the high order structure of Filgrastim by NMR. *Pharmaceutical research*, 32 (10), 3365–3375.

124 Quinternet, M., Starck, J. P., Delsuc, M. A., and Kieffer, B. (2013) Heteronuclear NMR provides an accurate assessment of therapeutic insulin's quality. *Journal of pharmaceutical and biomedical analysis*, 78–79,252–254.

125 Arbogast, L. W., Brinson, R. G., and Marino, J. P. (2015) Mapping monoclonal antibody structure by 2D 13C NMR at natural abundance. *Analytical chemistry*, 87 (7), 3556–3561.

126 Poppe, L., Jordan, J. B., Rogers, G., and Schnier, P. D. (2015) On the analytical superiority of 1D NMR for fingerprinting the higher order structure of protein therapeutics compared to multidimensional NMR methods. *Analytical chemistry*, 87 (11), 5539–5545.

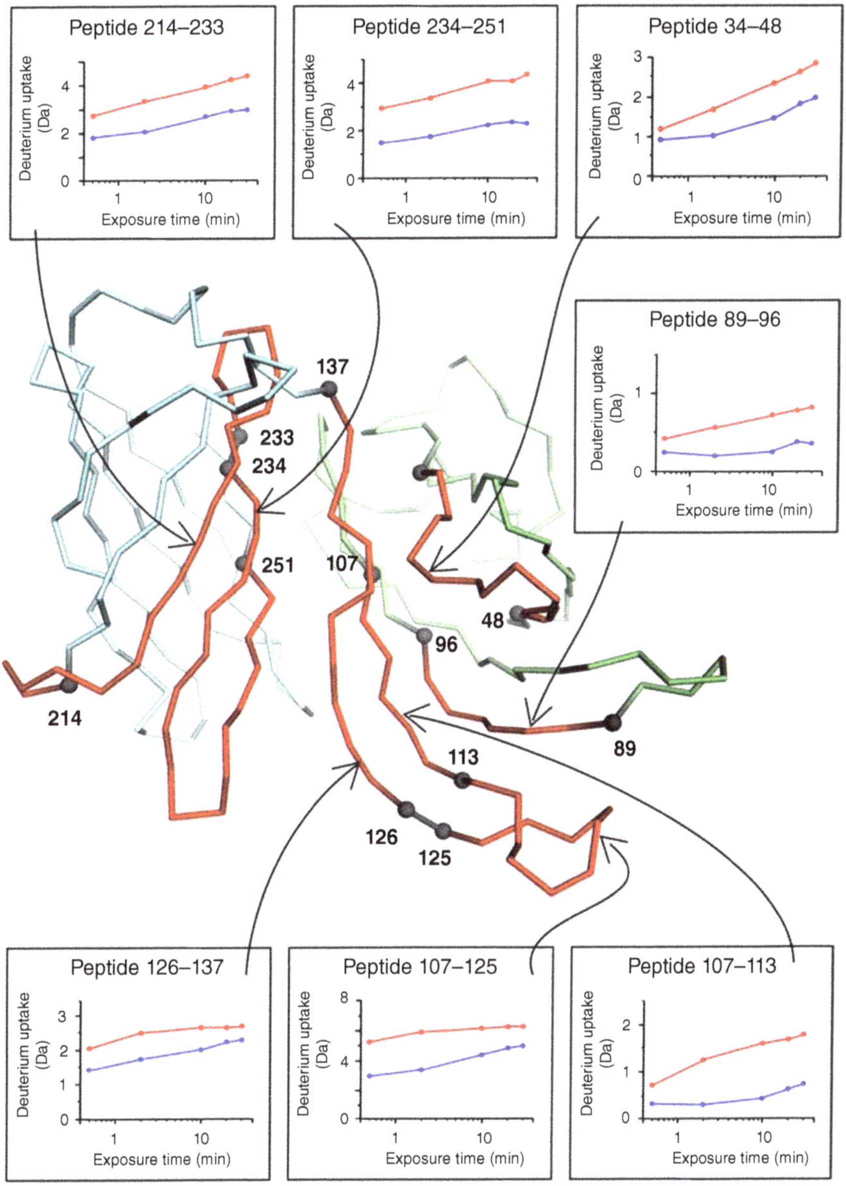

Figure 4.6 Epitope mapping by HDX. The inset show six different overlapping peptides that show reduced HDX over time in the presence of mAb 12C1. The segments of the protein corresponding to the protected peptides are highlighted in red. Source: Malito *et al.* [74]. Reproduced with permission of PNAS.

Analytical Characterization of Biotherapeutics, First Edition. Edited by Jennie R. Lill and Wendy Sandoval.

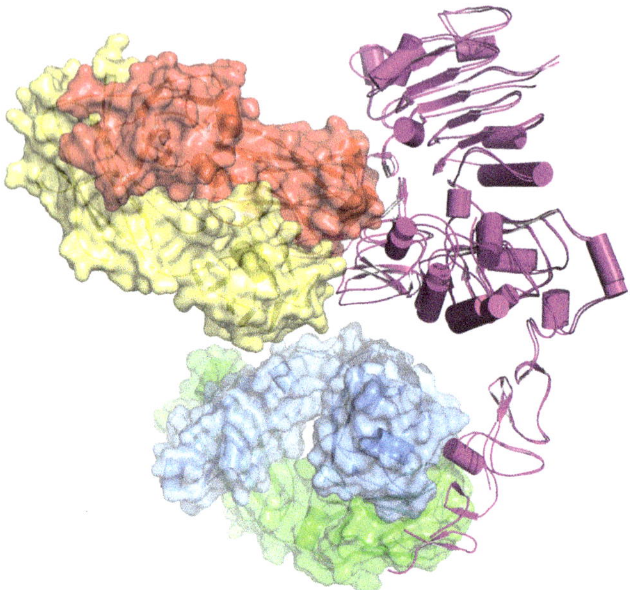

Figure 5.1 Perjeta (red/yellow) and Herceptin (blue/green) bound to the EGF family receptor HER2 (overlaid and colored in purple), illustrating the unique blocking epitopes bound by each of these marketed antibodies. PDB IDs are 1S78 [3] and 1NZ8 [2].

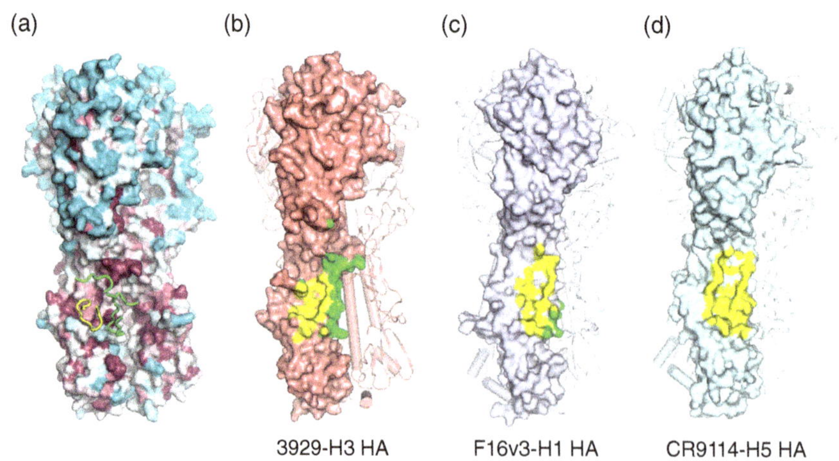

Figure 5.2 A stalk epitope on hemagglutinin (HA) that neutralizes most strains of influenza A. (a) Consurf analysis showing the conservation of surface residues in an H3 HA, displayed on a scale from purple (most conserved) to teal (least conserved). Heavy (yellow) and light (green) chains of the 39.29 antibody [5] are displayed to show the position of the epitope. (b–d) Epitopes of broadly neutralizing antibodies 39.29 bound to H3 HA [5] (b), FI6v3 bound to H1 HA [6] (c) and CR9114 bound to H5 HA [7] (d). Heavy chain epitope is shown as yellow surface and light chain epitope as green surface.

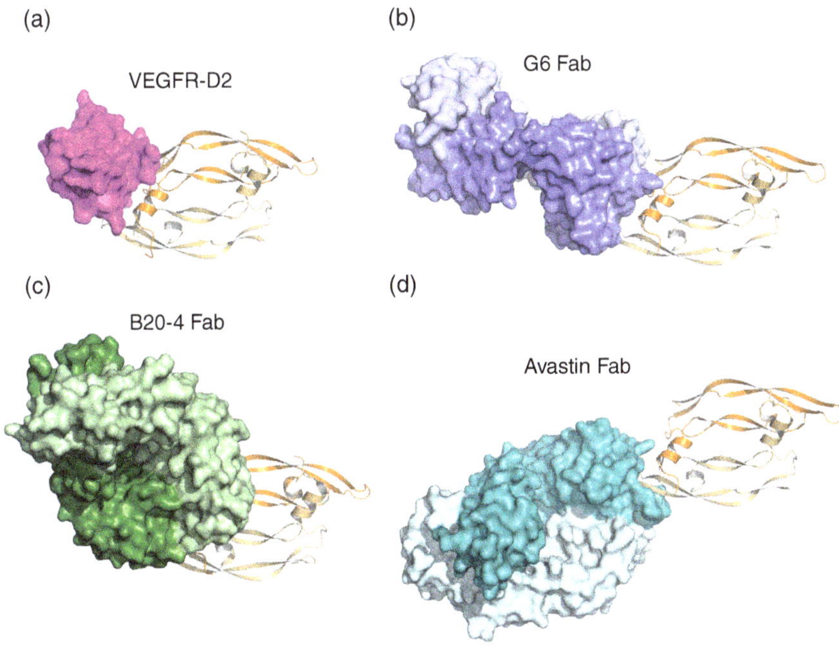

Figure 5.3 Comparison of antibody epitopes on VEGF. Surface representation shown for (a) VEGFR1 (magenta) bound to VEGF (orange) [92], (b) the G6 Fab (blue) bound to VEGF [9], (c) the B20-4 Fab (green) bound to VEGF [9], and (d) Avastin Fab bound to VEGF [10].

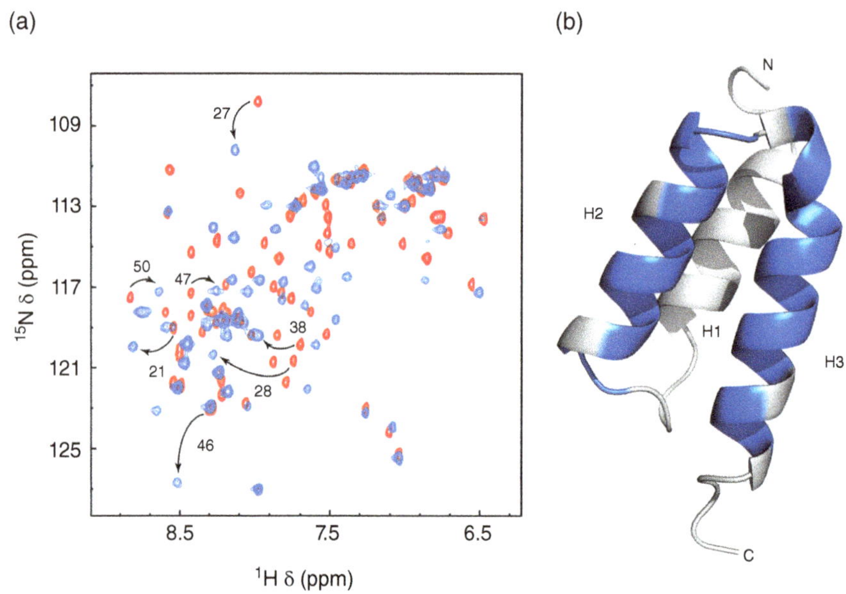

Figure 5.5 Epitope mapping by NMR. (a) Superposition of 2D ^{1}H–^{15}N–HSQC spectra of free (red) and Fv-bound ^{15}N–^{13}C–^{2}H-labeled EZ4, a mutant of the *Staphylococcal* protein A (SpA) domain E, in the free (red) and Fv-bound (blue) forms. Peaks undergoing a large change in chemical shifts are labeled. (b) Residues that experience significant chemical shift changes upon complex formation are colored in blue on the structure of the E-domain. Source: Meininger *et al.* [41]. Reproduced with permission of American Chemical Society.

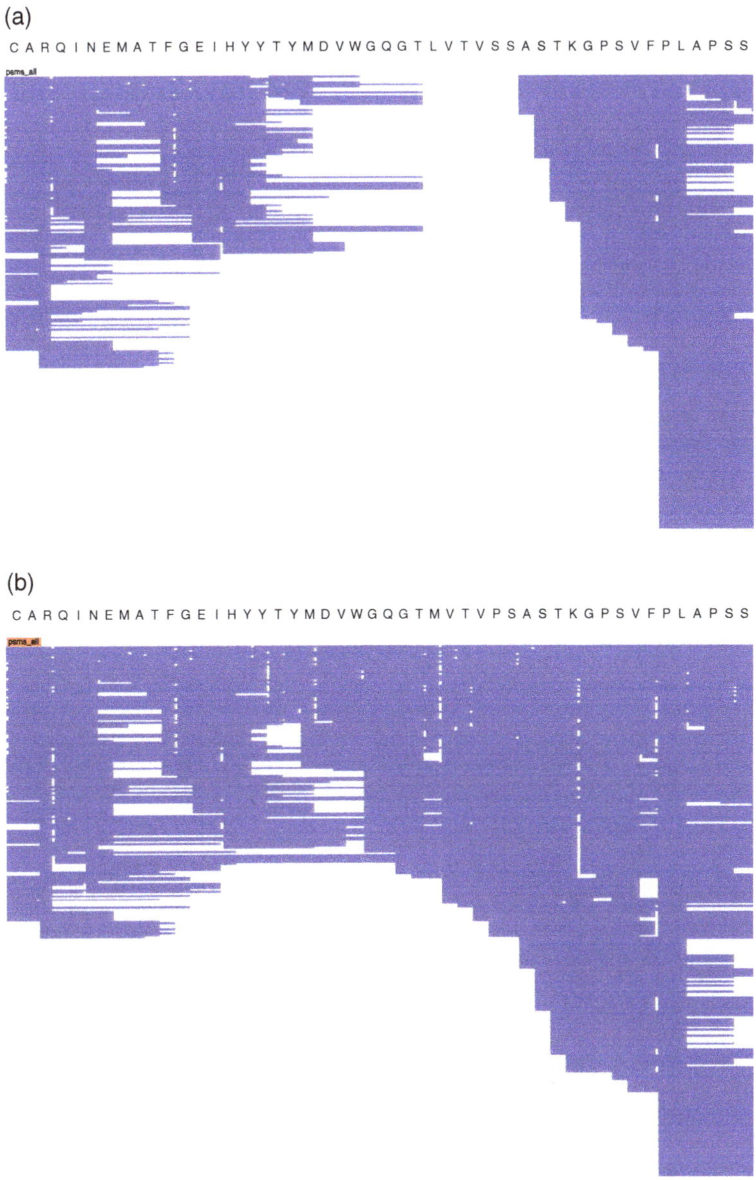

Figure 6.1 Zoomed-in views of the C-terminal end of the primary heavy chain variable region of a hybridoma sequenced with next generation sequencing (NGS) (a) and MS/MS (b) technologies. The sequence is shown at the top of each image. Each blue line denotes a peptide identified by a database search tool (MSGFDB) [2] (at <0.1% false discovery rate (FDR)) against three MS/MS spectra from the same precursor (CID, HCD, and ETD). Without MS/MS sequencing efforts, two sequencing errors in the framework 4 (FR4) region would have gone unnoticed: GQGTLVTVSSASTK → GQGTMVTVPSASTK. These figures were generated with the pViz software [3].

(a)

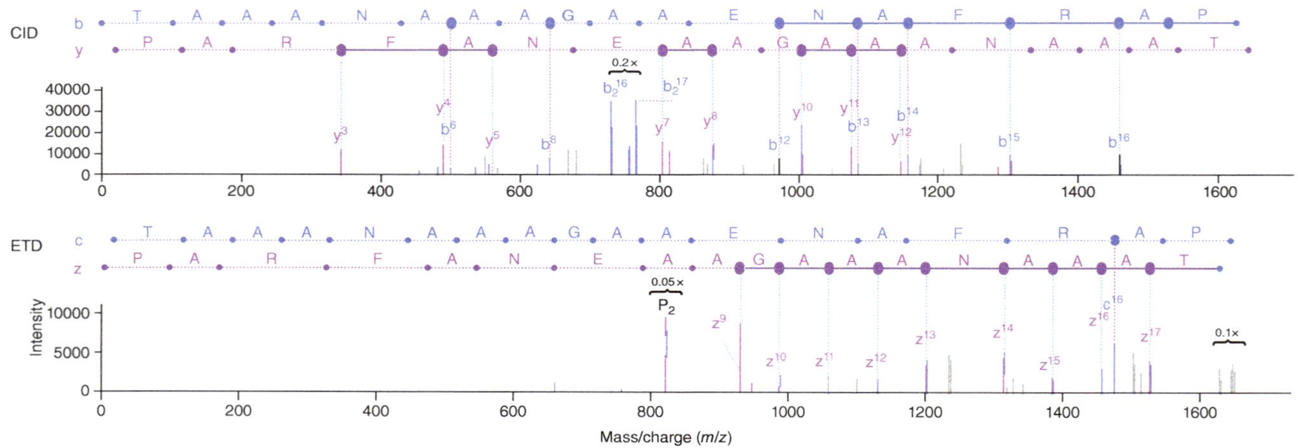

(b)

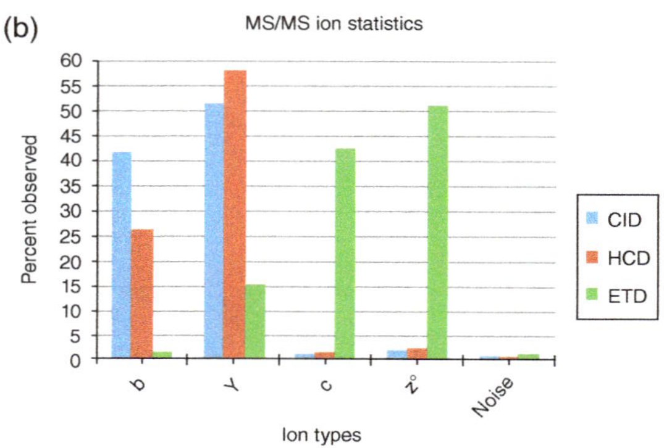

	Peptide breaks				
Prec. charge	2	3	4	5	All
# Unique pep.	4361	2873	712	87	8037
CID	72.3	45.7	32.2	25.7	55.8
HCD	74.1	54.0	34.9	27.1	60.3
ETD	64.1	63.6	55.9	40.4	62.4
CID/HCD	83.9	61.9	42.5	33.4	69.0
CID/ETD	88.9	80.7	70.7	56.4	82.8
HCD/ETD	89.1	84.3	72.1	57.4	84.4
CID/HCD/ETD	93.2	86.8	75.4	61.1	87.8

Figure 6.2 (a) Complementary fragmentation in CID and ETD for peptide TAAANAAAGAAENAFRAP. CID and ETD spectra were separately identified against this C-terminal tryptic peptide at 1% FDR. Enough ions were separately detected in each spectrum to identify the peptide (65% of breaks in CID, 53% in ETD). But combining the two yields full coverage of all possible breaks, thus giving higher confidence to breaks observed in both spectra and possibly enabling full-length *de novo* sequencing. (b) (Left) Peptide MS2 ion statistics for alternative fragmentation modes—This shows the percentage of breaks observed by each ion type over all identified MS2 spectra with precursor charge 2 or 3 for each fragmentation method. z° corresponds to peaks at offset +H from z ions. Ions were counted if observed peak masses were within 20 ppm of expected ion masses. The "noise" ion corresponded to offset $b + 0.5$, which was counted to show the level of noise in each type of MS2 spectra. (Right) Peptide break statistics for combinations of alternative fragmentation modes—peptide breaks were counted for all unique peptides identified by all three fragmentation modes. The six columns show the percentage of breaks detected by each fragmentation mode and combination of fragmentation modes per precursor charge state. In CID and HCD spectra, the presence of breaks was indicated by the presence of b- or y-ions. For ETD, c, z°, or z°+H ions indicated the presence of a break. Multiply charged ions (up to the spectrum's precursor charge) were also considered in each spectrum. Prior to this analysis, peak filtering was applied to all CID, HCD, and ETD spectra such that each peak was retained only if its intensity was ranked fifth or higher over all neighboring peaks in a ±56 Da radius. If a peptide was identified by more than one CID, HCD, or ETD spectrum, a single representative spectrum was randomly chosen for each fragmentation mode.

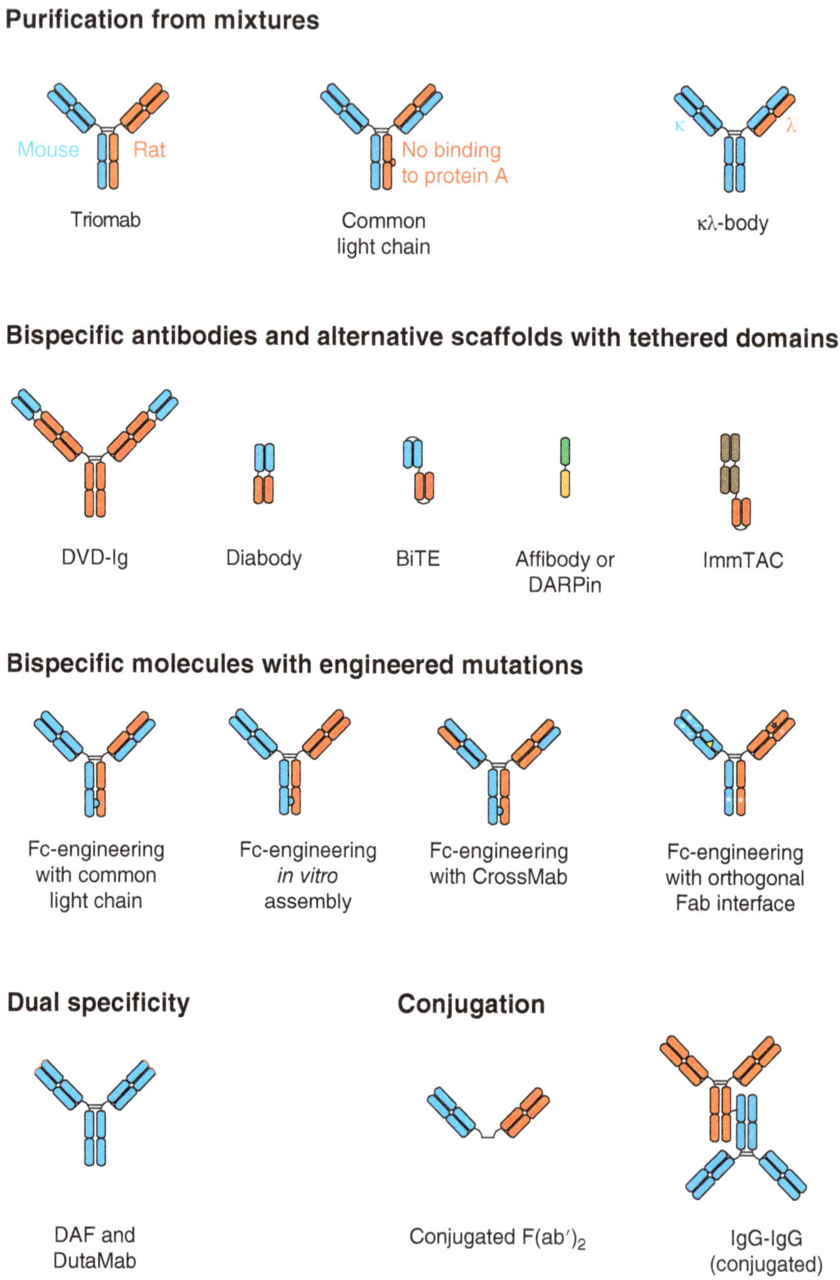

Figure 8.3 Examples of bispecific antibodies classified by their mode of production. Antibody domains are colored in red and blue, alternative scaffold domains in green and orange, and the T-cell receptor domains in brown.

5

Structural Characterization of Recombinant Proteins and Antibodies

Paola Di Lello and Patrick Lupardus

Department of Structural Biology, Genentech Inc., South San Francisco, CA, USA

Abbreviations

μg	microgram
Å	angstrom
AMPA	α-amino-3-hydroxy-5-methyl-4-isoxazolepropionic acid
Asn	asparagine
CDR	complementarity-determining region
CHO	Chinese Hamster Ovary
CSPs	chemical shift perturbations
Da	daltons
E. coli	*Escherichia coli*
EGF	epidermal growth factor
EM	electron microscopy
Endo	endoglycosidase
ER	endoplasmic reticulum
Fab	fragment antigen binding
Fc	fragment crystallizable
FcγR	Fc gamma receptor
Fv	fragment variable
GlcNAc	N-acetylglucosamine
GNT1	*N*-acetylglucosaminyltransferase I
Gor	glutathione reductase
GPCR	G-protein coupled receptor
HA	hemagglutinin
HEK	human embryonic kidney
HIV	human immunodeficiency virus

Analytical Characterization of Biotherapeutics, First Edition. Edited by Jennie R. Lill and Wendy Sandoval.

HOS	high order structure
HSQC	heteronuclear single quantum correlation
kDa	kilodaltons
ml	milliliter
NB-DNJ	*N*-butyldeoxynojirimycin
NMDA	*N*-methyl-D-aspartic acid
NMR	nuclear magnetic resonance
PDB	Protein Data Bank
PEG	polyethylene glycol
PEI	polyethylenimine
RSV	respiratory syncytial virus
scFv	single chain Fragment variable
SRP	signal recognition particle
TROSY	transverse relaxation optimized spectroscopy
TrxA	thioredoxin A
TrxB	thioredoxin reductase

5.1 Introduction

Defining the target epitope of a candidate clinical antibody is a critical aspect of the validation and characterization of a promising antibody therapeutic. Epitope mapping can help decipher the mechanism of action, decode species cross-reactivity, guide affinity improvement efforts, and protect intellectual property generated in the antibody discovery process. Epitope mapping studies can be as simple as competition binding experiments to find candidates with a desired binding profile, while other methods such as alanine scanning mutagenesis, proteolytic mapping, or mass spectrometry-based methods such as hydrogen–deuterium exchange and radiolytic footprinting can also provide useful epitope data quickly. Unfortunately, the inherent low resolution of these techniques does not provide a true molecular picture of how the antibody binds its target. Higher-resolution structural methods such as nuclear magnetic resonance (NMR) spectroscopy or X-ray crystallography can provide information about the exact amino acids involved in the interaction, which can be crucial for improving the properties of the candidate antibody or understanding how a molecule achieves *in vivo* activity. Electron microscopy (EM) can also provide a lower-resolution model of antibody binding quickly and with smaller sample requirements than crystallography or NMR. In this chapter we provide an overview of epitope mapping by these three structural methods. In addition, we include a section on techniques to express and purify antibody/antigen complexes quickly and efficiently, as well as tips for crystallization, structure determination, and NMR analysis. After reading this chapter, one will have the tools and understanding to begin deciphering their chosen epitope at an atomic level.

5.2 Antigens, Epitopes, and Paratopes

Given the complicated terminology of the antibody discovery and characterization process, it is important to first clearly define a few terms that describe antibody/target interactions. The first term is *antigen*, which is defined as the target molecule that a given antibody recognizes. In most cases the antigen is a purified protein used to stimulate an antibody response, or immunize, in a lab animal such as a mouse or rabbit. Importantly, antibodies can also be raised against peptides, polysaccharides, or small molecule haptens. For the purposes of this discussion, we will restrict the definition of an antigen to a peptide or protein. The second term that requires definition is *epitope*, which is the specific part of the antigen that the antibody CDR (complementarity-determining region) loops interact with. Epitopes can be either linear as in an unfolded peptide or three-dimensional and conformationally specific, as in the surface of a folded protein. Linear epitopes are generally continuous—meaning that the entire epitope consists of several amino acids in sequence. Three-dimensional epitopes are generally discontinuous—in that amino acids within the epitope can be far apart in sequence order but are arranged closely in the context of the tertiary structure of a folded protein. This is an important concept because most therapeutic antibodies target a native, folded protein, and therefore linear epitopes are not preferred. The third term needing a definition is a *paratope*, or the antibody amino acids that contact the epitope on the antigen. Paratope residues are almost always confined to the hypervariable CDR regions of an antibody, although it is often the case that not all the antibody CDRs will contact the antigen.

5.2.1 Rationale for Structural Characterization of Epitopes

There are a number of reasons a researcher might want to structurally determine an epitope during the antibody development process. One reason is to understand the molecular mechanism by which the antibody has its effect on the target, whether that is inhibition or activation of the target signaling pathway. Yet there are other important reasons to characterize an epitope, including understanding cross-reactivity, to inform engineering of antibody CDRs for affinity maturation, or to understand issues such as CDR cleavage, oxidation, or other modifications that are detrimental to the manufacturing of a therapeutic. In addition, epitope characterization can be extremely important for protection of the intellectual property generated during the development of antibody therapies. In this section we describe several published examples of structural studies informing the development of a therapeutic antibody.

A common rationale for structural characterization of a therapeutic candidate antibody with its target is to understand the molecular mechanism by which the antibody achieves *in vivo* activity. A noteworthy example of how epitope discovery can help inform mechanism is the case of the

HER2-inactivating therapeutics Herceptin (trastuzumab) and Perjeta (pertuzumab). The structures of each of these marketed antibodies bound to the extracellular domain of HER2 have been determined by X-ray crystallography. These data have been extremely informative for understanding the mechanisms by which each antibody works *in vivo*. Both molecules block HER2 signaling, but *via* different mechanisms, with trastuzumab primarily blocking ligand-independent receptor dimerization and pertuzumab sterically blocking ligand-mediated dimerization [1]. Structural analysis revealed that trastuzumab binds to the membrane proximal epidermal growth factor (EGF) domain of HER2 [2], while pertuzumab binds the receptor dimerization interface in HER2 domains I/II/III [3]. Based on the different mechanisms of action and structural characterization that revealed unique binding epitopes, trastuzumab and pertuzumab were developed into a combination therapy with better efficacy than the single agent therapies alone [4] (Figure 5.1).

Understanding mechanisms of cross-reactivity for a candidate antibody is another important reason to characterize an epitope. An excellent example of how structural analysis can help understand cross-reactivity can be found in the characterization of influenza A neutralizing antibodies. A class of antibodies has been discovered that neutralizes most clinical strains of influenza A by

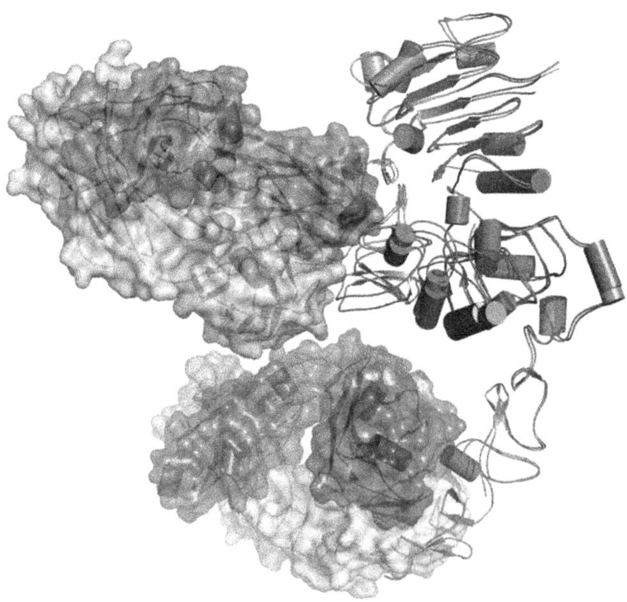

Figure 5.1 Perjeta (red/yellow) and Herceptin (blue/green) bound to the EGF family receptor HER2 (overlaid and colored in purple), illustrating the unique blocking epitopes bound by each of these marketed antibodies. PDB IDs are 1S78 [3] and 1NZ8 [2]. (*See insert for color representation of the figure.*)

binding to the stalk region of the viral surface protein hemagglutinin (HA) [5–7]. The epitope bound by this subset of antibodies is structurally and chemically conserved, likely due to a role in the pH-induced conformational change in the HA protein that occurs upon viral fusion [7] (Figure 5.2). In this case, structural analysis proposed a mechanism of action for the antibody class, as well as a rationale for cross-reactivity. Validation of this conserved epitope has also provided a target for unique therapeutic scaffolds [8].

Differences in species cross-reactivity are a common problem encountered in the therapeutic antibody discovery process. Practically, it is desirable for a candidate to have cross-reactivity with human, cynomolgous monkey, and mouse at the minimum in order to allow for preclinical mouse model utilization as well as pharmacokinetic and toxicology studies in the nonhuman primate cyno species all with the same antibody. Yet given the sequence diversity between human, nonhuman primates and mice, it is often difficult to find an antibody with this species cross-reactivity profile, along with all the other desired properties of a therapeutic antibody (pK, stability, solubility, etc.). To help understand differences in species cross-reactivity, a structural approach can be used to identify the epitope and the sequence differences that contribute to differences in binding affinity. Armed with this information, a development team can choose to reengineer the antibody or pick another candidate

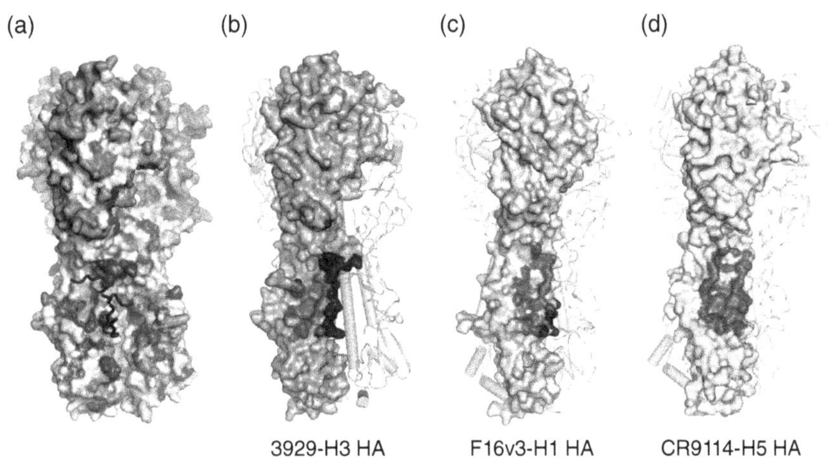

Figure 5.2 A stalk epitope on hemagglutinin (HA) that neutralizes most strains of influenza A. (a) Consurf analysis showing the conservation of surface residues in an H3 HA, displayed on a scale from purple (most conserved) to teal (least conserved). Heavy (yellow) and light (green) chains of the 39.29 antibody [5] are displayed to show the position of the epitope. (b–d) Epitopes of broadly neutralizing antibodies 39.29 bound to H3 HA [5] (b), Fl6v3 bound to H1 HA [6] (c) and CR9114 bound to H5 HA [7] (d). Heavy chain epitope is shown as yellow surface and light chain epitope as green surface. (*See insert for color representation of the figure.*)

for development. An example of structural characterization answering species cross-reactivity questions has been described for vascular endothelial growth factor (VEGF)-binding antibodies [9, 10]. This structural study of Avastin and two other human/mouse cross-reactive monoclonal antibodies targeted to VEGF showed that the phage-derived cross-reactive antibodies targeted the receptor binding site on VEGF, while Avastin bound a nearby epitope that only minimally overlaps with the receptor (Figure 5.3). The Avastin epitope is poorly conserved between mouse and human, illuminating the reason that Avastin is not cross-reactive with mouse VEGF. Studies such as this can help inform selection of preclinical efficacy and toxicology models, which is critical for efficient development and transition of a molecule into the clinic.

Understanding the structure of a candidate antibody bound to an antigen can also be valuable for the design of the best possible molecule to take forward into manufacturing and subsequent clinical trials. Once a candidate molecule is decided upon, several "polishing" steps are still needed before the candidate can be manufactured. Antibody affinity, sequence, pharmacokinetics, and effector function properties all may need to be tuned in order to achieve

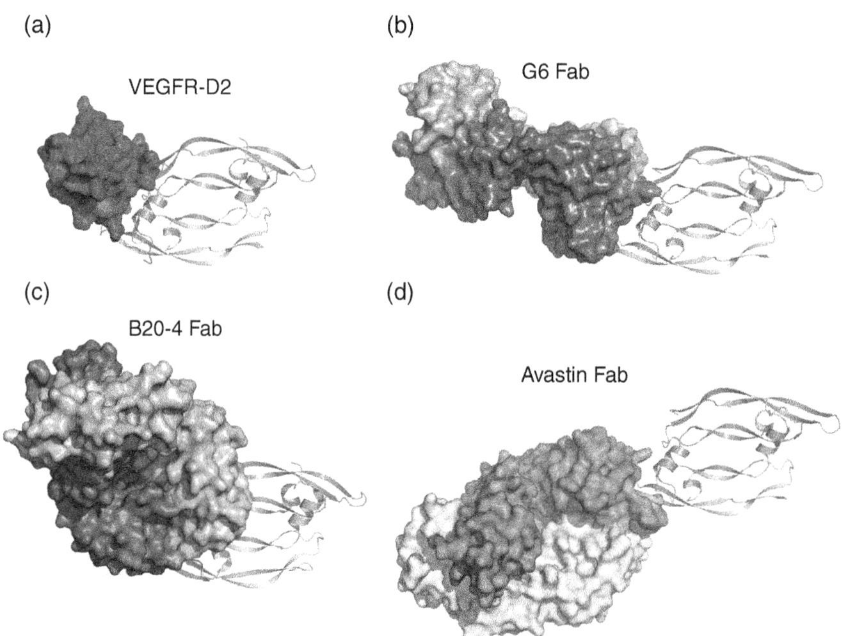

Figure 5.3 Comparison of antibody epitopes on VEGF. Surface representation shown for (a) VEGFR1 (magenta) bound to VEGF (orange) [92], (b) the G6 Fab (blue) bound to VEGF [9], (c) the B20-4 Fab (green) bound to VEGF [9], and (d) Avastin Fab bound to VEGF [10]. (*See insert for color representation of the figure.*)

optimal antibody stability and antigen targeting [11]. Importantly, all of these processes can be influenced by an available atomic structure. Knowledge of paratopic residues that contact an antigen can accelerate and improve the affinity maturation process [12]. For antibody humanization, a structure-guided approach can be useful, particularly for understanding which human framework to use for grafting CDRs [13, 14]. Crystal structures of antibody fragment crystallizable (Fc) interactions with Fc receptors have also revealed ways to engineer and modulate FcRn-mediated recycling and Fc gamma receptor (FcγR)-mediated effector functions [11].

Lastly, structural approaches can be utilized to improve antibody stability during the manufacturing, formulation, and storage procedures. Antibodies in general are extremely stable *in vitro*, yet can undergo modification during expression and purification steps or over time while in storage. Modifications include peptide cleavage or chemical changes to amino acid side chains like deamidation, isomerization, and oxidation [15]. These liabilities in antibody sequence can affect both long-term stability and affinity for the target molecule. One example of structural investigation of an antibody modification can be found in a study of aspartate isomerization in anti-IgE antibodies [16]. A second study utilized structural information to investigate a lysine glycation site near the paratope of a candidate antibody [17]. Structural knowledge can help a development team design changes to the antibody to help mitigate sequence deficiencies.

5.3 Choice of Analytical Method for Epitope Mapping

Structural mapping of an antibody epitope has many advantages, but it would be misleading to ignore the disadvantages of these methods as well. While an atomic model of the antibody/antigen complex of interest may be nice to have, crystallography, NMR, and EM can be time-consuming and expensive endeavors, and sometimes using a structural method is overkill for the question that is to be answered. Before committing resources to a structural effort, it is important to determine whether structural analysis is on the critical path for the development of a molecule. If not, a project may be better served using a simpler method for epitope mapping. If structural analysis is critical, the advent of contract structural biology groups in industry has put these methods within reach of those who do not have the time or budget to set up an in-house structural biology group.

5.3.1 EM for Epitope Analysis

EM has long been an important technique in structural biology, particularly for large protein complexes (i.e., the ribosome) and symmetric protein superstructures such as viral capsids. Technological advances in hardware and software have begun to make EM more accessible to researchers interested in understanding how their antibodies bind to a target protein quickly and with efficient use of sample (reviewed in [18, 19]). A typical single particle EM experiment in negative stain involves placing a dilute sample (μg/ml) onto a carbon grid, and staining that grid with a heavy atom such as uranium, which adheres to the carbon grid but leaves the protein unstained (hence the name "negative stain"). EM images of this grid are therefore inverted images in which the electrons transit through the protein but absorbed by the heavy atoms covering the surrounding grid. A number of EM images are taken from a single grid, and individual particles picked and "averaged" in two dimensions by a computer algorithm. These images, called "class averages," can provide a significant amount of information about the epitope of an antibody given a knowledge of what the antigen looks like. Further data collection and analysis can provide a three-dimensional envelope, which can be used for placement of a low-resolution model. For most negative stain EM experiments, a resolution of >15 Å is expected. Yet these data can be collected and analyzed in several days, rather than the weeks to months it might take for crystallography or NMR studies. EM has been used extensively in the study of antiviral antibodies, in particular those against the human immunodeficiency virus (HIV), respiratory syncytial virus (RSV), and influenza glycoproteins involved in viral fusion [7, 20, 21].

EM methods have two significant advantages over NMR and crystallography when it comes to epitope determination. The first advantage is that very small samples can yield a structure. Only micrograms of a pure, homogeneous sample are needed given that the structure is generated from images of individual particles. The second advantage is that EM has a fast turnaround time. Grid preparation and imaging for negative staining experiments can be done in an afternoon, and data analysis can be completed in a few days. The turnaround time increases for higher resolution (and more difficult) cryo-EM methods. The major disadvantage of EM is the resolution, which is the lowest among the three methods discussed here. However, detector technology and software has advanced cryo-EM techniques to the point that given the right sample, microscope, and detector, subnanometer resolution is achievable [22]. A second disadvantage is the need for the target complex to have a molecular weight greater than 100 kilodaltons (kDa), with 200–300 kDa being preferred [18]. Given these drawbacks, EM can be an extraordinarily powerful way to obtain a quick read on the location of an epitope, in particular for large targets such as viral surface glycoproteins like influenza HA [7].

5.3.2 Epitope and Paratope Mapping by NMR

NMR spectroscopy is a biophysical technique that provides information on the structural features and the dynamic properties of a protein in solution. It can also be applied to the study of protein–protein interactions, including the characterization of the binding interface in antibody/antigen complexes. The advantages of using NMR for these types of studies are manifold. First of all, this technique studies proteins in the solution state, thus in an environment where they are most likely to adopt their biologically relevant conformations. Moreover, NMR is quite sensitive to detect weak interactions ($K_d \sim 10^{-6}$ to 10^{-3} M) that might not be observable by other biophysical techniques. Another advantage of using NMR, especially for epitope mapping, is that one can apply this technique not only to well-folded antigens [23–26] but also to intrinsically disordered proteins (IDPs), like in the case of α-synuclein [27], a neuronal protein linked to Parkinson's disease [28], or the merozoite surface protein 2 (MSP2), an antigen from the malaria parasite *Plasmodium falciparum* [29, 30].

Although NMR is a versatile technique it has limitations, particularly when applied to large molecules like antibodies. Intact antibodies are not only large in size (~150 kDa) but they are also quite flexible, two characteristics that makes them problematic for NMR studies. Due to their size and dynamic properties, antibodies tumble slowly in solution and this feature gives rise to short relaxation times that, in return, cause significant line broadening of the NMR signals, often beyond detection. This is also observed in the case of antigen/antibody complexes. Formation of a tight complex between a full antibody and a smaller, well-structured antigen will induce a dramatic change in the tumbling rate of the antigen, leading to an overall line broadening of the NMR signals for the antigen, thus making it impossible to distinguish the regions of the antigens that are interacting with the antibody from the regions that are not.

The general strategy to counteract the detrimental effects of short relaxation times, typical of large proteins, is to study fragments rather than full antibodies, to use perdeuterated protein samples [31] and to apply the [^{1}H–^{15}N]–TROSY (transverse relaxation optimized spectroscopy) experiment [32], a variant of the [^{1}H–^{15}N]–HSQC (heteronuclear single quantum correlation) (see Section 5.3.2.1). These approaches, which improve the signal-to-noise ratio greatly and therefore lead to better sensitivity, have been successfully applied to the structural study of fragment variable (Fv) and Fc domains produced in *Escherichia coli* [33], as well as in CHO cells [34]. They also have been shown to be beneficial in mapping the epitope of tight binding antibody fragments to well-folded proteins, like in the case of canakinumab and gevokizumab Fabs to interleukin-1β [26]. NMR offers a variety of experiments to map protein–protein binding interfaces [35]; however, the most widely applied method is based on chemical shift perturbations (CSPs) mapping using [^{1}H–^{15}N]–HSQC or [^{1}H–^{15}N]–TROSY experiments.

5.3.2.1 Epitope/Paratope Mapping by Chemical Shift Perturbations

The [^{1}H–^{15}N]–HSQC experiment [36] is one of the key tools in protein NMR. This experiment correlates the resonances of directly bonded ^{1}H and ^{15}N nuclei. For a protein enriched in ^{15}N, one would expect to observe a signal for each of the backbone NH groups (i.e., for every amino acid in the protein, except prolines). The tryptophan and arginine side chain HNε groups as well as the glutamine and asparagine (Asn) side chain NH_2 groups will also be observed in this type of experiment.

Due to the high sensitivity of the ^{1}H and ^{15}N nuclei to the local chemical environment, well-folded proteins in aqueous solution will yield spectra with large dispersion of signals in both ^{1}H and ^{15}N dimensions (Figure 5.4). For these reasons the two-dimensional [^{1}H–^{15}N]–HSQC spectrum of a protein is an important tool; in fact, it provides structural information on a per-residues basis and in the form of a unique "fingerprint" that can be used for epitope and/or paratope mapping. This fingerprint can also be very useful in a variety of other applications, like in assessing and comparing the active conformations of biotherapeutics and biosimilars [37, 38] or in monitoring the effect of mutations and formulation components on the high order structure (HOS) of protein therapeutics [37].

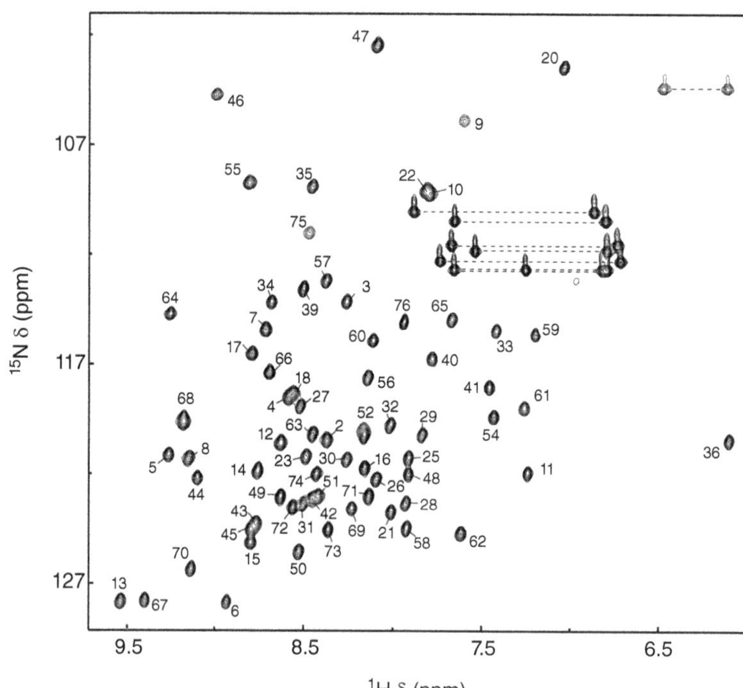

Figure 5.4 2D [^{1}H–^{15}N]–HSQC spectrum of uniformly ^{15}N–^{13}C-labeled and 50% deuterated ubiquitin. Peaks are labeled according to the corresponding amino acid number in the protein sequence. Resonances belonging to the asparagine and glutamine side chain amide protons are connected by dashed lines.

As noted, the ^{1}H and ^{15}N nuclei are very sensitive to the local chemical environment and this makes them excellent probes to detect perturbations arising from the interactions with any type of ligands. The typical workflow for epitope mapping using [^{1}H–^{15}N]–HSQC experiments consists in recording two spectra of the ^{15}N-labeled antigen, one in the absence and one in the presence of the interacting antibody fragment. Because the antibody is not enriched in ^{15}N, it will not interfere with the detection of the signals from the ^{15}N-labeled antigen. Upon addition of the antibody and formation of the ^{15}N-labeled antigen/antibody complex, select signals of the ^{15}N-labeled antigen will undergo significant changes in position (i.e., ^{1}H and ^{15}N chemical shift) and/or intensity (Figure 5.5, panel a). In general, the antigen residues experiencing the most significant changes upon complex formation will tend to cluster within a distinct surface of the antigen itself, thus defining the binding epitope (Figure 5.5, panel b). Therefore, by comparing the [^{1}H–^{15}N]–HSQC spectra collected for the ^{15}N-labeled antigen in the free and bound forms one can quickly identify the residues of the antigen involved in the interaction with the antibody.

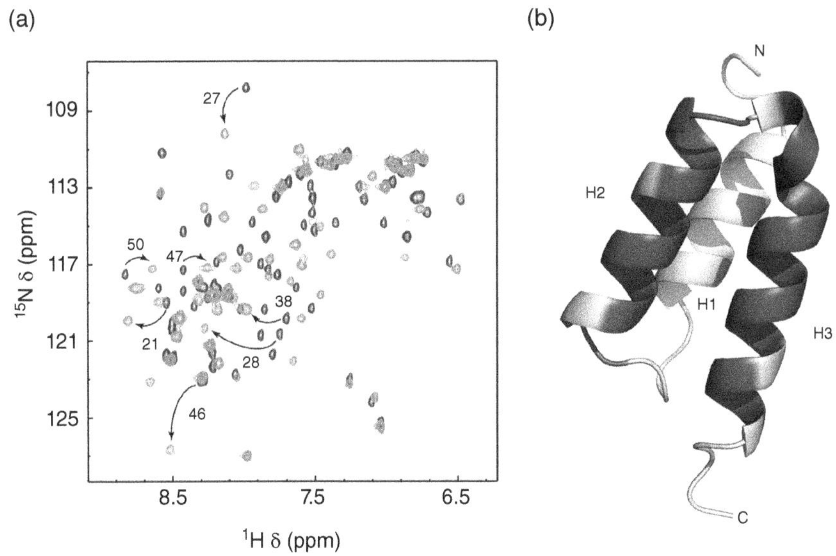

Figure 5.5 Epitope mapping by NMR. (a) Superposition of 2D ^{1}H–^{15}N–HSQC spectra of free (red) and Fv-bound ^{15}N–^{13}C–^{2}H-labeled EZ4, a mutant of the *Staphylococcal* protein A (SpA) domain E, in the free (red) and Fv-bound (blue) forms. Peaks undergoing a large change in chemical shifts are labeled. (b) Residues that experience significant chemical shift changes upon complex formation are colored in blue on the structure of the E-domain. Source: Meininger *et al.* [41]. Reproduced with permission of American Chemical Society. (*See insert for color representation of the figure.*)

One aspect to keep in mind when using the CSPs approach is that the changes in the NMR signals induced by the binding antibody could also arise from allosteric effects. The nuclei observed by NMR are sensitive to the local chemical environment and this can be perturbed by direct interaction with the antibody surface as well as by local conformational changes. To help distinguish between these two effects one can use additional NMR experiments like the ones based on cross-saturation [39] or deuterium exchange [40]. Of course, one can apply the CSPs method to carry on paratope mapping and this has been shown to work well for a variety of isotopically labeled Fv fragments [30, 41, 42].

5.3.2.2 Final Considerations

Although epitope/paratope mapping by CSPs is widely used, it has some drawbacks that need to be carefully considered before embarking in this type of studies. An essential requirement of this approach is to have a full or nearly full assignment of the resonances of the isotopically labeled antigen or antibody fragment—in other words one needs to know which signal belongs to which residue in the protein sequence. This can be achieved through the application of a suite of three-dimentional NMR experiments [43], which often requires production of different protein samples with various labeling schemes, making the process time-consuming if the assignments are not already known. Another aspect to take into consideration are the costs associated with the production of ^{15}N-, ^{13}C-, and ^{2}H-labeled protein samples, which for some expression systems, like insect cells, are still quite expensive. In spite of the disadvantages, this NMR approach can still offer a valuable method to gain low-resolution information on antigen/antibody interactions. For instance, even in the absence of resonance assignment, one could use this technique in a comparative analysis to distinguish between the binding fingerprints of different antibodies on the same antigen.

In the last decade, the technical development of cryogenic probes and the implementation of various methods for fast data acquisition have opened new avenues to tackle more and more complex biological systems, and new NMR approaches have started to emerge that are making possible to obtain structural fingerprints of systems as large as a full IgG1κ antibody, without the need to produce isotopically enriched samples [44].

5.3.3 Epitope Mapping by X-ray Crystallography

Since the first antibody–antigen complex was crystallized [45], X-ray crystallography has been a gold standard method for epitope determination. In a typical crystallography experiment, a complex of highly purified monoclonal fragment antigen binding (Fab) fragment and antigen is prepared and concentrated to >10 mg/ml. The concentrated protein solution is set up in drops containing an equal volume of "mother liquor" solutions that contain varied buffers, pH values, precipitants, salts, and additives. These precipitant solutions are organized into 96-well screens (sold by a number of different vendors) and the drops are set in specially designed crystallization plates by a robot liquid handler. Drops are monitored for crystal formation over days to weeks, and once a crystallization condition is found, it is optimized to provide well-diffracting crystals of a size large enough to be harvested for data collection. Diffraction images are then collected by placing the crystal in a focused X-ray beam, from either a commercially available X-ray generator or more often at one of the high-energy synchrotron radiation sources run by a handful of governmental labs around the world. Images are continually collected during rotation of the crystal around a single axis, with anywhere from 45° to 360° of data needed for a complete dataset. The spots found in these images are then indexed based on the crystal lattice properties, with the location and intensity of each spot measured and catalogued in a single data file. It is from this "reduced" data the electron density maps can be obtained. A three-dimensional polypeptide model can then be built into the electron density maps, with the model improving in accuracy as the resolution of the data increases. While a discussion of in-depth methods for X-ray data collection and structure solution are beyond the scope of this chapter, there are many textbooks, reviews, and online resources available for those interested in the details of protein crystallography [46, 47].

Attempting crystallization of an antigen in complex antibody Fab fragment has a number of important benefits. First, Fabs often improve the crystallization properties of an antigen. Monoclonal Fab fragments crystallize readily by themselves and are often used as chaperones to improve the crystallization of difficult structural biology targets [48]. Second, the Fab fragment provides a known structure for overcoming the "phase problem" with X-ray diffraction. Due to the wealth of Fab crystal structures available in the Protein Data Bank (PDB), Fab models from the database can generally be used for molecular replacement into the diffraction data to start the model building process. Despite these advantages, crystallization is an inherently unpredictable experimental method and sometimes a "perfect" antibody/antigen complex will yield no crystals at all. To improve crystallization properties, several "tricks" can be utilized to produce an antibody/antigen complex with the highest chance of crystallization. We have therefore devoted the next section in this chapter to the protein expression and engineering methods that can be employed to improve crystallization properties of an antibody/antigen complex.

5.4 Recombinant Antigen Generation

A critical step in the process of epitope characterization is the production of an antigen to be used for structural studies. Often the antigen used for structural characterization is the same antigen used for the antibody discovery campaign (i.e., immunization), without proper consideration of the consequences for structural experiments. Protein reagents perfectly acceptable for immunization, *in vivo* or cell-based experiments are often impractical for use in structural studies. One must think about the structural methods that will be used before making a decision on how to generate the antigen. For example, extracellular proteins expressed in CHO or 293 cells are often used for immunization for reasons such as "native" glycosylation and readily available cell culture capacity. While proteins expressed in mammalian cells with native glycosylation are appropriate for EM, they are only rarely amenable to crystallization or isotopic labeling for NMR studies. It is advisable to spend some time thinking about what experiments will be performed before choosing an expression system, or if possible, trying multiple expression systems to ensure a better chance of success. While the following sections are geared toward generation of antibody/antigen complexes for crystallography, the methods are universal and can be adapted for the structural biologist using NMR or EM for epitope discovery.

5.4.1 *E. coli* Expression of Antigens

Protein expression in *E. coli* bacteria has been a mainstay of biochemistry and structural biology labs for decades due to a number of advantages that *E. coli* protein production has over more recently developed eukaryotic expression systems [49]. The first advantage is speed—one can start with a cloned construct, transform and grow the bacteria, and purify the protein all within a few days. This turnaround time allows for iterative construct testing and design with a speed impossible for eukaryotic expression systems. The second advantage is cost—*E. coli* expression media and equipment is relatively inexpensive compared to insect or mammalian cell media and the requisite tissue culture facilities. Finally, expression in *E. coli* can be easily optimized to allow production of proteins that are enriched in ^{15}N, ^{13}C, ^{2}H isotopes, as required for NMR studies [49]. The downsides to *E. coli* expression, which unfortunately loom large, come primarily from a problem that has particular ramifications for the production of protein antigens. Extracellular proteins, which are often the targets of antibody discovery, commonly have folding and posttranslational requirements not supported by prokaryotic protein expression machinery [50]. These proteins are usually still produced by the bacterial ribosomal machinery, but, instead of being properly folded, they are shunted in an insoluble form to inclusion bodies within the bacterial cytoplasm. While several methods (discussed later) have been developed to overcome these issues, it is common practice in structural biology labs to concurrently test eukaryotic expression systems in order to have more "shots on goal" for successful expression and only use *E. coli* if these early tests show the bacteria make acceptable levels of soluble protein.

Extracellular proteins often contain disulfide bonds that require a proper oxidative environment as well as chaperones to properly form. Therefore secreted proteins expressed within the *E. coli* cytoplasm are often driven as misfolded intermediates into inclusion bodies. A classical way of dealing with proteins in inclusion bodies is to purify them, denature the inclusion bodies, and then refold at a low concentration in a buffer containing oxidized and reduced glutathione or cysteine as a redox pair [51]. The presence of the redox pair allows reduction of improperly formed disulfide bonds, while proper disulfide bonds remain stably paired. If refolding is successful, it can be a fantastic way to obtain large amounts of antigen for multiple types of structural studies. Unfortunately, refolding can be difficult for many proteins, in particular those dependent on glycosylation for stability.

A second effective method for *E. coli* expression of extracellular proteins is to engineer a prokaryotic secretion signal into the sequence, which targets the protein for transport into the periplasm *via* the Sec or signal recognition particle (SRP) translocation systems [52]. A protocol specific for periplasmic protein purification can then be used to isolate the protein, which has the added benefit of removing any cytoplasmic components that may interfere with purification. This method works particularly well for antibody fragments [53], but can be tested for any extracellular protein.

A third way to make extracellular proteins in *E. coli* is using a strain or tag that helps with efficient disulfide bond formation in the bacterial cytoplasm. Using the thioredoxin A (TrxA protein) fusion tag is one such method, as TrxA can help catalyze disulfide bond formation in the cytoplasm [54]. Another method (which can be combined with the TrxA tag) is the use of a modified BL21 strains (Origami or SHuffle) with mutations in thioredoxin reductase (TrxB) and glutathione reductase (Gor) that shift the cytoplasmic environment to more oxidizing conditions [55]. This method has been successfully paired with a low temperature expression protocol to produce engineered extracellular proteins such as the ligand binding domains from α-amino-3-hydroxy-5-methyl-4-isoxazolepropionic acid (AMPA) and *N*-methyl-D-aspartic acid (NMDA) receptors [56, 57].

5.4.2 Insect Cell Expression of Antigens

Over the last several decades insect cell expression using baculovirus as a DNA delivery vector has become one of the most widely used methods for expressing extracellular antigens [49]. This method involves the preparation of a baculovirus transfer vector containing the target gene of interest, followed by cotransfection with a viral bacmid that recombines with the transfer vector insert to generate a baculovirus genome containing the desired gene to be expressed. This virus is produced by the host cell and is subject to several rounds of amplification to obtain high-titer viral stocks that can be used for large-scale insect cell infection. After a 60–72 h expression period, the protein is purified either from the cell-free supernatant or from the cell paste itself. There are many advantages to using the baculovirus system, including reasonably fast turnaround times, relatively simple cell culture requirements, eukaryotic protein folding, and posttranslational modification machinery. The use of virus-containing supernatant as the infective agent also makes coexpression of binding partners as simple as adding aliquots of each separate virus to the flask of cells. For the purposes of structural biology, insect cell expression is a desirable system because of its flexibility. Proteins expressed in insect cells can be expressed fully glycosylated and used for EM, or they can be deglycosylated just as easily for use in crystallization experiments (discussed later). In addition, the advent of insect cell media that contains isotopically labeled amino acids means that insect cell–expressed proteins can be used for NMR experiments [58]. In rare cases one may find an extracellular protein that does not express or fold properly in insect cells—and in those cases mammalian methods may be required. But for the vast majority of extracellular proteins the insect cell system provides the simplest means to high-quality protein expression.

5.4.3 Mammalian Expression of Antigens

Mammalian cell protein expression is a highly developed method because of its use in manufacturing of biotherapeutics [49, 59, 60]. The primary cell lines used for large-scale production are Chinese hamster ovary (CHO) cells, and properly selected stable cell lines can be engineered to produce gram quantities of protein per liter of cell culture. Unfortunately, development of cell lines with this kind of yield is resource-intensive and takes months of development time. For the purposes of antigen generation for structural biology, expression is usually induced by polyethylenimine (PEI)-mediated transient transfection. Other cell lines can also be used for expression, such as the human embryonic kidney (HEK) 293 cell line or the murine myeloma NS0 cell line [59]. All three have their own advantages and disadvantages, which can include surprisingly large differences in expression as well as different glycosylation patterns. A normal workflow for antigen production would take advantage of CHO and 293 expression, followed up by expression in a specialty cell like NS0 if issues arise. While not common, proper expression and folding of a protein from CHO or 293 can be an issue in particular with proteins normally expressed in immune cells.

Another more recent development in mammalian cell expression is the use of baculovirus as a DNA transfection vector. This system, called BACMAM, uses high-titer stocks of baculovirus containing a mammalian promoter in front of the gene of interest to deliver DNA into the mammalian cells [61]. Stocks of baculovirus for mammalian expression are produced in the same manner as stocks used for insect cell expression, although much more virus (generally 10–100×) is needed to induce high expression in mammalian cells due to reduced virus tropism for mammalian cells. Fortunately, high-titer baculovirus can be made in large amounts in flasks, making the virus preparation a trivial step in the process of BACMAM expression. This method has been used effectively for expression of extracellular and membrane proteins for structural studies [62, 63].

5.5 N-linked Glycosylation

While purified extracellular proteins are often easy to work with because of their "hardiness" (stability at room temperature, no need for detergents or reductants), extracellular proteins nearly always contain at least one (and often many) Asn- or N-linked glycosylation sites that can make crystallization extremely difficult. N-linked glycosylation occurs at Asn sites followed by a serine or threonine at the +2 position (NxS/T motif) [64] and is catalyzed by the glycosylation machinery found in the endoplasmic reticulum (ER), with final processing continuing upon transit into the Golgi apparatus [65]. While each N-linked glycan starts out as a common dolichol-linked precursor oligosaccharide (glucose$_3$–mannose$_9$–*N*-acetylglucosamine$_2$), these initial sugars are trimmed as well as added to by enzymatic machinery that is both species and cell type specific [64, 65]. For example, CHO cells have different glycosylation patterns than 293S cells, and the pattern of sugars attached to a single protein can also vary. Therefore glycosylation introduces heterogeneity to the surface of the target protein that often interferes with highly ordered crystal packing required for strong crystal diffraction.

Given the complications that glycosylation creates for the crystallographer, several techniques have been developed to minimize glycosylation on the target protein. First, expression systems where glycosylation can be better controlled like *E. coli* or insect cells are generally more desirable for a crystallization target. *E. coli*, for example, do not contain glycosylation machinery, yet as discussed earlier in the chapter this lack of posttranslational processing is often detrimental to expression of normally glycosylated proteins. In insect cells, the simplest method for removal of N-linked glycans is the mutation of the Asn in the NxS/T motifs to the related amino acid glutamine, which blocks transfer of the precursor oligosaccharide to the target protein. Unfortunately, mutation of glycosylation consensus sites often reduces protein expression, and each protein often contains several glycosylated NxS/T motifs. It can be an exponentially complicated effort to find the optimal combination of mutations that gives maximum expression yield with minimum glycosylation. Therefore much effort in technology development has been put forward to find ways to remove or minimize N-linked glycans in a wholesale manner—methods which are described in the next sections.

5.5.1 *E. coli* Expression to Remove Glycosylation as a Factor

A key benefit as well as drawback of expressing extracellular antigens in *E. coli* is the lack of a glycosylation pathway in prokaryotes. This can be a huge benefit for proteins that are well-behaved in the absence of their N-linked glycans, but it can also mean that the *E. coli* system is not a tractable alternative if the protein needs surface sugars for folding or stability. In general, extracellular antigens are best expressed as inclusion bodies and then refolded, or secreted into the periplasm. Good examples of refolded proteins used for structural biology are IL-13 [66] and VEGF [67]. Secretion has been successful for T-cell receptors [68] and antibody variable domains [53, 69, 70].

5.5.2 Manipulating N-linked Glycans on Antigens

For production of highly glycosylated antigens, eukaryotic expression systems are the traditional choice given the presence of appropriate posttranslational machinery. Glycans attached to proteins expressed in insect cells are generally 1–6 and 1–3 fucosylated paucimannosidic structures of ~1000 Da [71]. Glycans attached by the mammalian glycosylation machinery are normally more complex, with biantennary and hybrid structures dominating [72]. As previously discussed, the heterogeneity introduced by N-linked oligosaccharides frequently obstructs crystallization [73]. Several methods have been developed for removing N-linked glycans from proteins expressed in eukaryotic cells, which have been successful in a number of studies.

The first method is simple digestion of the sugar off of the protein using a bacterially derived endoglycosidic enzyme. The endoglycosidase (Endo) family of proteins (including EndoH, F1, F2, and F3) digest between the N-linked N-acetylglucosamine (GlcNAc) residue and the second GlcNAc residue, leaving the initial GlcNAc attached to the protein [74–76]. The "Endo" proteins have varying specificity for N-linked sugars, with the activity of EndoH and F1 high against high mannose, afucosylated sugars, and EndoF2 and F3 more active against complex biantennary structures with or without fucosylation on the core-bound *N*-acteylglucosamine [75]. A cocktail of these enzymes is available for purchase and can sometimes be used for digestion, in particular if the protein has only one or two glycosylation sites. In addition, there is also the possibility of PNGase digestion, which cuts the N-linked sugar off the Asn residue and in the process converts the Asn to aspartate [77]. Importantly, proteins that contain a terminal GlcNAc linked to the Asn glycosylation site are often better behaved than those fully deglycosylated by PNGase or with mutations removing the glycosylation site, likely due to shielding of hydrophobic patches evolved to interact with an attached glycan [73].

An alternate method by which glycosylation can be controlled in eukaryotic expression systems is with small molecule inhibitors of the enzymatic pathway. Several molecules have been characterized for this purpose, with the end point being a recombinant protein that contains afucosylated glycans sensitive to Endo H or F1 digestion. The first described method utilized a Lec3.2.8.1 CHO cell line along with the alpha-glucosidase inhibitor *N*-butyldeoxynojirimycin (NB-DNJ) to produce afucosylated glycoproteins for crystallization [78, 79]. Other studies have utilized the mannosidase inhibitors swainsonine or kifunensine to block mannose removal and trap the glycans in an afucosylated state [73]. Swainsonine blocks mannosidase II, resulting in high mannose/hybrid structures that are largely EndoH sensitive [73, 80]. Kifunensine is a potent mannosidase I inhibitor that retains N-linked glycans in a high mannose, EndoH-sensitive state [81]. Proteins expressed in the presence of kifunensine can also be coexpressed with EndoH to allow for glycan processing in the ER. Insect cell expression in the presence of kifunensine in combination with EndoH coexpression has been used for deglycosylation successfully in a number of cases [82, 83]. On occasion a protein will not express when coexpressed with EndoH, and in those cases a rescue strategy can be to express the protein with kifunensine alone and follow with EndoH or F1 digestion *in vitro* after purification [77].

Another method which has been recently utilized with great success is the use of a 293S cell line that has an inactivating mutation in *N*-acetylglucosaminyltransferase I (GNT1) [84]. Proteins expressed in this cell line only contain high mannose, afucosylated glycans that can be readily digested by EndoH. Coupled with a BACMAM transfection approach, this expression system has been used for generation of crystallization-quality membrane proteins such as the NMDA receptor [85], G-protein coupled receptors (GPCRs) [86], and extracellular receptors that could not be generated in insect cells [87].

5.6 Antibody Generation for Crystallography

While intact antibody IgG molecules are often available for structural studies during the development process, the complete IgG containing two Fab fragments connected to an Fc region is not compatible with most structural studies. Full IgGs are often too flexible for crystallization and class averaging in EM studies, and IgGs are generally too large for NMR analysis. Therefore generation of the isolated Fab fragment is a key step for enabling epitope mapping studies.

The simplest method for obtaining a Fab fragment is by digestion of the intact IgG molecule. Several methods are available for Fab generation, depending on the source of the IgG. Mouse IgG can be efficiently cleaved with Ficin [88] while human IgG can be cleaved with Lys-C protease [89]. As a third alternative, IdeS can be used for generation of the Fab'2 fragment, which is a bivalent molecule containing both Fabs connected by uncleaved disulfide-bonded antibody hinge residues [90].

A second method for generation of Fab or single chain Fv (scFv) is by expression of the protein in the periplasm of *E coli* [53, 69]. *E. coli* secretion can provide high yields of the variable domain fragment of interest and has the benefit of not needing a cleavage step. For antibody/antigen complexes with stability issues, coexpression of the Fab with the protein of interest can be tested as a rescue approach. In this case, the antibody heavy and light chains are cloned individually and cotransfected with the antigen construct. This method was used for the RSV F protein in complex with a neutralizing antibody D25 [20]. The advantages of Fab coexpression are twofold—the antigen is stabilized during the expression phase, and only one purification is necessary rather than purification of individual components followed by purification of the antibody/antigen complex.

5.7 Crystallization of Antibody/Antigen Complexes

A protein crystallization experiment begins by incubating a molecule of interest (or antibody/antigen complex in this case) in a large number of "precipitant" solutions in an attempt to coax the protein to form an ordered array of molecules, held together by noncovalent contacts [46]. The protein is driven from solution into this ordered array, otherwise known as a crystal, by the precipitant, which is often a salt or polyethylene glycol (PEG) derivative. Protein crystallization is an inherently stochastic process, involving hundreds of miniature crystallization "experiments" with different precipitants, additives, at a range of pH values performed in a high-throughput manner. Some have called crystallography a "dark art" given the randomness and uncontrolled nature of the crystallization process. Yet a number of tricks can be utilized to improve the chances at successful crystallization of an antibody/antigen complex of interest.

A first suggestion is to work with the highest affinity antibody clone available. A general rule of thumb is that if the antibody has an affinity (K_d) weaker than 100 nM, there is a reasonable chance of the formation of crystals of the Fab fragment by itself, without the bound antigen. As the affinity improves, there is less free antibody available in the solution to crystallize at any one time, raising the odds that any crystals that form are complex crystals. In addition, crystals that form at low pH values (<5) are often Fab crystals, due to a reduction in the affinity of most antibody/antigen interactions at low pH. This leaves more free Fab in the solution to crystallize. Although all crystals should be checked, it is wise not to assume a crystal grown at low pH is a complex crystal until the structure is solved. A second important step to improve crystallization chances is to ensure high purity of both the Fab and the antigen, and perform a final purification of the complex using a size exclusion column. This final sizing column is an important quality control step to confirm absolute purity of the sample destined for crystallization as well as confirm formation of a complex between the Fab fragment and antigen. This step also ensures a 1 : 1 antibody/antigen ratio in the crystallization drop.

If after extensive crystallization trials, no crystals form or X-ray diffraction of the crystals is not adequate for structure solution, several rescue strategies are available to reinvigorate the crystallization process. As discussed earlier in this chapter, deglycosylation of the antigen is very important and can make an enormous difference in the crystallization process. In addition to deglycosylation, one can also produce protein with universally homogenous high-mannose N-linked glycans by expression in the presence of kifunensine or in the GNT1-/- cell line. Proteins with homogeneous, high-mannose sugars are more amenable to crystallization than those with a diversity of N-linked glycans. Another protein modification step that can be very successful is methylation of surface lysines [91]. This method uses dimethylaminoborane and formaldehyde to convert all surface lysines to dimethyl-lysine. This does not alter the charge of the lysine, but the addition of two methyl groups significantly modifies the surface chemistry of both the antibody and the antigen. On occasion this method will break up an antibody/antigen complex if either the epitope or the paratope contains key lysine-mediated interactions, but the methylation reaction can be performed on the Fab or antigen separately if the complex formation is disrupted.

5.8 Conclusion

Methods for structurally mapping an antibody epitope have expanded tremendously over the last two decades. EM, NMR, and X-ray crystallography have all become more routine, and these methods are now readily accessible to drug development teams with an interest in understanding how their drug interacts with its target. We foresee these methods becoming even more accessible in the future, with EM in particular undergoing a renaissance and likely becoming an even bigger part of the repertoire for epitope discovery. Given the importance of understanding mechanism and cross-reactivity for a potential antibody therapeutic, this progression is great news for both academic and industrial researchers interested in drug development.

References

1 Stern, H M (2012) Improving treatment of HER2-positive cancers: opportunities and challenges. *Sci Transl Med*, 4 (127), 127rv2–127rv2.
2 Cho, H-S, Mason, K, Ramyar, K X, *et al.* (2003) Structure of the extracellular region of HER2 alone and in complex with the Herceptin Fab. *Nature*, 421 (6924), 756–760.

3 Franklin, M C, Carey, K D, Vajdos, F F, *et al.* (2004) Insights into ErbB signaling from the structure of the ErbB2-pertuzumab complex. *Cancer Cell*, 5 (4), 317–328.
4 Baselga, J, Cortés, J, Kim, S-B, *et al.* (2012) Pertuzumab plus trastuzumab plus docetaxel for metastatic breast cancer. *N Engl J Med*, 366 (2), 109–119.
5 Nakamura, G, Chai, N, Park, S, *et al.* (2013) An in vivo human-plasmablast enrichment technique allows rapid identification of therapeutic influenza a antibodies. *Cell Host Microbe*, 14 (1), 93–103.
6 Corti, D, Voss, J, Gamblin, S J, *et al.* (2011) A neutralizing antibody selected from plasma cells that binds to group 1 and group 2 influenza A hemagglutinins. *Science*, 333 (6044), 850–856.
7 Dreyfus, C, Laursen, N S, Kwaks, T, *et al.* (2012) Highly conserved protective epitopes on influenza B viruses. *Science*, 337 (6100), 1343–1348.
8 Fleishman, S J, Whitehead, T A, Ekiert, D C, *et al.* (2011) Computational design of proteins targeting the conserved stem region of influenza hemagglutinin. *Science*, 332 (6031), 816–821.
9 Fuh, G, Wu, P, Liang, W-C, *et al.* (2006) Structure-function studies of two synthetic anti-vascular endothelial growth factor Fabs and comparison with the Avastin Fab. *J Biol Chem*, 281 (10), 6625–6631.
10 Muller, Y A, Chen, Y, Christinger, H W, *et al.* (1998) VEGF and the Fab fragment of a humanized neutralizing antibody: crystal structure of the complex at 2.4 A resolution and mutational analysis of the interface. *Struct Fold Des*, 6 (9), 1153–1167.
11 Carter, P J (2006) Potent antibody therapeutics by design. *Nat Rev Immunol*, 6 (5), 343–357.
12 Levin, A M, Weiss, G A (2006) Optimizing the affinity and specificity of proteins with molecular display. *Mol Biosyst*, 2 (1), 49–57.
13 Co, M S, Queen, C (1991) Humanized antibodies for therapy. *Nature*, 351 (6326), 501–502.
14 Queen, C, Schneider, W P, Selick, H E, *et al.* (1989) A humanized antibody that binds to the interleukin 2 receptor. *Proc Natl Acad Sci U S A*, 86 (24), 10029–10033.
15 Wang, W, Singh, S, Zeng, D L, *et al.* (2007) Antibody structure, instability, and formulation. *J Pharm Sci*, 96 (1), 1–26.
16 Wakankar, A A, Borchardt, R T, Eigenbrot, C, *et al.* (2007) Aspartate isomerization in the complementarity-determining regions of two closely related monoclonal antibodies. *Biochemistry*, 46 (6), 1534–1544.
17 Zhang, B, Yang, Y, Yuk, I, *et al.* (2008) Unveiling a glycation hot spot in a recombinant humanized monoclonal antibody. *Anal Chem*, 80 (7), 2379–2390.
18 Orlova, E V, Saibil, H R (2011) Structural analysis of macromolecular assemblies by electron microscopy. *Chem Rev*, 111 (12), 7710–7748.
19 Cheng, Y, Grigorieff, N, Penczek, Pawel A, Walz, Thomas (2015) A primer to single-particle cryo-electron microscopy. *Cell*, 161 (3), 438–449.

20 McLellan, J S, Chen, M, Joyce, M G, *et al.* (2013) Structure-based design of a fusion glycoprotein vaccine for respiratory syncytial virus. *Science*, 342 (6158), 592–598.
21 Pejchal, R, Doores, K J, Walker, L M, *et al.* (2011) A potent and broad neutralizing antibody recognizes and penetrates the HIV glycan shield. *Science*, 334 (6059), 1097–1103.
22 Cheng, Y (2015) Single-particle Cryo-EM at crystallographic resolution. *Cell*, 161 (3), 450–457.
23 Grinstead, J S, Schuman, J T, Campbell, A P (2003) Epitope mapping of antigenic MUC1 peptides to breast cancer antibody fragment B27.29: a heteronuclear NMR study. *Biochemistry*, 42 (48), 14293–14305.
24 Razzera, G, Gadermaier, G, de Paula, V, *et al.* (2010) Mapping the interactions between a major pollen allergen and human IgE antibodies. *Structure*, 18 (8), 1011–1021.
25 Blech, M, Seeliger, D, Kistler, B, *et al.* (2012) Molecular structure of human GM-CSF in complex with a disease-associated anti-human GM-CSF autoantibody and its potential biological implications. *Biochem J*, 447 (2), 205–215.
26 Blech, M, Peter, D, Fischer, P, *et al.* (2013) One target-two different binding modes: structural insights into gevokizumab and canakinumab interactions to interleukin-1β. *J Mol Biol*, 425 (1), 94–111.
27 Sasakawa, H, Sakata, E, Yamaguchi, Y, *et al.* (2007) Ultra-high field NMR studies of antibody binding and site-specific phosphorylation of alpha-synuclein. *Biochem Biophys Res Commun*, 363 (3), 795–799.
28 Stefanis, L (2012) α-Synuclein in Parkinson's disease. *Cold Spring Harb Perspect Med*, 2 (2), a009399.
29 Adda, C G, MacRaild, C A, Reiling, L, *et al.* (2012) Antigenic characterization of an intrinsically unstructured protein, *Plasmodium falciparum* merozoite surface protein 2. *Infect Immun*, 80 (12), 4177–4185.
30 Morales, R A V, MacRaild, C A, Seow, J, *et al.* (2015) Structural basis for epitope masking and strain specificity of a conserved epitope in an intrinsically disordered malaria vaccine candidate. *Sci Rep*, 5, 10103.
31 Garrett, D S, Seok, Y J, Liao, D I, *et al.* (1997) Solution structure of the 30 kDa N-terminal domain of enzyme I of the *Escherichia coli* phosphoenolpyruvate:sugar phosphotransferase system by multidimensional NMR. *Biochemistry*, 36 (9), 2517–2530.
32 Pervushin, K, Riek, R, Wider, G, Wüthrich, K (1997) Attenuated T2 relaxation by mutual cancellation of dipole-dipole coupling and chemical shift anisotropy indicates an avenue to NMR structures of very large biological macromolecules in solution. *Proc Natl Acad Sci U S A*, 94 (23), 12366–12371.
33 Liu, D, Ren, D, Huang, H, *et al.* (2008) Structure and stability changes of human IgG1 Fc as a consequence of methionine oxidation. *Biochemistry*, 47 (18), 5088–5100.

34 Yamaguchi, Y, Nishimura, M, Nagano, M, *et al.* (2006) Glycoform-dependent conformational alteration of the Fc region of human immunoglobulin G1 as revealed by NMR spectroscopy. *Biochim Biophys Acta*, 1760 (4), 693–700.
35 Zuiderweg, E R P (2002) Mapping protein-protein interactions in solution by NMR spectroscopy. *Biochemistry*, 41 (1), 1–7.
36 Bodenhausen, G, Ruben, D J (1980) Natural abundance nitrogen-15 NMR by enhanced heteronuclear spectroscopy. *Chem Phys Lett*, (69), 185–189.
37 Aubin, Y, Hodgson, D J, Thach, W B, *et al.* (2015) Monitoring effects of excipients, formulation parameters and mutations on the high order structure of filgrastim by NMR. *Pharm Res*, 32 (10), 3365–3375.
38 Quinternet, M, Starck, J-P, Delsuc, M-A, Kieffer, B (2013) Heteronuclear NMR provides an accurate assessment of therapeutic insulin's quality. *J Pharm Biomed Anal*, 78–79, 252–254.
39 Takahashi, H, Nakanishi, T, Kami, K, *et al.* (2000) A novel NMR method for determining the interfaces of large protein-protein complexes. *Nat Struct Biol*, 7 (3), 220–223.
40 Mueller, G A, Smith, A M, Chapman, M D, *et al.* (2001) Hydrogen exchange nuclear magnetic resonance spectroscopy mapping of antibody epitopes on the house dust mite allergen Der p 2. *J Biol Chem*, 276 (12), 9359–9365.
41 Meininger, D P, Rance, M, Starovasnik, M A, *et al.* (2000) Characterization of the binding interface between the E-domain of Staphylococcal protein A and an antibody Fv-fragment. *Biochemistry*, 39 (1), 26–36.
42 Wilkinson, I C, Hall, C J, Veverka, V, *et al.* (2009) High resolution NMR-based model for the structure of a scFv-IL-1beta complex: potential for NMR as a key tool in therapeutic antibody design and development. *J Biol Chem*, 284 (46), 31928–31935.
43 Kay, L E, Ikura, M, Tschudin, R, Bax, A (1990) Three-dimensional triple-resonance NMR spectroscopy of isotopically enriched proteins. *J Magn Reson*, 89, 496–514.
44 Arbogast, L W, Brinson, R G, Marino, J P (2015) Mapping monoclonal antibody structure by 2D 13C NMR at natural abundance. *Anal Chem*, 87 (7), 3556–3561.
45 Amit, A G, Mariuzza, R A, Phillips, S E, Poljak, R J (1985) Three-dimensional structure of an antigen-antibody complex at 6 A resolution. *Nature*, 313 (5998), 156–158.
46 McPherson, A, Gavira, J A (2014) Introduction to protein crystallization. *Acta Crystallogr F Struct Biol Commun*, 70 (Pt 1), 2–20.
47 Shi, Y (2014) A glimpse of structural biology through X-ray crystallography. *Cell*, 159 (5), 995–1014.
48 Bukowska, M A, Grütter, M G (2013) New concepts and aids to facilitate crystallization. *Curr Opin Struct Biol*, 23 (3), 409–416.

49 Assenberg, R, Wan, P T, Geisse, S, Mayr, L M (2013) Advances in recombinant protein expression for use in pharmaceutical research. *Curr Opin Struct Biol*, 23 (3), 393–402.

50 Baneyx, F, Mujacic, M (2004) Recombinant protein folding and misfolding in *Escherichia coli*. *Nat Biotechnol*, 22 (11), 1399–1408.

51 Burgess, R R (2009) Refolding solubilized inclusion body proteins. *Meth Enzymol*, 463, 259–282.

52 de Marco, A (2009) Strategies for successful recombinant expression of disulfide bond-dependent proteins in *Escherichia coli*. *Microb Cell Fact*, 8, 26.

53 Carter, P, Kelley, R F, Rodrigues, M L, *et al.* (1992) High level *Escherichia coli* expression and production of a bivalent humanized antibody fragment. *Biotechnology (NY)*, 10 (2), 163–167.

54 LaVallie, E R, DiBlasio, E A, Kovacic, S, *et al.* (1993) A thioredoxin gene fusion expression system that circumvents inclusion body formation in the *E. coli* cytoplasm. *Biotechnology (NY)*, 11 (2), 187–193.

55 Prinz, W A, Aslund, F, Holmgren, A, Beckwith, J (1997) The role of the thioredoxin and glutaredoxin pathways in reducing protein disulfide bonds in the *Escherichia coli* cytoplasm. *J Biol Chem*, 272 (25), 15661–15667.

56 Furukawa, H, Singh, S K, Mancusso, R, Gouaux, E (2005) Subunit arrangement and function in NMDA receptors. *Nature*, 438 (7065), 185–192.

57 Chen, G Q, Sun, Y, Jin, R, Gouaux, E (1998) Probing the ligand binding domain of the GluR2 receptor by proteolysis and deletion mutagenesis defines domain boundaries and yields a crystallizable construct. *Protein Sci*, 7 (12), 2623–2630.

58 Gossert, A D, Jahnke, W (2012) Isotope labeling in insect cells. *Adv Exp Med Biol*, 992, 179–196.

59 Zhu, J (2012) Mammalian cell protein expression for biopharmaceutical production. *Biotechnol Adv*, 30 (5), 1158–1170.

60 Dalton, A C, Barton, W A (2014) Over-expression of secreted proteins from mammalian cell lines. *Protein Sci*, 23 (5), 517–525.

61 Boyce, F M, Bucher, N L (1996) Baculovirus-mediated gene transfer into mammalian cells. *Proc Natl Acad Sci U S A*, 93 (6), 2348–2352.

62 Dukkipati, A, Park, H H, Waghray, D, *et al.* (2008) BacMam system for high-level expression of recombinant soluble and membrane glycoproteins for structural studies. *Protein Expr Purif*, 62 (2), 160–170.

63 Goehring, A, Lee, C-H, Wang, K H, *et al.* (2014) Screening and large-scale expression of membrane proteins in mammalian cells for structural studies. *Nat Protoc*, 9 (11), 2574–2585.

64 Gahmberg, C G, Tolvanen, M (1996) Why mammalian cell surface proteins are glycoproteins. *Trends Biochem Sci*, 21 (8), 308–311.

65 Davis, S J, Crispin, M (2010) *Functional and Structural Proteomics of Glycoproteins*. Springer, Dordrecht, pp. 127–158.

66 Moy, F J, Diblasio, E, Wilhelm, J, Powers, R (2001) Solution structure of human IL-13 and implication for receptor binding. *J Mol Biol*, 310 (1), 219–230.
67 Christinger, H W, Muller, Y A, Berleau, L T, *et al.* (1996) Crystallization of the receptor binding domain of vascular endothelial growth factor. *Proteins*, 26 (3), 353–357.
68 Golden, A, Khandekar, S S, Osburne, M S, *et al.* (1997) High-level production of a secreted, heterodimeric alpha beta murine T-cell receptor in *Escherichia coli*. *J Immunol Methods*, 206 (1–2), 163–169.
69 Skerra, A, Plückthun, A (1988) Assembly of a functional immunoglobulin Fv fragment in *Escherichia coli*. *Science*, 240 (4855), 1038–1041.
70 Bird, R E, Hardman, K D, Jacobson, J W, *et al.* (1988) Single-chain antigen-binding proteins. *Science*, 242 (4877), 423–426.
71 Harrison, Robert L, Jarvis, Donald L (2006) Protein N-glycosylation in the baculovirus-insect cell expression system and engineering of insect cells to produce "mammalianized" recombinant glycoproteins. *Adv Virus Res*, 68, 159–191.
72 Nettleship, J E, Assenberg, R, Diprose, J M, *et al.* (2010) Recent advances in the production of proteins in insect and mammalian cells for structural biology. *J Struct Biol*, 172 (1), 55–65.
73 Chang, V T, Crispin, M, Aricescu, A R, *et al.* (2007) Glycoprotein structural genomics: solving the glycosylation problem. *Structure*, 15 (3), 267–273.
74 Trimble, R B, Tarentino, A L (1991) Identification of distinct endoglycosidase (endo) activities in *Flavobacterium meningosepticum*: endo F1, endo F2, and endo F3. Endo F1 and endo H hydrolyze only high mannose and hybrid glycans. *J Biol Chem*, 266 (3), 1646–1651.
75 Tarentino, A L, Plummer, T H (1994) Enzymatic deglycosylation of asparagine-linked glycans: purification, properties, and specificity of oligosaccharide-cleaving enzymes from *Flavobacterium meningosepticum*. *Meth Enzymol*, 230, 44–57.
76 Tarentino, A L, Quinones, G, Changchien, L M, Plummer, T H (1993) Multiple endoglycosidase F activities expressed by *Flavobacterium meningosepticum* endoglycosidases F2 and F3. Molecular cloning, primary sequence, and enzyme expression. *J Biol Chem*, 268 (13), 9702–9708.
77 Plummer, T H, Tarentino, A L (1981) Facile cleavage of complex oligosaccharides from glycopeptides by almond emulsin peptide: N-glycosidase. *J Biol Chem*, 256 (20), 10243–10246.
78 Butters, T D, Sparks, L M, Harlos, K, *et al.* (1999) Effects of N-butyldeoxynojirimycin and the Lec3.2.8.1 mutant phenotype on N-glycan processing in Chinese hamster ovary cells: application to glycoprotein crystallization. *Protein Sci*, 8 (8), 1696–1701.
79 Davis, S J, Puklavec, M J, Ashford, D A, *et al.* (1993) Expression of soluble recombinant glycoproteins with predefined glycosylation: application to the crystallization of the T-cell glycoprotein CD2. *Protein Eng*, 6 (2), 229–232.

80 Crispin, M, Aricescu, A R, Chang, V T, *et al.* (2007) Disruption of alpha-mannosidase processing induces non-canonical hybrid-type glycosylation. *FEBS Lett*, 581 (10), 1963–1968.
81 Elbein, AD, Tropea, JE, Mitchell, M, Kaushal, GP (1990) Kifunensine, a potent inhibitor of the glycoprotein processing mannosidase I. *J Biol Chem*, 265 (26), 15599–15605.
82 Lupardus, P J, Garcia, K C (2008) The structure of interleukin-23 reveals the molecular basis of p40 subunit sharing with interleukin-12. *J Mol Biol*, 382 (4), 931–941.
83 Lupardus, P J, Birnbaum, M E, Garcia, K C (2010) Molecular basis for shared cytokine recognition revealed in the structure of an unusually high affinity complex between IL-13 and IL-13Ralpha2. *Structure*, 18 (3), 332–342.
84 Reeves, P J, Callewaert, N, Contreras, R, Khorana, H G (2002) Structure and function in rhodopsin: high-level expression of rhodopsin with restricted and homogeneous N-glycosylation by a tetracycline-inducible N-acetylglucosaminyltransferase I-negative HEK293S stable mammalian cell line. *Proc Natl Acad Sci U S A*, 99 (21), 13419–13424.
85 Lee, C-H, Lü, W, Michel, J C, *et al.* (2014) NMDA receptor structures reveal subunit arrangement and pore architecture. *Nature*, 511 (7508), 191–197.
86 Burg, J S, Ingram, J R, Venkatakrishnan, A J, *et al.* (2015) Structural biology. Structural basis for chemokine recognition and activation of a viral G protein-coupled receptor. *Science*, 347 (6226), 1113–1117.
87 Ely, LK, Fischer, S, Garcia, KC (2009) Structural basis of receptor sharing by interleukin 17 cytokines. *Nat Immunol*, 10 (12), 1245–1251.
88 Mariani, M, Camagna, M, Tarditi, L, Seccamani, E (1991) A new enzymatic method to obtain high-yield F(ab)2 suitable for clinical use from mouse IgGl. *Mol Immunol*, 28 (1–2), 69–77.
89 Gadgil, H S, Bondarenko, P V, Pipes, G D, *et al.* (2006) Identification of cysteinylation of a free cysteine in the Fab region of a recombinant monoclonal IgG1 antibody using Lys-C limited proteolysis coupled with LC/MS analysis. *Anal Biochem*, 355 (2), 165–174.
90 Pawel-Rammingen, von, U, Johansson, B P, Björck, L (2002) IdeS, a novel streptococcal cysteine proteinase with unique specificity for immunoglobulin G. *EMBO J*, 21 (7), 1607–1615.
91 Walter, T S, Meier, C, Assenberg, R, *et al.* (2006) Lysine methylation as a routine rescue strategy for protein crystallization. *Struct Fold Des*, 14 (11), 1617–1622.
92 Wiesmann, C, Fuh, G, Christinger, H W, *et al.* (1997) Crystal structure at 1.7 A resolution of VEGF in complex with domain 2 of the Flt-1 receptor. *Cell*, 91 (5), 695–704.

6

Antibody *de novo* Sequencing

Natalie Castellana[1,2] *and Adrian Guthals*[1,2]

[1] *Digital Proteomics LLC, La Jolla, CA, USA*
[2] *Mapp Biopharmaceutical, Inc., San Diego, CA, USA*

Abbreviations

CID	collision-induced disassociation
ETD	electron-transfer disassociation
Da	Dalton
PTM	posttranslational modification

6.1 Introduction

Accurate primary sequence characterization is crucial in the development of an antibody therapeutic. While this sequence may be known *a priori*, it may also be determined by genomic sequencing the source cell line [1]. However, when the source cell is not available, direct proteomic sequencing of the antibody is the only option. Proteomic sequencing may be required when the hybridoma is no longer producing or is unavailable, if the antibody was a gift, or in cases where genomic sequencing efforts fail. In many cases genomic sequencing can yield a mAb sequence with amino acid mutations that are not reflected in the primary hybridoma product, often a result of novel mAb framework domains that are not recognized by the DNA primers used in the real time polymerase chain reaction (rtPCR) reaction during RNA expression investigations (Figure 6.1).

In addition to the primary sequence, chemical modifications to the protein (either intentionally attached or otherwise) can significantly impact therapeutic activity [4–7]. To accurately determine the types of modification, their sites, and relative abundance, proteomic technology is required.

Analytical Characterization of Biotherapeutics, First Edition. Edited by Jennie R. Lill and Wendy Sandoval.

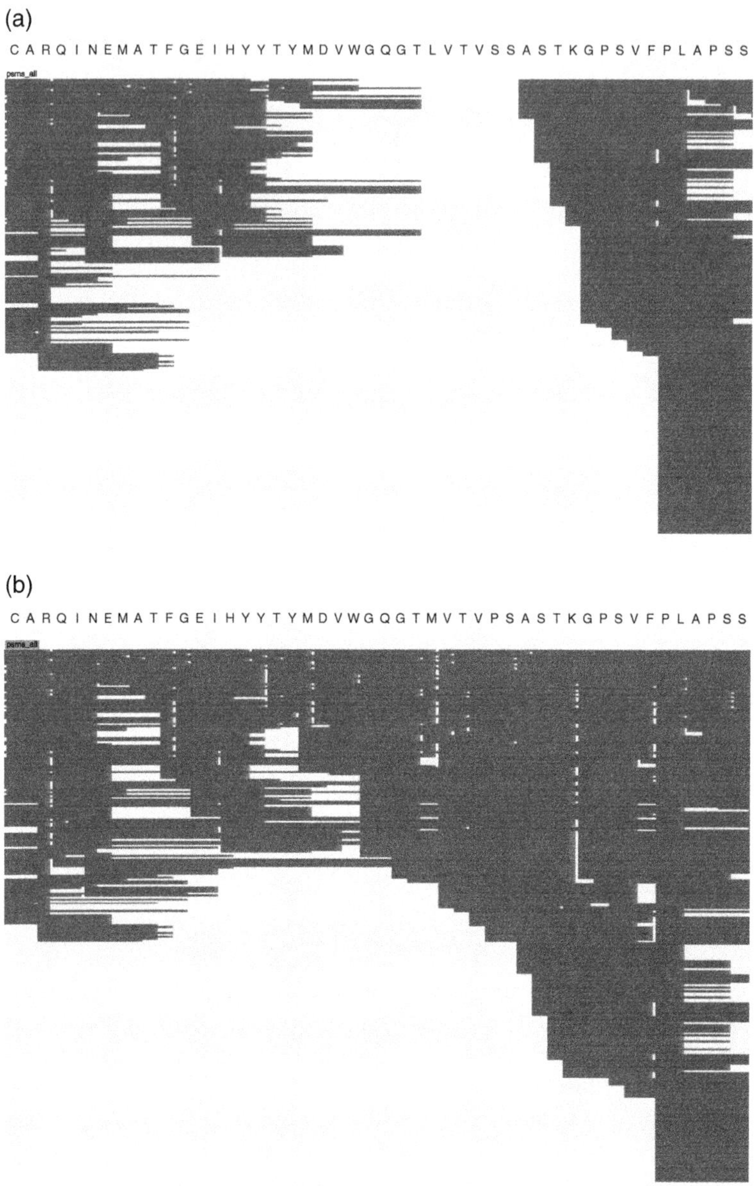

Figure 6.1 Zoomed-in views of the C-terminal end of the primary heavy chain variable region of a hybridoma sequenced with next generation sequencing (NGS) (a) and MS/MS (b) technologies. The sequence is shown at the top of each image. Each blue line denotes a peptide identified by a database search tool (MSGFDB) [2] (at <0.1% false discovery rate (FDR)) against three MS/MS spectra from the same precursor (CID, HCD, and ETD). Without MS/MS sequencing efforts, two sequencing errors in the framework 4 (FR4) region would have gone unnoticed: GQGTLVTVSSASTK → GQGTMVTVPSASTK. These figures were generated with the pViz software [3]. (*See insert for color representation of the figure.*)

For many years, Edman sequencing [8] has been a widely accepted method for primary sequence determination of proteins. In each cycle of Edman sequencing, the amino-terminal residue is labeled and cleaved, yielding the protein sequence one residue at a time. The technique is limited to sequencing molecules of no more than 50–60 amino acids. Digestion of the antibody using a restriction enzyme can yield shorter molecules that can then be subjected to sequencing. Another major drawback is the difficulty in sequencing beyond modified amino acids. For antibodies, pyroglutamate formation is common on the N-terminus, making Edman degradation challenging.

Mass spectrometry-based protein characterization has grown in both speed and accuracy in recent years [9]. However, many of the available tools require the protein sequence to be present in a sequence database [2, 10, 11]. For bottom-up mass spectrometry, in which the protein(s) are digested into shorter peptides, *de novo* sequencing tools can be used to determine the peptide sequence without reference to a sequence database [12–14].

De novo peptide sequencing algorithms typically report a scored list of candidate peptides for each spectrum where top-scoring peptides have an accuracy of ~80–90% for low-resolution collision-induced disassociation (CID) spectra [13, 14] and ~90–92% for high-resolution CID spectra [12]. But these levels of accuracy are rarely achieved in the real world as *de novo* sequencing tools face a significant trade-off between sequencing accuracy and protein sequence coverage. Tandem mass spectra exhibiting complete peptide fragmentation are required to reconstruct accurate peptide sequences, yet such spectra are rarely observed covering entire proteins. As a result, these tools don't achieve the high level of accuracy needed for antibody characterization, and they only act on mass spectra derived from interpretable peptides (<20 AA long).

Several successful attempts at *de novo* protein sequencing [15–18] using bottom-up mass spectrometry have been made. These techniques harness the strengths of the database-mediated proteomic tools as well as *de novo* sequencing techniques. As a result of these efforts, protein sequencing *via* mass spectrometry has become very accurate, and can often be completed in less time than is required for Edman sequencing. This chapter will focus on methods available for protein sequencing using bottom-up mass spectrometry.

6.2 Technical Details on Antibody *de novo* Sequencing

6.2.1 Achieving Complete Protein Coverage

Bottom-up mass spectrometry first requires the protein sample to be subjected to restriction enzymes to digest the protein(s) into peptides. Trypsin is a popular choice for digestion since it reliably cleaves C-terminal to lysine and arginine, producing peptides roughly 10–30 amino acids long. Peptides of this length are ideal for analysis with bottom-up mass spectrometry. However, trypsin digestion of a typical antibody only results in partial coverage of the sequence by identifiable spectra. The rest of the sequence produces peptides which are too short, too long, or unsuitable in composition for analysis by mass spectrometry. In order to recover at least one peptide covering every amino acid in the antibody, all protein sequencing methods utilize multiple enzymes for digestion.

While the ideal set and number of enzymes differs depending on antibody sequence composition, most methods suggest between three and six distinct enzymes. Bandeira *et al.* [16] suggested six separate digests by Lys-C, Glu-C, Asp-N, chymotrypsin, pepsin, and trypsin. Castellana *et al.* [15] omitted the Lys-C digest. Liu *et al.* [18] suggests that three enzymes (Glu-C, trypsin, and Lys-C) may be sufficient.

The particular loading columns used to elute digested peptides into the mass spectrometer is also an important consideration for achieving full MS/MS coverage. In most cases, 1-h liquid chromatography (LC) gradient columns from established vendors are suitable. But in rare cases this may not yield enough peptide separation (usually when the hybridoma line is contaminated with multiple mAbs), and a 2-h LC gradient would address the issue by doubling the number of peptides being analyzed. It is important, however, to match these gradients with optimized MS/MS dynamic exclusion settings and fragmentation settings. Guthals *et al.* [17] provide unadulterated RAW files with such optimized settings for an ETD-enabled Thermo Orbitrap Velos Mass Spectrometer (or later model) with a 1- or 2-h LC gradient (these files are freely available at the University of California at San Diego (UCSD)'s MassIVE data repository: ftp://massive.ucsd.edu/MSV000078436). Similar RAW files with settings optimized for earlier model Orbitrap instruments are also available at this repository [15, 16, 19].

In cases where MS/MS coverage of hydrophobic complementarity determining region (CDR) domains is insufficient while other domains are highly covered, use of a trapping column may be to blame. These columns are routinely used to automatically load sample onto the LC over a 1- or 2-h gradient without requiring an operator to continuously load small aliquots. However, in some cases, trapping columns have been known to scatter the signal of hydrophobic peptides. Thus, the extra effort involved with manually loading sample onto the column may pay off by filling gaps in CDR coverage.

6.2.2 Achieving High Sequencing Accuracy

Mass spectrometry analysis of peptides yields higher fragmentation efficiency than intact protein analysis. However, fragmentation is often far from perfect, introducing the possibility of short sequence reversals or substitutions during sequencing. To address this, Guthals *et al.* [20] and Kim *et al.* [2] suggest the use of alternate fragmentation modes to achieve higher fragmentation efficiency and therefore higher sequencing accuracy. CID and its variant higher-energy collisional dissociation (HCD) produce b- and y-ions and due to the fragmentation pathways, favor breaking certain peptide bonds more than others. In contrast, electron-transfer dissociation (ETD) produces primarily c- and z-ions, and exhibits complementary peptide bond breakage. In addition, CID and HCD produce more informative spectra for lowly charged peptides. ETD fragmentation favors highly charge peptides, which often include longer peptides. By combining both CID/HCD and ETD in an antibody sequencing project, the range of identifiable peptide lengths and charges is increased as more peptide fragmentation points are observed in the union of both types of spectra (Figure 6.2). In addition, by selecting the same precursor ion for fragmentation with both CID or HCD and ETD, *de novo* sequencing accuracy of each individual peptide improves as fragmentation points are corroborated in the intersection of both types of spectra [17, 20, 21].

6.2.3 Handling Protein Modifications

Certain modifications on the antibody can confound sequencing by mass spectrometry. The modifications cause the spectra to be indecipherable or too complex to be confidently identified. While progress has been made on identifying disulfide-bonded peptides [22], these have yet to become a standard part of antibody sequencing pipelines. Prior to analysis with mass spectrometry, antibodies should be reduced to break the disulfide bonds and then alkylated to prevent bond reformation [23].

Glycosylation is another modification that can significantly impact the interpretability of the spectrum. Most methods recommend deglycosylation prior to mass spectrometry with enzymes such as PNGase.

Many common, small molecular mass modifications can be acceptable during sequencing. These include oxidation, deamidation, and pyroglutamate formation. Large molecular mass modifications, such as drugs conjugated to the antibody, impede the ability to sequence, unless the modification is of very low frequency in the sample.

6.2.4 Handling Sample Purity

Most protein sequencing methods require a high level of protein purity. Spectra from common contaminating proteins, such as trypsin or keratin, can be easily identified; however, sequence variants of the antibody can confound sequencing efforts. Sequencing methods used today require only a single antibody to be present in the sample, and often require that the heavy and light chains be separated prior to mass spectrometry analysis.

(a)

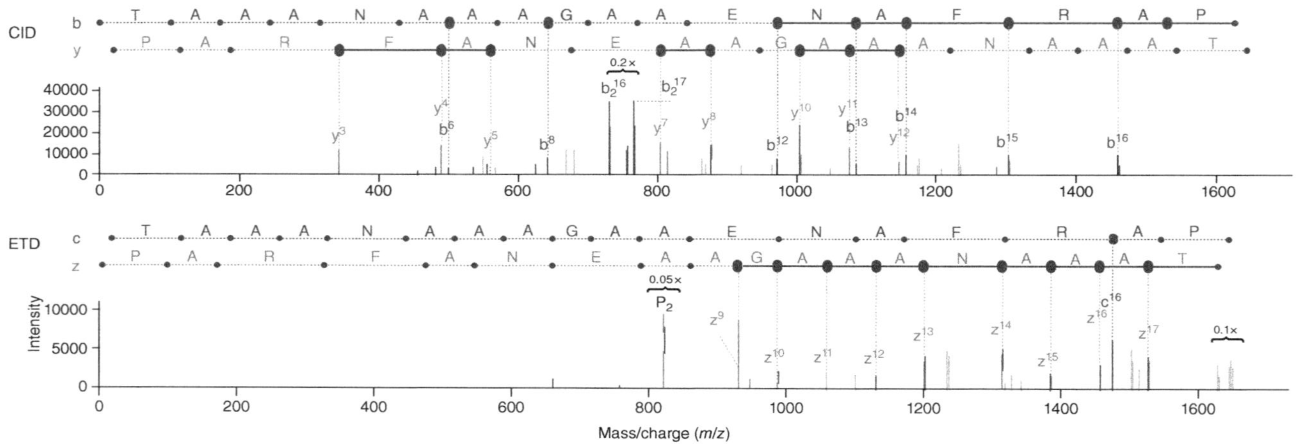

(b)

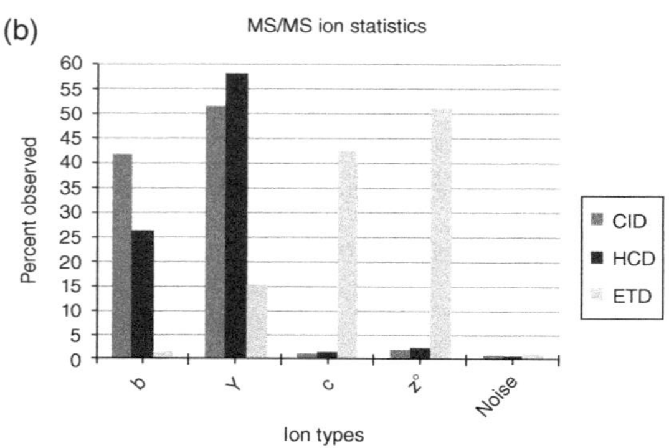

	Peptide breaks				
Prec. charge	2	3	4	5	All
# Unique pep.	4361	2873	712	87	8037
CID	72.3	45.7	32.2	25.7	55.8
HCD	74.1	54.0	34.9	27.1	60.3
ETD	64.1	63.6	55.9	40.4	62.4
CID/HCD	83.9	61.9	42.5	33.4	69.0
CID/ETD	88.9	80.7	70.7	56.4	82.8
HCD/ETD	89.1	84.3	72.1	57.4	84.4
CID/HCD/ETD	93.2	86.8	75.4	61.1	87.8

Figure 6.2 (a) Complementary fragmentation in CID and ETD for peptide TAAANAAAGAAENAFRAP. CID and ETD spectra were separately identified against this C-terminal tryptic peptide at 1% FDR. Enough ions were separately detected in each spectrum to identify the peptide (65% of breaks in CID, 53% in ETD). But combining the two yields full coverage of all possible breaks, thus giving higher confidence to breaks observed in both spectra and possibly enabling full-length *de novo* sequencing. (b) (Left) Peptide MS2 ion statistics for alternative fragmentation modes—This shows the percentage of breaks observed by each ion type over all identified MS2 spectra with precursor charge 2 or 3 for each fragmentation method. z° corresponds to peaks at offset +H from z ions. Ions were counted if observed peak masses were within 20 ppm of expected ion masses. The "noise" ion corresponded to offset b + 0.5, which was counted to show the level of noise in each type of MS2 spectra. (Right) Peptide break statistics for combinations of alternative fragmentation modes—peptide breaks were counted for all unique peptides identified by all three fragmentation modes. The six columns show the percentage of breaks detected by each fragmentation mode and combination of fragmentation modes per precursor charge state. In CID and HCD spectra, the presence of breaks was indicated by the presence of b- or y-ions. For ETD, c, z°, or z°+H ions indicated the presence of a break. Multiply charged ions (up to the spectrum's precursor charge) were also considered in each spectrum. Prior to this analysis, peak filtering was applied to all CID, HCD, and ETD spectra such that each peak was retained only if its intensity was ranked fifth or higher over all neighboring peaks in a ±56 Da radius. If a peptide was identified by more than one CID, HCD, or ETD spectrum, a single representative spectrum was randomly chosen for each fragmentation mode. (*See insert for color representation of the figure.*)

6.3 Bioinformatics Workflow

6.3.1 Spectral Preprocessing

Preprocessing steps can be used to improve the success of antibody sequencing. While many mass spectrometers make use of dynamic exclusion lists, many experiments result in several redundant spectra being acquired for the same precursor. While each spectrum alone does not contain complete information, combining the spectra into a consensus spectrum not only reduces the number of spectra that must be considered, but also improves the signal-to-noise ratio. Clustering identifies spectra from the same precursor based on parent mass and charge as well as high-intensity fragment peaks [24]. Each cluster is then reduced to a single consensus spectrum that can then be used in the downstream sequencing pipeline.

High-charge peptides produce fragment ions with charge up to the charge of the entire peptide. As the charge gets higher, the spectrum can become complicated by the multiple charge states present for each fragment. Spectral deconvolution identifies fragment ions by identifying isotopic envelopes. The peak can then be decharged, reducing the spectrum to all peaks of charge 1. Reducing the complexity of the spectra improves the sequencing accuracy [19].

Tandem mass spectra can be further simplified by converting fragment ion peaks to *prefix residue mass peaks* [13]. In a tandem mass spectrum, the peptide backbone breakage produces a variety of ions including ions for the peptide prefix (often b-ions or c-ions), ions for the peptide suffix (often y-ions or z-ions), and neutral losses from these ions. The variety of ions complicates the spectrum. In contrast, a prefix residue mass spectrum consists of only peptide prefix masses. Tandem mass spectra can be converted to prefix residue mass spectra by identifying the set of complementary ions produced by a particular breakage and replacing them with the mass of the peptide prefix. This approach is primarily adopted by spectrum alignment-based sequencing approaches (see next section).

6.3.2 Spectral Alignment-based Approach

While per-spectrum *de novo* sequencing is plagued by low accuracy, merging together several spectra can improve sequencing accuracy. For example, in the case of spectral clustering described previously, merging the spectra for the same peptide together improves the signal-to-noise ratio of peaks. The multiple enzymes used to ensure full sequence coverage produce overlapping peptides. The spectra from overlapping peptides can be merged together, thus sharing the peak information in the overlapping regions. The merged spectrum represents a longer sequence than any one peptide, contains more complete fragmentation than any one spectrum, and has a higher signal-to-noise ratio. The merged spectrum is called a spectral contig and this approach is called shotgun protein sequencing [19, 25, 26]. One can draw an analogy between shotgun protein sequencing, whereby peptide mass spectra are stitched together to recover a spectrum for the entire protein, and shotgun genome sequencing, whereby short sequencing reads are stitched together to recover a complete genome sequence.

Identifying the spectra from overlapping peptides can be performed without knowledge of the peptides that produced the spectra. Spectrum alignment [25, 27, 28] has long been used to identify spectra from similar molecules. In shotgun protein sequencing, an all-to-all alignment is performed between input spectra, thus identifying a "network" of spectra. Each node in the network corresponds to a spectrum and an edge between two nodes indicates that the spectra are from overlapping peptides. Connected components in the network are consolidated into a spectral contig.

Once the contigs have been assembled, sequencing can be performed with very high accuracy due to the abundance of information provided by merging multiple spectra together. Shotgun protein sequencing attempts to *de novo* sequence these contigs without the aid of a reference antibody sequence. Another benefit of the shotgun protein sequencing approach is that spectral alignment allows identification of unknown posttranslational modifications (PTMs) from unidentified spectra (Figure 6.3). Castellana *et al.* [15] make use of the nature of the antibody construction to identify conserved regions of the antibody (e.g., the constant region) using a sequence database of homologous sequences, and focus *de novo* sequencing efforts on the variable region.

The spectral alignment-based approach relies heavily on high peptide density, meaning the method fails if there are insufficient spectra from overlapping peptides. This can result from too few enzymes being used for digestion.

Spectral alignment-based approaches are ideally suited to situations in which no homologous protein sequences are available. While spectral alignment-based sequencing of protein mixtures has been demonstrated [17, 19], extending this method to mixtures of antibodies or polyclonal antibodies remains an open problem.

6.3.3 Sequence Homology-based Approaches

While spectrum alignment-based approaches merge together spectra prior to *de novo* sequencing, sequence homology-based approaches *de novo* sequence spectra first and then assemble the sequences into contigs. Liu *et al.* [18] proposed the use of a homologous sequence to correct errors introduced during *de novo* sequencing of peptide mass spectra. First, spectra are sequenced, producing a collection of peptide sequences. These sequences are mapped to the homolog sequence. By looking at multiple peptide sequences simultaneously, Liu *et al.* were able to distinguish between *de novo* sequencing errors (which occur in only a few peptides overlapping a residue) and mutations (which should occur in all peptides overlapping a residue).

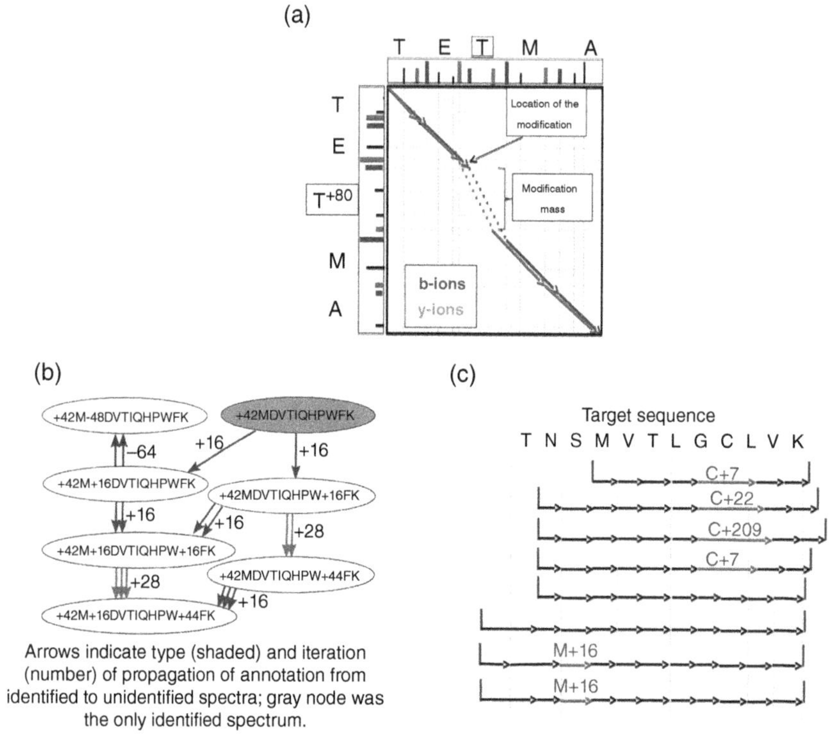

Figure 6.3 Discovery and identification of posttranslational modifications through spectral networks; (a) spectral alignment between modified and unmodified variants of the peptide TETMA (*b*-ions shown in light gray, *y*-ions in dark gray, black lines track consecutively matched *b*/*y*-ions); (b) grouped modification states of the peptide MDVTIQHPWFK from a sample of cataractous lenses. Nodes in the spectral network represent individual MS^2 spectra and edges between nodes represent significant spectral alignments such as that shown in part (a); (c) spectra assembled in the spectral network, or *contig*, for TNSMVTLGCLVK with diverse cysteine modifications on a monoclonal antibody. Each arrow corresponds to the propagation of a sequence and/or PTM from an identified spectrum to an unidentified spectrum (repeated arrows are iterative propagations). Shaded arrows correspond to types of modifications transferred.

An obvious benefit of this approach is that it does not require the density of overlapping spectra needed for the spectrum alignment-based approaches. Liu *et al.* were able to use only three enzymes to achieve full coverage of the protein. Since a reference sequence is used to determine proximal peptides, the peptides do not need to be significantly overlapping. The need for a reference protein can limit the utility of this approach, however. First a reference protein must be correctly chosen for each project. Second, the reference protein must be sufficiently similar to the antibody of interest so that peptide sequences can be confidently mapped to it.

6.3.4 Semi-automated and Manual *de novo* Sequencing

In rare cases insufficient tandem mass spectra are acquired to enable fully automated *de novo* sequencing of the target antibody. If there is not enough sample available to attempt alternative digestion protocols, this can be problematic, especially when gaps in sequence coverage occur over CDR regions where homology to known antibodies cannot be used to resolve the gaps. It is often the case that spectra which do cover these gaps are acquired, but either there are too few of them or they do not contain sufficient peptide fragmentation for automated sequencing.

Prior to the advent of automated tools for processing tandem mass spectra, manual *de novo* sequencing was the only viable method for interpreting tandem mass spectra [29]. Manual sequencing protocols have since been updated and are still in use today [30]. But due to the efficiency of the latest generation of mass spectrometers, too many tandem mass spectra can be acquired (>20 000) from a single run to manually sequence all of them. Thus, computational techniques are often used to "filter" spectra down to a few candidates for manual interpretation that might fill gaps in automated *de novo* sequencing coverage.

Many tools are available that can generate "tags" of 3–5 AA from unidentified spectra [13, 14, 31–33]. Without having the constraint of predicting full-length peptide sequences, these tools can generate sets of shorter tags at high accuracy from spectra with poorer peptide fragmentation. These tags can then be efficiently matched to sequence fragments adjoining gaps. Manual sequencing efforts can then be focused on spectra containing one or more high-confidence tags that match the flanking ends of *de novo* sequences generated by homology-based or alignment-based tools.

6.4 Sequence Validation

6.4.1 Mass Spectrometry-based Statistics

Once a putative antibody sequence has been determined, verification of the sequence is required. A number of statistics indicating sequencing quality can be determined from the mass spectrometry data. *Spectral count* is the number of spectra supporting a given amino acid. While high spectral count is often a good indicator for sequence correctness, low spectral count may not always indicate incorrectness. Low coverage regions of the antibody may simply have yielded few peptides measureable by mass spectrometry. High spectral count regions may in fact be the result of biases in the informatics workflow.

The *digest count* is a complementary statistic to the *spectral count*. It indicates the number of distinct enzymatic digestions that produced a peptide matching a given residue. If six enzymes were used in an experiment, then each predicted amino acid would have a digest count between 0 and 6. The more independent digests supporting an amino acid the better. In general, two or more digests supporting an amino acid are desirable. Again, low digest count may not indicate incorrectness.

The spectral count and digest count have an obvious flaw. While two adjacent amino acids may have high spectra counts and high digest counts, nothing indicates that there may not be missing amino acids between them. For this, the *overlap count* can be used to identify possible gaps in the prediction. The overlap count for a given residue is the number of spectra supporting the residue as well as the next residue.

6.4.2 Intact Mass Comparison

Non-tandem mass spectrometry (non-MS/MS)-based verification can also be useful, providing an orthogonal means of determining sequencing correctness. The intact mass measurement of the heavy and light chains can be compared to the mass of the predicted sequence. Depending on the technology used, the intact mass measurement may exhibit errors of up to 2 Da providing only rough evidence of correctness. Modifications may also confound the measurement. Only a handful of deamidations can cause the mass measurement to deviate by a few Daltons. Similarly, surviving disulfide bonds can cause the intact mass measurement to be off by approximately 2 Da for the light chain or 4 Da for the heavy chain.

6.4.3 Synthetic Peptides

Synthetic peptides are the most accepted way to verify sequence correctness. Once the putative sequence has been determined, peptide sequences covering regions of uncertainty are identified. A synthetic peptide is then produced and analyzed by MS/MS. The spectrum for the synthetic peptide is compared to the experimental spectra matching the peptide. While visual interpretation can be used, a simple cosine can be computed between the two spectra. The spectral intensities must first be normalized to one. If the experimental spectrum was correctly interpreted (and thus the region of the antibody in question is correct), then it should have a cosine score close to one when matched to the synthetic spectrum. If the cosine score is low between the two spectra, then the experimental spectrum was likely incorrectly interpreted and the sequence contains errors.

6.5 Conclusions

When lack of hybridoma necessitates the *de novo* sequencing of an antibody, mass spectrometry plays a critical role. Once the antibody in question has been determined to be monoclonal through a combination of analytical techniques including N-terminal sequencing and intact mass measurement, a strategy utilizing enzymatic digestions followed by LC–MS/MS may proceed. Proteomic sequence determination requires careful spectral interpretation followed by sequence assembly. Homology searches provide antibody framework sequence surrounding the CDR regions. Through the use of sequence tag search or sequence extension via sophisticated software tools, the complete *de novo* sequence may be determined. Finally, synthetic isotopically labeled peptides containing the putative CDR sequences may be spiked into the antibody digest for validation.

References

1 Koren, S., Kosmač, M., Colja Venturini, A., Montanič, S., and Čurin Šerbec, V. (2008) Antibody variable-region sequencing as a method for hybridoma cell-line authentication. *Appl. Microbiol. Biotechnol.* 78, 1071–1078.
2 Kim, S., Mischerikow, N., Bandeira, N., Navarro, J. D., Wich, L., Mohammed, S., Heck, A. J. R., and Pevzner, P. A. (2010) The generating function of CID, ETD, and CID/ETD pairs of tandem mass spectra: applications to database search. *Mol. Cell. Proteomics* 9, 2840–2852.
3 Mukhyala, K. and Masselot, A. (2014) Visualization of protein sequence features using JavaScript and SVG with pViz.js. *Bioinformatics* 30(23), 3408–3409.
4 Huang, L., Lu, J., Wroblewski, V. J., Beals, J. M., and Riggin, R. M. (2005) In vivo deamidation characterization of monoclonal antibody by LC/MS/MS. *Anal. Chem.* 77, 1432–1439.
5 Vlasak, J., Bussat, M. C., Wang, S., Wagner-Rousset, E., Schaefer, M., Klinguer-Hamour, C., Kirchmeier, M., Corvaïa, N., Ionescu, R., and Beck, A. (2009) Identification and characterization of asparagine deamidation in the light chain CDR1 of a humanized IgG1 antibody. *Anal. Biochem.* 392, 145–154.
6 Chumsae, C., Gaza-Bulseco, G., Sun, J., and Liu, H. (2007) Comparison of methionine oxidation in thermal stability and chemically stressed samples of a fully human monoclonal antibody. *J. Chromatogr. B Anal. Technol. Biomed. Life Sci.* 850, 285–294.
7 Liu, H., Gaza-Bulseco, G., Xiang, T., and Chumsae, C. (2008) Structural effect of deglycosylation and methionine oxidation on a recombinant monoclonal antibody. *Mol. Immunol.* 45, 701–708.
8 Pham, V., Tropea, J., Wong, S., Quach, J., and Henzel, W. J. (2003) High-throughput protein sequencing. *Anal. Chem.* 75, 875–882.
9 Aebersold, R., and Mann, M. (2003) Mass spectrometry-based proteomics. *Nature* 422, 198–207.
10 Eng, J. K., McCormack, A. L., and Yates, J. R. (1994) An approach to correlate tandem mass spectral data of peptides with amino acid sequences in a protein database. *J. Am. Soc. Mass Spectrom.* 5, 976–989.

11 Perkins, D. N., Pappin, D. J., Creasy, D. M., and Cottrell, J. S. (1999) Probability-based protein identification by searching sequence databases using mass spectrometry data. *Electrophoresis* 20, 3551–3567.
12 Frank, A. M., Savitski, M. M., Nielsen, M. L., Zubarev, R. A., and Pevzner, P. A. (2007) De novo peptide sequencing and identification with precision mass spectrometry. *J. Proteome Res.* 6, 114–123.
13 Frank, A., and Pevzner, P. A. (2005) PepNovo: de novo peptide sequencing via probabilistic network modeling. *Anal. Chem.* 77, 964–973.
14 Ma, B., Zhang, K., Hendrie, C., Liang, C., Li, M., Doherty-Kirby, A., and Lajoie, G. (2003) PEAKS: powerful software for peptide de novo sequencing by tandem mass spectrometry. *Rapid Commun. Mass Spectrom.* 17, 2337–2342.
15 Castellana, N. E., McCutcheon, K., Pham, V. C., Harden, K., Nguyen, A., Young, J., Adams, C., Schroeder, K., Arnott, D., Bafna, V., Grogan, J. L., and Lill, J. R. (2011) Resurrection of a clinical antibody: template proteogenomic de novo proteomic sequencing and reverse engineering of an anti-lymphotoxin-α antibody. *Proteomics* 11, 395–405.
16 Bandeira, N., Pham, V., Pevzner, P., Arnott, D., and Lill, J. R. (2008) Automated de novo protein sequencing of monoclonal antibodies. *Nat. Biotechnol.* 26, 1336–1338.
17 Guthals, A., Clauser, K. R., Frank, A. M., and Bandeira, N. (2013) Sequencing-Grade De novo Analysis of MS/MS Triplets (CID/HCD/ETD) From Overlapping Peptides. *J. Proteome Res.* 12, 2846–2857.
18 Liu, X., Han, Y., Yuen, D., and Ma, B. (2009) Automated protein (re)sequencing with MS/MS and a homologous database yields almost full coverage and accuracy. *Bioinformatics* 25, 2174–2180.
19 Guthals, A., Clauser, K. R., and Bandeira, N. (2012) Shotgun protein sequencing with meta-contig assembly. *Mol. Cell. Proteomics* 10, 1084–1096.
20 Guthals, A., and Bandeira, N. (2012) Peptide identification by tandem mass spectrometry with alternate fragmentation modes. *Mol. Cell. Proteomics* 11, 550–557.
21 Datta, R., and Bern, M. (2009) Spectrum fusion: using multiple mass spectra for de novo peptide sequencing. *J. Comput. Biol.* 16, 1169–1182.
22 Tsai, P. L., Chen, S. F., and Huang, S. Y. (2013) Mass spectrometry-based strategies for protein disulfide bond identification. *Rev. Anal. Chem.* 32, 257–268.
23 Gundry, R. L., White, M. Y., Murray, C. I., Kane, L. A., Fu, Q., Stanley, B. A., and Van Eyk, J. E. (2009) *Current Protocols in Molecular Biology* (John Wiley & Sons, Inc., Hoboken), pp 10.25.1–10.25.23.
24 Frank, A. M., Bandeira, N., Shen, Z., Tanner, S., Briggs, S. P., Smith, R. D., and Pevzner, P. A. (2008) Clustering millions of tandem mass spectra. *J. Proteome Res.* 7, 113–122.

25 Bandeira, N., Tang, H., Bafna, V., and Pevzner, P. A. (2004) Shotgun protein sequencing by tandem mass spectra assembly. *Anal. Chem.* 76, 7221–7233.

26 Bandeira, N., Clauser, K. R., and Pevzner, P. A. (2007) Shotgun protein sequencing: assembly of peptide tandem mass spectra from mixtures of modified proteins. *Mol. Cell. Proteomics* 6, 1123–1134.

27 Bandeira, N., Tsur, D., Frank, A., and Pevzner, P. A. (2007) Protein identification by spectral networks analysis. *Proc. Natl. Acad. Sci. U. S. A.* 104, 6140–6145.

28 Tsur, D., Tanner, S., Zandi, E., Bafna, V., and Pevzner, P. A. (2005) Identification of post-translational modifications by blind search of mass spectra. *Nat. Biotechnol.* 23, 1562–1567.

29 Johnson, R. S., and Biemann, K. (1987) The primary structure of thioredoxin from Chromatium vinosum determined by high-performance tandem mass spectrometry. *Biochemistry* 26, 1209–1214.

30 Seidler, J., Zinn, N., Boehm, M. E., and Lehmann, W. D. (2010) De novo sequencing of peptides by MS/MS. *Proteomics* 10, 634–649.

31 Shen, Y., Tolić, N., Hixson, K. K., Purvine, S. O., Anderson, G. A., and Smith, R. D. (2008) De novo sequencing of unique sequence tags for discovery of post-translational modifications of proteins. *Anal. Chem.* 80, 7742–7754.

32 Dasari, S., Chambers, M. C., Slebos, R. J., Zimmerman, L. J., Ham, A.-J. L., and Tabb, D. L. (2010) TagRecon: high-throughput mutation identification through sequence tagging. *J. Proteome Res.* 9, 1716–1726.

33 Bern, M., Cai, Y., and Goldberg, D. (2007) Lookup peaks: a hybrid of de novo sequencing and database search for protein identification by tandem mass spectrometry. *Anal. Chem.* 79, 1393–400.

7

Characterization of Antibody–Drug Conjugates

Yichin Liu

Department of Biochemical and Cellular Pharmacology, Genentech Inc., South San Francisco, CA, USA

Abbreviations

ADC	antibody–drug conjugate
CCS	collision cross section
DAR	drug-to-antibody ratio
HCD	higher-energy collisional dissociation
HDX	hydrogen–deuterium exchange
HIC	hydrophobic interaction chromatography
IM-MS	ion mobility mass spectrometry
LC-MS	liquid chromatography mass spectrometry
MGBA	minor groove-binding alkylating
MS	mass spectrometry
nano-ESI	nanoelectrospray ionization
NHS	*N*-hydroxysuccinimide
TMT	tandem mass tag
TOF	time-of-flight
TWIM-MS	traveling wave ion mobility MS
VC-PABC	valine-citrulline-*p*-aminobenzylcarbamate
WT	wild type

Analytical Characterization of Biotherapeutics, First Edition. Edited by Jennie R. Lill and Wendy Sandoval.

7.1 Introduction

Target-specific antibodies armed with potent cytotoxic agents to form antibody–drug conjugates (ADCs) have shown promise as novel therapeutic agents in fighting human cancer. An ADC consists of three major components: antibody, linker (either cleavable or noncleavable), and cytotoxic payload. The rationale behind using ADCs is to bring specificity of a cytotoxic agent by coupling it to an antibody that selectively targets tumor cells with the goal of enhancing the therapeutic index (TI), the window between maximal tolerated dose (MTD) and minimal efficacious dose (MED). In principle the TI should be greatly enhanced for ADCs compared to the direct treatment with the cytotoxic agent, which brings damage to normal cells and tissues. However, in practice, ADCs have not lived up to the initial promise of complete tumor specificity. The limited success of ADCs is due to numerous challenges in designing a molecule that selectively releases its toxic payload only in the tumor environment while sparing normal tissues. These challenges include (i) identifying appropriate tumor antigens and their respective epitopes to trigger efficient internalization of ADC to the lysosomal compartment for degradation and (ii) designing an effective intracellular payload release mechanism in tumor cells while maintaining its stability in circulation.

To optimize an ADC to achieve tumor specific uptake, internalization, degradation, and drug release without damaging the normal healthy tissues requires significant ADC characterization and validation using various bioanalytical, cellular, and biochemical methods that may not be typically applied to either the conventional biologics or small molecule therapeutics [1]. With the increasing complexity of the ADC designs, conventional immunoassays are not sufficient in providing the high content information required to understand the pharmacology of the ADC both *in vitro* and *in vivo*. The advances in mass spectrometry (MS) in the past few years have provided research scientists alternative solutions in profiling these therapeutic agents and in obtaining details of the molecules that are essential to validate their on-target efficacy and potentially to identify the off-target toxicity liabilities. However, additional improvements in future mass spectrometers are still much needed to support the development of ADCs. For example, due to the high potency of an ADC, it is often administered at a much lower dose *in vivo* relative to other conventional antibody therapeutics. As a result, the characterization of the sample from *in vivo* studies can potentially reach the limit of bioanalytical detection sensitivity including those observed with MS.

The conventional approach of analyzing intact proteins in MS directly without digestion (referred hereafter as intact mass analyses) has typically been performed under denaturing conditions to provide drug-to-antibody ratio (DAR) information. However, this technique cannot be applied to ADCs conjugated through their interchain disulfide cysteine residues. The recent advances in native MS, using nondenaturing conditions to preserve noncovalent interactions, are able to address this limitation. In addition, these technologies, which include ion mobility, have provided alternative new biophysical approaches in profiling the three-dimensional structure of ADCs. Finally, with the growing complexity of new linker–drug designs that may be reactive and prone to modifications in biological matrices both *in vitro* and *in vivo*, MS can play a critical role in elucidating ADC catabolism and metabolism. The profiles of catabolites in circulation and in tumor cells will provide key data to understand the stability of the linked drug conjugated at the antibody and to confirm the on-target mechanism (e.g., DNA damage) of cell killing. The focus of this review is to highlight the applications of the increasingly diversified MS techniques in characterizing the integrity of ADCs from production to *in vitro* and *in vivo* testing as well as in elucidating the mechanisms of their biotransformation.

7.2 Characterization of DAR Utilizing MS

To describe ADC integrity, a DAR is typically calculated to assess the conjugation efficiency or the stability of the conjugate in biological matrices, either *in vivo* or *in vitro*. In this section, the most advanced MS techniques for characterizing the stability of conjugation chemistry, as well as the linkers and payloads of ADCs, are reviewed.

7.2.1 The Stability of Conjugation Chemistry and the Cleavable Linker of ADC

Historically, ADCs were generated through conjugation of a linker–drug moiety to endogenous nucleophilic residues such as lysine or cysteine [2]. This type of conjugation typically resulted in a distribution of DAR ranging from 0 to 8; but with conjugation using interchain disulfide bonds, the DAR should exist only as even numbers. In order to achieve ADCs with more homogenous DAR with minimal production-dependent variation, exploration of site-specific conjugation has become the primary focus for the next generation ADCs. For example, THIOMAB technology utilizes site-specific mutagenesis to introduce one additional cysteine pair to an antibody for conjugation [3]. This type of ADC should yield a maximum of DAR 2 [3]. Unlike the stable lysine conjugation which typically uses *N*-hydroxysuccinimide (NHS), THIOMABs utilize thiol conjugation with maleimide through the cysteine side chain to form succinimide, which is relatively unstable in biological matrices. The stability of the cysteine conjugate, however, can be enhanced when located in site-specific residues [3, 4]. Others have also reported using microbial transglutaminase or the incorporation of nonnatural amino acid for alternative site-specific conjugations that potentially can provide good stability of ADCs *in vivo* [5, 6]. Recently, Pfizer has reported that the stability of the protease-cleavable linkers, such as the valine-citrulline-*p*-aminobenzylcarbamate (VC-PABC), can also be impacted by the sites of antibody conjugation [7]. The authors have hypothesized that either the local environment or steric hindrance by proximal antibody domains may play a role in protecting the linker from proteolytic cleavage. In summary, the pharmacokinetic profile of the ADC is the aggregate of the total antibody clearance and the stability of conjugation chemistry as well as the linker–drug moiety. Any of these components can lead to changes of tumor exposure to the intact active ADC and thus likely affect efficacy and toxicity.

7.2.2 Historical Usage of Hydrophobic Interaction Chromatography in ADC Characterization

Hydrophobic interaction chromatography (HIC) paired with UV/VIS spectroscopy was one of the first bioanalytical methods implemented to characterize DAR [8–10]. Taking advantage of the hydrophobic nature of a small molecule toxic payload that can increase the overall lipophilicity of the antibody surface, HIC separates different DAR species under native conditions. As a result, this technique is advantageous for maintaining intact ADCs that conjugate through interchain cysteinyl residues [3]. Because a salt gradient is generally used for analyte elution, HIC is usually coupled to UV/VIS detection measuring absorbance. Several limitations, however, are known to be associated with DAR profiling using HIC. First, in targeted monitoring of conjugated drug where high sensitivity of the payload is desired, a high extinction coefficient is required. Second, for spectroscopic absorbance specificity, it is crucial that the conjugated payload has a nonoverlapping wavelength from aromatic amino acid residues such as tryptophan. The relatively high limit of detection for UV absorbance and the lack of specificity also make it impractical to apply to *in vivo* sample analyses. In addition, both HIC and spectroscopy provide relatively low-resolution information that is often incapable of distinguishing minor chemical changes to the linker–drug on the ADC. Characterizing these modifications is crucial because they may dramatically impact the potency of the payload. Finally, monitoring alterations of the linker–drug is critical in assessing the integrity of the ADC either during storage or when dosed *in vivo*. An example of a common minor chemical change includes the hydrolysis of the succinimide ring for the maleimide-based linker–drugs, which can lead to an increase in the stability of the conjugation chemistry [4]. Consequently, there is a growing recognition that MS is needed to overcome the limitations of HIC by providing more complete information with unambiguous protein and linker–drug characterization at higher sensitivity.

7.2.3 Intact MS Detection under Denaturing Condition

DAR characterization using intact MS was first reported by Xu *et al.* in 2011 [11] where they described a sensitive approach on characterizing THIOMAB drug conjugates by coupling affinity capture with reverse-phase liquid chromatography mass spectrometry (LC-MS). This intact MS approach can be more versatile than the antibody-conjugated drug method reported for drug load assessment which involves affinity capture of the target ADC, linker cleavage using recombinant enzyme, followed by LC-MS/MS quantitation of the released payload [12]. The latest time-of-flight (TOF) mass analyzer used by Xu allowed the direct detection of intact ADCs with sufficient resolution to observe succinimide ring hydrolysis [4]. They further applied the method to pharmacokinetic analyses of ADCs [13] and have been widely cited from publications that characterize ADCs that do not use interchain cysteine for conjugation. One of those publications described further optimization of intact ADC MS detection sensitivity by incorporating ion mobility mass spectrometry (IM-MS). The increase in signal to noise (by at least 10 fold) was afforded after separating the deglycosylation enzyme, PNGaseF, in the gas phase from the ADC, which would otherwise coelute under conventional conditions using reverse-phase liquid chromatography [14]. While intact MS is now a mature technique for providing routine bioanalytical characterization of ADCs, data analysis is still largely a manual process and has emerged as a major bottleneck. The data analysis process involved with intact ADCs and catabolites requires identifying, matching, and labeling deconvoluted mass peaks, which can be highly complex. Moreover, different ADC catabolisms can occur that often result in a complex mass spectrum, especially for ADCs with a starting DAR > 1. To address these challenges, many ADC MS laboratories have started exploring custom-made software. One example presented by Genentech in the 2014 Dotmatics Day Symposium (San Francisco, CA) utilizes the existing analytical tools within Dotmatics Vortex software to automate the entire process in four steps. The software begins, first, with reconstructing deconvoluted intact mass spectra in batch; second, identifying and marking peaks above a prespecified signal threshold; third, determining the mass differences between the marked peaks; and, fourth, automatically matching and labeling the marked peaks with mass differences (within a specified tolerance) that correspond to a list of known chemical modifications and degradations (Figure 7.1). As additional intact protein MS techniques become more mature for ADC characterization, there is still much to be done to improve and streamline the data processing, interpretation and reporting.

7.2.4 Intact MS Characterization under Native Conditions

With the growing interest in using intact protein MS in characterizing antibody therapeutics including ADC molecules, the diversity of these approaches has also increased. As mentioned previously, the disadvantage of the conventional

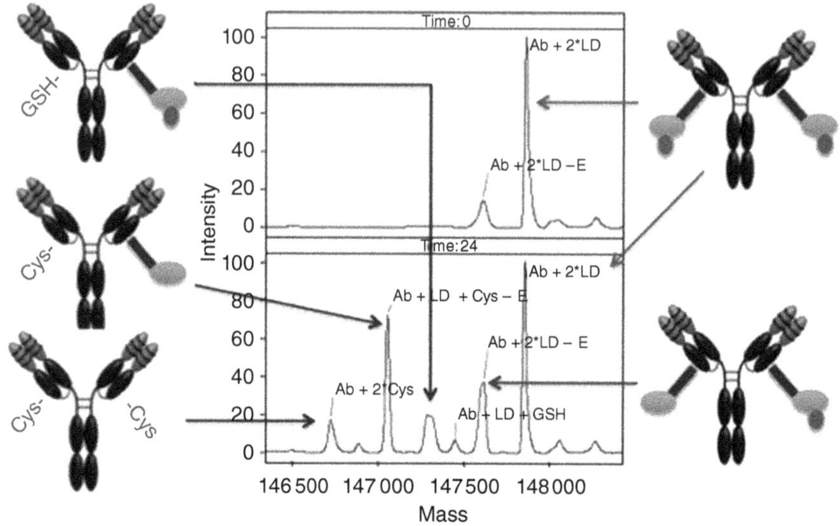

Figure 7.1 Dotmatics Vortex processing of the deconvoluted intact protein mass spectra of a mixture of ADC and its catabolites. Features were added to the software to reconstruct the deconvoluted mass spectra and automatically label the observed antibody, linker–drug moieties, as well as the possible modifications. Antibody graphics have been added manually for illustrative purposes.

reverse-phase LC-MS methodology in characterizing ADCs is the inability to maintain the integrity of the antibody when the linker–drugs are conjugated at the cysteine residues involved in interchain disulfide bonds. As a result, the mass spectra can only reveal the average DAR by comparing the conjugated versus nonconjugated heavy and light chains of the antibody instead of measuring the distribution across a range of DAR. In 2012, Valliere-Douglass and coworkers reported the implementation of native MS to enable the analyses of interchain cysteinyl-linked ADCs [15]. In subsequent years, additional publications have emerged to describe various methods or instrumentations operated at native conditions for improving accuracy and sensitivity [16–20]. For example, Chen *et al.* reported the usage of a more gentle nanoelectrospray ionization (nano-ESI) source operated at a lower temperature and voltage, compared to high-flow electrospray ionization (ESI), to further prevent the disintegration of the noncovalent ADC complex [16]. In recent years, high-resolution native MS was also applied to characterize the sites of conjugation containing interchain disulfide cysteine residues. By isolating the intact ADC under native conditions in the quadrupole followed by fragmenting in a higher-energy collisional dissociation (HCD) cell, the subunits of antibody with drug conjugated can be readily identified [19]. Application of the analysis of an ADC complex with its antigen using native MS has also been described [19].

In addition to the application of characterizing ADCs conjugated through interchain disulfide bonds resulting in a noncovalent antibody complex, native MS has been shown to provide great utility in profiling ADCs with lysine conjugates as well. Applying tandem mass tag (TMT) labeling and detecting with high resolution under native conditions, Gautier *et al.* were able to show that 69 out of 88 total available amine groups (87 lysines plus N-terminus) on deglycosylated IgG1 were reactive and the reactivity of some sites (presumably near the glycosylation sites) was reduced in IgG1 when glycosylation was still intact [21]. Using the combination of TMT labeling and native MS analyses, they were also able to show that lysine-labeled IgG4, another common mAb employed in ADC design, can potentially impair the reassociation of two symmetrical monomeric subunits to form a full antibody even though there was no difference in the labeling rates between subunit versus full antibody [21]. In another recent example, Marcoux *et al.* described the application of native MS to characterize T-DM1 (Kadcyla®), which is trastuzumab conjugated to emtansine through lysine residues. The authors were able to show that charge state reduction of the ADC under native versus denaturing conditions has allowed different levels of DAR species to be more clearly resolved with fewer overlapping ion distributions even for glycosylated samples [20].

7.2.5 Middle-down and Bottom-up MS Approach in Mapping Drug Conjugates

Prior to the introduction of site-specific conjugation such as engineered cysteine mutations or the site-specific incorporation of nonnatural amino acids, ADCs were typically produced as a heterogeneous population with a distribution of DAR as well as different conjugation site combinations. In order to understand the potential lot-to-lot variations in ADC production, it is apparent that identifying the common sites of labeling as well as the level of reactivity on each site has become critical in ensuring quality control in manufacturing. To complement the ADC analyses at the intact level, approaches using middle-down and bottom-up MS [22] have been described over the past few years aiming to define the conjugation levels in each antibody subunit (light chain versus heavy chain or Fab versus Fc). These types of analyses can also provide the exact locations of conjugation [21, 23–25]. For middle-down analysis, the two most commonly described methods are (i) reducing interchain disulfide bonds to release light chain and heavy chain subunits and (ii) proteolytic digestion with IdeS to separate Fab and Fc domains [26–30]. The products of these two approaches will likely still contain positional isomers and bottom-up MS can assist in identifying the exact locations of the conjugation within the antibody subunit to differentiate the isomers [23, 25]. While applying TMT analyses to assess the global level of lysine reactivity for a mAb using native MS, Gautier *et al.* have also coupled the analysis with quantitative bottom-up MS to elucidate the spectrum of lysine reactivity on IgG1 toward the NHS ester functional group of TMT [21]. It is unclear, however, if the relative reactivity of different lysine residues can be influenced by linkers and payloads due to their unique physicochemical properties including potential steric hindrance from the shape of the linker–drugs. Future studies with diverse conjugates will likely provide additional insights.

7.3 Structural Characterization of ADC

With the advances of the MS techniques, structural characterization of protein therapeutics including mAb has also emerged as a popular research topic in both the pharmaceutical industry and academic institutes. Structural analyses of antibodies using MS that employ IM-MS and hydrogen–deuterium exchange MS (HDX-MS) have been reported [31–33]. For ADCs, the structural characterization has been focused on the impact of the conjugated payloads on the overall conformation of the antibody. In this section, the recent progress of techniques in IM-MS and HDX-MS is discussed.

7.3.1 Ion-Mobility Mass Spectrometry

Various biophysical methods have been employed over the past couple of decades in characterizing the structures or conformations of different mAb isoforms. In recent years, IM-MS showed great promise in elucidating global conformational changes of mAb. IM-MS has been applied to enhance sensitivity of detection by further separating molecules with the same mass per charge based on their different shapes. The principle is analogous to gel electrophoresis in separating macromolecules (e.g., oligonucleotides or proteins) with different dimensions or curvatures. One of the first examples in applying IM-MS characterization on a mAb was demonstrated by Bagal *et al.* The authors were able to differentiate two IgG2 isoforms that showed disulfide heterogeneity at the hinge region using IM-MS [32]. A few years later, Debaene *et al.* showed how IM-MS under native conditions can monitor Fab exchange in the formation of a bispecific antibody [34]. In both cases, traveling wave ion mobility MS (TWIM-MS) was used to determine the collision cross section (CCS) of the analyte, which is essentially the average diameter of the molecule [35]. This application provides great utility in quality control of antibody therapeutic production complementary to the standard LC-MS techniques.

The characterization of ADC conformation under native conditions using IM-MS was first reported in 2014 by Debeane *et al.* [34]. In this work, the authors showed that a higher DAR from brentuximab vedotin has led to a larger CCS. In addition, the higher charge states of the same DAR also resulted in a larger CCS consistent with the slight unfolding of the ADC by Coulombic charge repulsions [34]. The same group later reported the IM-MS characterization of TDM-1, which is an ADC conjugated through lysine residues [20]. While the potential of IM-MS in ADC characterization is immense, this field of research will benefit from further improvement in instrumentation such as better resolution in the IM cell. Understanding the potential difference between gas phase folding in MS and solid phase conformation (e.g., crystallography) of the large protein therapeutics will also be critical in cross-comparing the structural analyses among different biophysical techniques.

7.3.2 Hydrogen–Deuterium Exchange Mass Spectrometry

In addition to IM-MS, hydrogen–deuterium exchange mass spectrometry (HDX-MS) has also emerged as a practical technique in elucidating potential changes in mAb conformation [31]. However, unlike IM-MS, which provides folding information at the global level, HDX-MS is more informative of local changes in protein conformation. The principle behind HDX-MS is similar to other biophysical techniques such as nuclear magnetic resonance (NMR) that utilizes hydrogen–deuterium exchange (HDX) of the amide backbone of the protein to assess the solvent accessibility of individual residues. Collectively, the mapping of the solvent accessibility to the entire protein can yield crucial information on the protein's tertiary structure. For ADC, HDX-MS has been implemented to characterize interchain cysteine-linked ADC brentuximab vedotin [36] as well as site-directed conjugation to the mutant S239C [37]. The data on brentuximab vedotin showed that there was an increase in HDX only at the interchain disulfide region primarily due to the absence of the disulfide bond; the presence of payload conjugation, however, has no impact [36]. In contrast, the ADC utilizing the S239C site-specific mutation showed no impact on HDX kinetics compared to the wild type (WT) mAb, consistent with the molecular modeling that the serine 239 side chain is not involved in maintaining the protein tertiary structure [36]. Taken together, HDX-MS has great utility in understanding the impact of site-specific conjugation to the structural integrity of the ADC.

7.4 Characterization of ADC Catabolism by MS

As data have emerged from the increasing number of clinical trials of ADC molecules, there is a general realization that understanding ADC catabolism in both tumor and normal tissues as well as in circulation is critical for validating the on-target efficacy and predicting potential toxicity. These types of studies should also include profiling the metabolism of the small molecule payload released from the antibody. In addition to the conventional immunoassays, MS has recently become the mainstream technique in profiling ADC catabolism and small molecule payload metabolism [38]. For example, the characterization of catabolism of T-DM1 was recently reported to provide a potential mechanistic understanding of thrombocytopenia that resulted in dose-limiting toxicity of the ADC in clinical trials [39]. Saad and colleagues recently described a comprehensive workflow involving various qualitative and quantitative MS approaches to categorize different catabolites that arise from a single ADC [38]. The complexity and heterogeneity of ADC catabolism adds significant challenges to this process, however, especially in detecting the presence of potentially potent catabolites or metabolites that are in low abundance. Consequently, radioactive labeling of the ADC is often required to provide sufficient detection sensitivity especially for *in vivo* studies [38, 40]. For an ADC that is conjugated with potent covalent inhibitors such as DNA damaging payloads, potential DNA adducts need to be taken into consideration when monitoring the generation of targeted catabolites in cells. An example of this type of analysis utilizing MS has been demonstrated by Thevanayagam *et al.* in detecting the adenine adduct formed with a DNA minor groove-binding alkylating (MGBA) drug MDX-1203 released from an ADC both *in vitro* and *in vivo* [41]. The authors were able to show that in mice the nucleotide adduct was more abundant in tumor than other tissues or organs confirming the on-target delivery of the ADC payload. With continuous improvement of MS technology to assist characterization, catabolism and biotransformation analyses will greatly contribute to the validation and future design of the next generation ADCs.

7.5 Conclusions

With many ADCs currently in clinical trials, the learning from human efficacy and toxicity data will define the focus of the future design improvements for the next generation ADC molecules. Understanding the mechanisms behind the emerging results will require detailed biotransformation analyses of the ADCs in addition to the conventional pharmacokinetics characterization for biologics. DAR characterization of ADCs is recognized to be important for production quality control as well as pharmacokinetics analysis. Structural characterization of ADCs is also gaining attention as it can potentially provide information to predict the stability of the intact molecule. In recent years, MS has risen to become the mainstream technology platform to characterize ADCs. With the advances in various MS techniques, we have also witnessed growing success in characterizing ADC processing both *in vitro* and *in vivo*. There is no doubt that we will see more innovative approaches in the near future that may eventually reveal the full potential of ADCs as highly potent and specific drugs for the treatment of cancer and other diseases.

References

1 Stephan JP, Kozak KR, Wong WL. Challenges in developing bioanalytical assays for characterization of antibody-drug conjugates. *Bioanalysis*. 2011;3(6):677–700.

2 Panowksi S, Bhakta S, Raab H, Polakis P, Junutula JR. Site-specific antibody drug conjugates for cancer therapy. *MAbs*. 2014;6(1):34–45.

3 Junutula JR, Raab H, Clark S, Bhakta S, Leipold DD, Weir S, *et al.* Site-specific conjugation of a cytotoxic drug to an antibody improves the therapeutic index. *Nat Biotechnol.* 2008;26(8):925–932.

4 Shen BQ, Xu K, Liu L, Raab H, Bhakta S, Kenrick M, *et al.* Conjugation site modulates the in vivo stability and therapeutic activity of antibody-drug conjugates. *Nat Biotechnol.* 2012;30(2):184–189.

5 Strop P, Liu SH, Dorywalska M, Delaria K, Dushin RG, Tran TT, *et al.* Location matters: site of conjugation modulates stability and pharmacokinetics of antibody drug conjugates. *Chem Biol.* 2013;20(2):161–167.

6 Jackson D, Atkinson J, Guevara CI, Zhang C, Kery V, Moon SJ, *et al.* In vitro and in vivo evaluation of cysteine and site specific conjugated herceptin antibody-drug conjugates. *PLoS One*. 2014;9(1):e83865.

7 Dorywalska M, Strop P, Melton-Witt JA, Hasa-Moreno A, Farias SE, Galindo Casas M, *et al.* Effect of attachment site on stability of cleavable antibody drug conjugates. *Bioconjug Chem*. 2015;26(4):650–659.

8 Alley SC, Anderson KE. Analytical and bioanalytical technologies for characterizing antibody-drug conjugates. *Curr Opin Chem Biol.* 2013;17(3):406–411.

9 Beck A, Wagner-Rousset E, Ayoub D, Van Dorsselaer A, Sanglier-Cianferani S. Characterization of therapeutic antibodies and related products. *Anal. Chem.* 2013;85(2):715–736.

10 Beck A. Review of antibody-drug conjugates, methods in molecular biology series: a book edited by Laurent Ducry. *MAbs*. 2014;6(1):30–33.

11 Xu K, Liu L, Saad OM, Baudys J, Williams L, Leipold D, *et al.* Characterization of intact antibody-drug conjugates from plasma/serum in vivo by affinity capture capillary liquid chromatography-mass spectrometry. *Anal Biochem.* 2011;412(1):56–66.

12 Kaur S, Xu K, Saad OM, Dere RC, Carrasco-Triguero M. Bioanalytical assay strategies for the development of antibody-drug conjugate biotherapeutics. *Bioanalysis*. 2013;5(2):201–226.

13 Xu K, Liu L, Dere R, Mai E, Erickson R, Hendricks A, *et al.* Characterization of the drug-to-antibody ratio distribution for antibody-drug conjugates in plasma/serum. *Bioanalysis*. 2013;5(9):1057–1071.

14 Huang RY, Deyanova EG, Passmore D, Rangan V, Deshpande S, Tymiak AA, *et al.* Utility of Ion Mobility mass spectrometry for drug-to-antibody ratio

measurements in antibody-drug conjugates. *J Am Soc Mass Spectrom*. 2015;26(10):1791–1794.
15 Valliere-Douglass JF, McFee WA, Salas-Solano O. Native intact mass determination of antibodies conjugated with monomethyl Auristatin E and F at interchain cysteine residues. *Anal Chem*. 2012;84(6):2843–2849.
16 Chen J, Yin S, Wu Y, Ouyang J. Development of a native nanoelectrospray mass spectrometry method for determination of the drug-to-antibody ratio of antibody-drug conjugates. *Anal Chem*. 2013;85(3):1699–1704.
17 Debaene F, Boeuf A, Wagner-Rousset E, Colas O, Ayoub D, Corvaia N, *et al*. Innovative native MS methodologies for antibody drug conjugate characterization: high resolution native MS and IM-MS for average DAR and DAR distribution assessment. *Anal Chem*. 2014;86(21):10674–10683.
18 Hengel SM, Sanderson R, Valliere-Douglass J, Nicholas N, Leiske C, Alley SC. Measurement of in vivo drug load distribution of cysteine-linked antibody-drug conjugates using microscale liquid chromatography mass spectrometry. *Anal Chem*. 2014;86(7):3420–3425.
19 Dyachenko A, Wang G, Belov M, Makarov A, de Jong RN, van den Bremer ET, *et al*. Tandem native mass-spectrometry on antibody-drug conjugates and submillion Da antibody-antigen protein assemblies on an Orbitrap EMR equipped with a high-mass quadrupole mass selector. *Anal Chem*. 2015;87(12):6095–6102.
20 Marcoux J, Champion T, Colas O, Wagner-Rousset E, Corvaia N, Van Dorsselaer A, *et al*. Native mass spectrometry and ion mobility characterization of trastuzumab emtansine, a lysine-linked antibody drug conjugate. *Protein Sci*. 2015;24(8):1210–1223.
21 Gautier V, Boumeester AJ, Lossl P, Heck AJ. Lysine conjugation properties in human IgGs studied by integrating high-resolution native mass spectrometry and bottom-up proteomics. *Proteomics*. 2015;15(16):2756–2765.
22 Zhang Z, Pan H, Chen X. Mass spectrometry for structural characterization of therapeutic antibodies. *Mass Spectrom Rev*. 2009;28(1):147–176.
23 Farias SE, Strop P, Delaria K, Galindo Casas M, Dorywalska M, Shelton DL, *et al*. Mass spectrometric characterization of transglutaminase based site-specific antibody-drug conjugates. *Bioconj Chem*. 2014;25(2):240–250.
24 Wagner-Rousset E, Janin-Bussat MC, Colas O, Excoffier M, Ayoub D, Haeuw JF, *et al*. Antibody-drug conjugate model fast characterization by LC-MS following IdeS proteolytic digestion. *MAbs*. 2014;6(1):273–285.
25 Janin-Bussat MC, Dillenbourg M, Corvaia N, Beck A, Klinguer-Hamour C. Characterization of antibody drug conjugate positional isomers at cysteine residues by peptide mapping LC-MS analysis. *J Chromatogr B Analyt Technol Biomed Life Sci*. 2015;981–982:9–13.
26 von Pawel-Rammingen U, Johansson BP, Bjorck L. IdeS, a novel streptococcal cysteine proteinase with unique specificity for immunoglobulin G. *EMBO J*. 2002;21(7):1607–1615.

27 Vincents B, von Pawel-Rammingen U, Bjorck L, Abrahamson M. Enzymatic characterization of the streptococcal endopeptidase, IdeS, reveals that it is a cysteine protease with strict specificity for IgG cleavage due to exosite binding. *Biochemistry*. 2004;43(49):15540–15549.
28 Wenig K, Chatwell L, von Pawel-Rammingen U, Bjorck L, Huber R, Sondermann P. Structure of the streptococcal endopeptidase IdeS, a cysteine proteinase with strict specificity for IgG. *Proc Natl Acad Sci U S A*. 2004;101(50):17371–17376.
29 Johansson BP, Shannon O, Bjorck L. IdeS: a bacterial proteolytic enzyme with therapeutic potential. *PLoS One*. 2008;3(2):e1692.
30 An Y, Zhang Y, Mueller HM, Shameem M, Chen X. A new tool for monoclonal antibody analysis: application of IdeS proteolysis in IgG domain-specific characterization. *MAbs*. 2014;6(4):879–893.
31 Houde D, Arndt J, Domeier W, Berkowitz S, Engen JR. Characterization of IgG1 conformation and conformational dynamics by hydrogen/deuterium exchange mass spectrometry. *Anal Chem*. 2009;81(7):2644–2651.
32 Bagal D, Valliere-Douglass JF, Balland A, Schnier PD. Resolving disulfide structural isoforms of IgG2 monoclonal antibodies by ion mobility mass spectrometry. *Anal Chem*. 2010;82(16):6751–6755.
33 Pacholarz KJ, Porrini M, Garlish RA, Burnley RJ, Taylor RJ, Henry AJ, *et al*. Dynamics of intact immunoglobulin G explored by drift-tube ion-mobility mass spectrometry and molecular modeling. *Angew Chem Int Ed Engl*. 2014;53(30):7765–7769.
34 Debaene F, Wagner-Rousset E, Colas O, Ayoub D, Corvaia N, Van Dorsselaer A, *et al*. Time resolved native ion-mobility mass spectrometry to monitor dynamics of IgG4 Fab arm exchange and "bispecific" monoclonal antibody formation. *Anal Chem*. 2013;85(20):9785–9792.
35 Leary JA, Schenauer MR, Stefanescu R, Andaya A, Ruotolo BT, Robinson CV, *et al*. Methodology for measuring conformation of solvent-disrupted protein subunits using T-WAVE ion mobility MS: an investigation into eukaryotic initiation factors. *J Am Soc Mass Spectrom*. 2009;20(9):1699–1706.
36 Pan LY, Salas-Solano O, Valliere-Douglass JF. Conformation and dynamics of interchain cysteine-linked antibody-drug conjugates as revealed by hydrogen/deuterium exchange mass spectrometry. *Anal Chem*. 2014;86(5):2657–2664.
37 Pan LY, Salas-Solano O, Valliere-Douglass JF. Antibody structural integrity of site-specific antibody-drug conjugates investigated by hydrogen/deuterium exchange mass spectrometry. *Anal Chem*. 2015;87(11):5669–5676.
38 Saad OM, Shen BQ, Xu K, Khojasteh SC, Girish S, Kaur S. Bioanalytical approaches for characterizing catabolism of antibody-drug conjugates. *Bioanalysis*. 2015;7(13):1583–1604.
39 Uppal H, Doudement E, Mahapatra K, Darbonne WC, Bumbaca D, Shen BQ, *et al*. Potential mechanisms for thrombocytopenia development with trastuzumab emtansine (T-DM1). *Clin Cancer Res*. 2015;21(1):123–133.

40 Lai KC, Deckert J, Setiady YY, Shah P, Wang L, Chari R, *et al.* Evaluation of targets for maytansinoid ADC therapy using a novel radiochemical assay. *Pharm Res.* 2015;32:3593–3603.

41 Thevanayagam L, Bell A, Chakraborty I, Sufi B, Gangwar S, Zang A, *et al.* Novel detection of DNA-alkylated adducts of antibody-drug conjugates with potentially unique preclinical and biomarker applications. *Bioanalysis.* 2013;5(9):1073–1081.

8

Characterization of Bispecific or Other Hybrid Molecules

T. Noelle Lombana and Christoph Spiess

Department of Antibody Engineering, Genentech Research and Early Development, South San Francisco, CA, USA

Abbreviations

BiTE™	bispecific T-cell engager
BsAbs	bispecific antibodies
CDRs	complementarity determining regions
CE-SDS	capillary electrophoresis–sodium dodecyl sulfate
CEX	cation-exchange chromatography
DAF	dual-action Fab
DSC	differential scanning calorimetry
DSF	differential scanning fluorimetry
DVD-Ig™	dual variable domain immunoglobulin
ELISA	enzyme-linked immunosorbent assay
EpCAM	anti-epithelial cell adhesion molecule
ESI-Q-TOF	electrospray ionization and quadrupole time-of-flight
HIC	hydrophobic interaction chromatography
HPLC	high-performance liquid chromatography
MAb	monospecific antibody
MS	mass spectrometry
NK	natural killer
RP-HPLC	reverse-phase high-performance liquid chromatography
SDS-PAGE	sodium dodecyl sulfate–polyacrylamide gel electrophoresis
SEC	size-exclusion chromatography
SEC-HPLC	size-exclusion high-performance liquid chromatography
SPR	surface plasmon resonance
UPLC	ultra-high-performance chromatography

Analytical Characterization of Biotherapeutics, First Edition. Edited by Jennie R. Lill and Wendy Sandoval.

8.1 Introduction

Bispecific antibodies (BsAbs) and bispecific molecules have specificity for two distinct protein antigens compared to a monospecific antibody (MAb) that binds only a single antigen. The idea of using a BsAb as a therapeutic has been around for almost as long as its monospecific counterpart; however, during this time over 40 MAbs but only two BsAbs were approved. One of the main reasons why BsAbs lag behind is that they are far more complex to produce. In the past decade, significant advances have been made in the production and characterization of BsAbs. This has partially contributed to a revived interest in advancing BsAbs into clinical development, with over 30 under investigation covering a range of therapeutic indications. Over two-thirds of the BsAbs target the oncology disease area [1], while immunology, ocular, and neurobiology indications are also being explored.

8.1.1 Bispecific Antibody Applications

There are many therapeutic reasons to choose a BsAb although they can be more complex to produce and characterize than a MAb. Adding an additional layer of specificity has the advantage of bestowing antibodies with new modes of action, increasing potency or efficacy, or even combatting biological redundancy [2]. BsAbs function in a number of ways, namely, by bringing two cells together (retargeting effector cells to tumor cells), binding molecules on the same cell (to block or activate), neutralizing two soluble receptor ligands, or delivering antibodies to the brain (Figure 8.1). Additionally, BsAbs can be utilized to improve cancer diagnostics and imaging.

The classical application of BsAbs is the retargeting of effector cells to kill tumor cells, which is achieved by binding a receptor on each cell type. The retargeting BsAbs comprise about half of the roughly 30 bispecific molecules currently in clinical development [1, 3]. In their original application, these molecules retargeted natural killer (NK) cells and macrophages *via* CD16 to eradicate target cells [4], but now mainly focus on redirecting cytotoxic T-cells [5]. The only two approved bispecifics are in this category, and have anti-CD3 (T-cell receptor subunit)/anti-tumor activity. Another application of effector cell retargeting BsAbs, a fusion of T-cell receptor and anti-CD3 scFv named ImmTAC, expands their capabilities to include binding intracellular proteins as peptides displayed on the major histocompatibility complex of cells [6].

Most of the remaining BsAbs in clinical development block or inactivate two receptors or ligands. Many of these include human epidermal growth factor receptor (EGFR/HER/ErbB) family members such as EGFR, HER2, and HER3 or proinflammatory cytokine ligands including TNF, interleukin-1 (IL-1), and IL-13 [1, 3]. There are also preclinical stage biepitopic BsAbs that bind to two different epitopes on the same target, as this approach can have

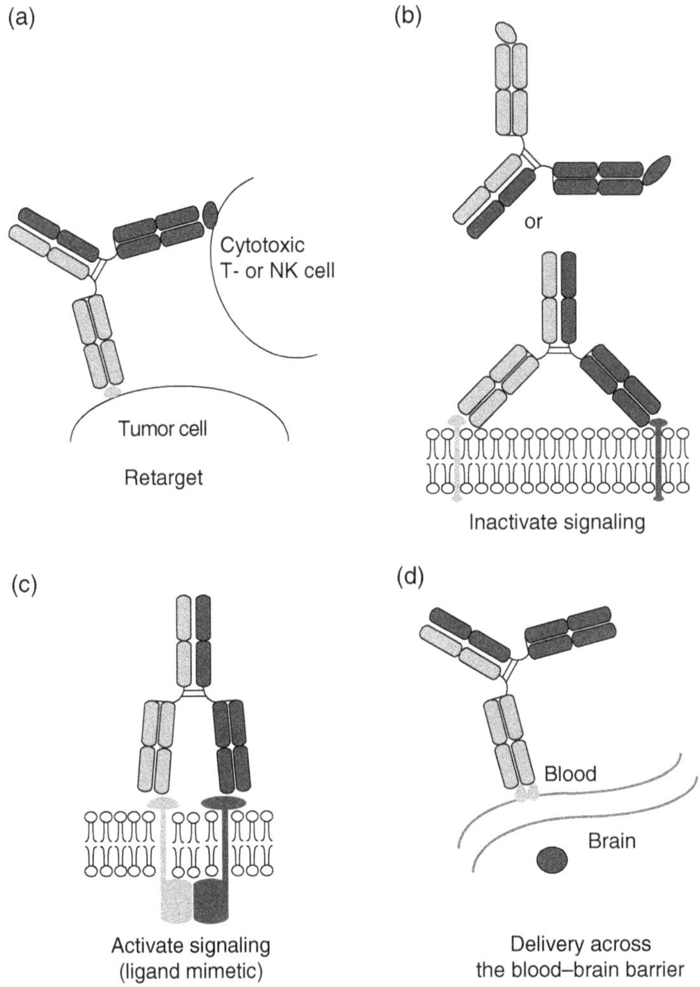

Figure 8.1 Bispecific antibody modes of action. (a) Retarget cytotoxic T-cells or NK cells to kill tumor cells, (b) inactivate signaling of two soluble ligands or receptors, (c) activate signaling by a ligand mimetic, and (d) delivery across the blood–brain barrier.

better antagonistic properties [7, 8]. Alternatively to inactivating signaling pathways, there is one BsAb in clinical development that acts as a Factor VIII ligand mimetic to dimerize Factor IXa and Factor X to induce signaling for the clotting disorder hemophilia A [9]. In addition, preclinical studies with a BsAb that binds a receptor complex, fibroblast growth factor receptor 1 and βKlotho, demonstrated selective activation of the pathway and a promising therapeutic route for the treatment of type 2 diabetes and other obesity-linked disorders [10].

BsAbs can also function as delivery vehicles to target previously inaccessible antigens in the brain. To combat the difficulties of getting therapeutic antibodies across the tight epithelial cell junctions of the blood–brain barrier, a receptor-mediated transport arm can be combined with a therapeutic arm. A BsAb that binds the transferrin receptor and either the enzyme β-secretase [11] or Aβ directly [12] can "hitchhike" on the transferrin-mediated endocytosis mechanism to increase brain uptake and decrease Aβ levels. A similar strategy was also employed to transmigrate a single domain antibody conjugated to blood–brain barrier–impermeable neuropeptides for inflammatory pain [13].

In addition to disease applications, BsAbs can be used as highly specific tools for infectious disease and cancer diagnostics as well as imaging. In a diagnostic assay, the first arm of the BsAb recognizes the pathogen or tumor cell while the second arm is specific for a detection enzyme, such as horseradish peroxidase [14]. This approach allows one-step detection with no need to label the antibody. BsAbs can also improve cancer cell imaging by pretargeting the BsAb to the cancer cell antigen and then binding a radioactive hapten peptide, for example, which is added at a later time. This imaging technology shows higher sensitivity, specificity, and lower background binding than radiolabeled antibodies [15].

8.2 Overview of the Various Bispecific Formats

There are more than 60 recombinant bispecific formats that have been created due to the modular nature of the immunoglobulin domain. Each format offers a distinct set of properties, such as size, antigen-binding valency, geometry of antigen-binding regions, flexibility, pharmacokinetic half-life, effector function, and biodistribution. The bispecific format and its properties should be carefully considered depending on the requirements of the therapeutic or diagnostic application. Another important consideration is the complexity of the bispecific molecule, as a more complex mode of production and more extensive characterization are required.

A BsAb can be made in a single cell by coexpression of the two different heavy chains and two different light chains; however, there are a few challenges to overcome. Due to the proposed random assembly of the four individual protein chains [16], there can be up to nine unwanted species with only one correct species containing the properly assembled heavy chain heterodimer (heavy chain-pairing problem) and the correct pairing of each light chain with its cognate heavy chain (light chain-pairing problem) [17, 18] (Figure 8.2). Although a BsAb molecule can be produced this way, the desired bispecific is found in low yield and it is difficult to purify away the unwanted species [19].

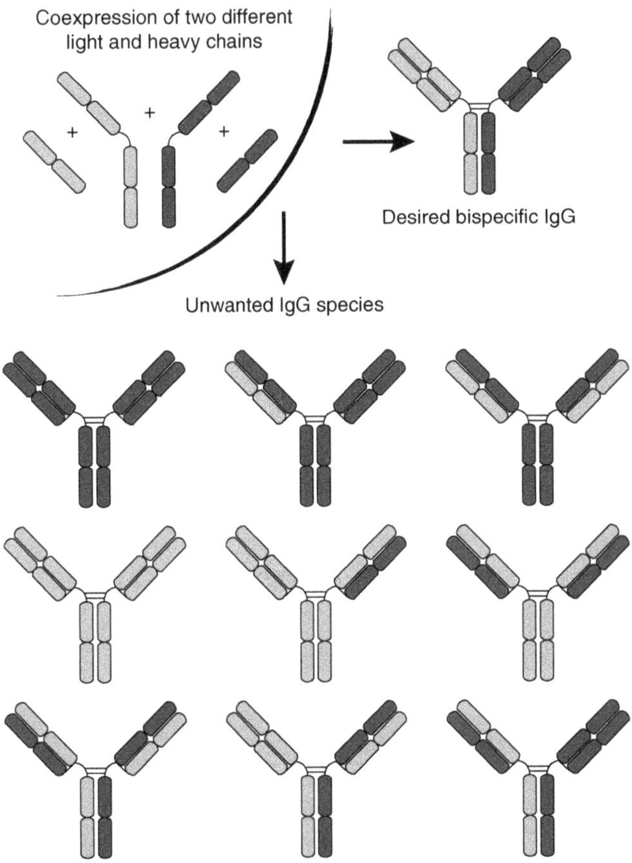

Figure 8.2 The bispecific antibody light and heavy chain pairing problems. Coexpression of two light and two heavy chains can result in nine unwanted species besides the desired bispecific antibody. An efficient bispecific platform requires correct light chain pairing with its cognate heavy chain, as well as heterodimeric heavy chain pairing.

To overcome these challenges, several solutions/formats were developed. Several BsAb formats are introduced here by the mode of production, as this has profound consequences on downstream analytical characterization (see Figure 8.3). Here, they are classified into five groups: (i) purification from mixtures, (ii) tethered bispecific molecules and alternative scaffolds, (iii) untethered bispecific molecules with engineered mutations, (iv) native bispecific IgG with dual binding behavior, and (v) BsAb conjugates. Their advantages and limitations are summarized in Table 8.1.

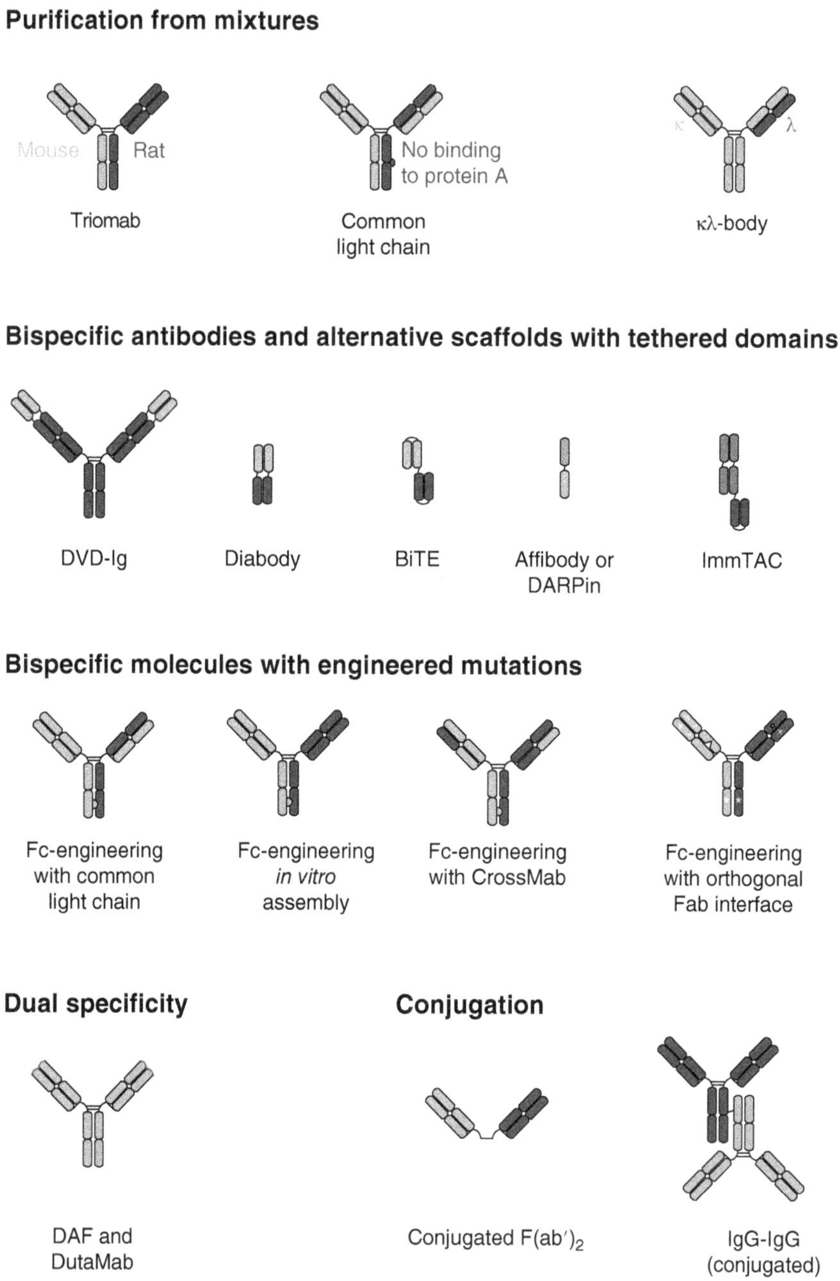

Figure 8.3 Examples of bispecific antibodies classified by their mode of production. Antibody domains are colored in red and blue, alternative scaffold domains in green and orange, and the T-cell receptor domains in brown. (*See insert for color representation of the figure.*)

Table 8.1 Production technologies of bispecific antibody formats and their advantages and limitations.

Production technology	Advantages	Limitations
Purification from mixtures 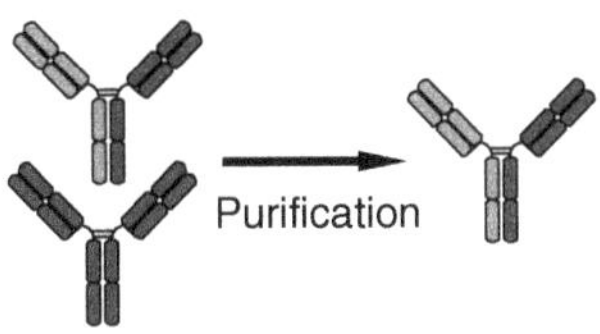 	BsAb without mutations	Complex purification Significant loss of material due to mispaired species Constraints on discovery technology for common light and heavy chain Characterization more complex (3 or 4 chains)
Tethered bispecific molecules 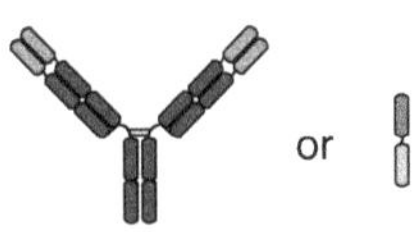 	Easy to purify and characterize (1 or 2 chains) Multivalency possible	Linkers between domains are nonnative Linker optimization may be necessary to retain affinity
IgG with engineered mutations 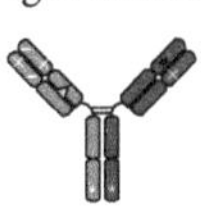	Native-like IgG architecture Plug and play from parental antibodies	Upfront engineering to identify Fc and/or Fab mutations Characterization more complex (3 or 4 chains) May require assembly
Dual specificity 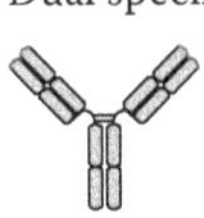	Native IgG Standard IgG purification and characterization	Significant upstream engineering efforts
Conjugation 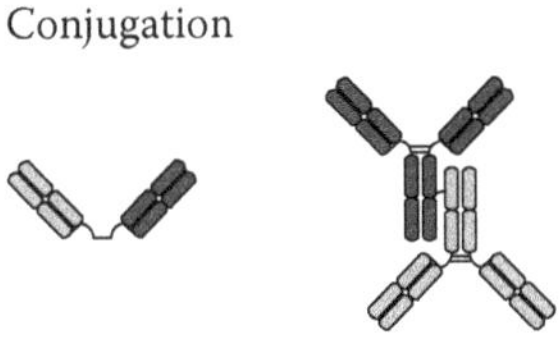	Multivalency possible	Lower yields and efficiency

8.2.1 Purification from Mixtures

A number of BsAbs have been purified from complex mixtures and have gone into clinical development. Catumaxomab (Removab®), an anti-CD3/anti-epithelial cell adhesion molecule (EpCAM) that retargets T-cells to treat malignant ascites in EpCAM-positive tumors was the first approved

BsAb in 2009. Catumaxomab is made by coexpression of a mouse and rat antibody (Triomab®) in a single cell (quadroma technology) [20]. Due to the preferential species-specific pairing of the light and heavy chains, the light chain-pairing problem is overcome [21]. To facilitate purification of the desired BsAb, protein A is used with a sequential pH elution to separate the BsAb from the mouse IgG, as only mouse antibodies can bind.

A common light chain approach combined with engineering distinct resin-binding specificities into each of the two heavy chains can be used to purify unwanted species away from the desired bispecific. In this approach, one light chain is engineered to pair with two different heavy chains and each pair has a different antigen-binding specificity. Using this approach, the light chain-pairing problem is avoided and the heavy chain-pairing problem is solved by differentially tagging the Fc region for purification with protein A [22].

An alternative way to isolate a BsAb from unwanted species is to use a common heavy chain, which eliminates the heavy chain-pairing problem. The κλ-body format has a common heavy chain with kappa and lambda light chains where each light chain–heavy chain pair is engineered to bind a distinct antigen [23]. Two sequential affinity purification steps that bind to either the kappa- or lambda-specific light chain produce the desired BsAb. This two-step process disfavors binding to antibodies that only contain a single type of light chain, elegantly solving a complex purification process.

Isolating the preferred BsAb from other unwanted species in a mixture does come with purification challenges. It is a process with low efficiency due to the production of at least two other species besides the desired bispecific. Engineered preferential binding can streamline purification of the wanted BsAb [22, 23]; however, there is significant upfront engineering to use a common light or heavy chain, and this may not always be amenable for certain antigen targets. Therefore, alternative modes of production may be more preferred.

8.2.2 Bispecific Antibodies and Alternative Scaffolds with Tethered Domains

One way to eliminate the pairing problem and to simplify characterization of the desired BsAb format is to fuse an additional antibody-variable region to the N- or C-terminus of the light chain or heavy chain. These molecules are full-length antibodies with additional domains added. One such molecule in clinical development is the dual variable domain immunoglobulin (DVD-Ig™). This molecule has V_L and V_H domains having specificity to a second antigen added N-terminally by a short peptide linker onto the V_L and V_H domains, respectively, of the full-length antibody molecule. In this format, the DVD-Ig™ is bivalent for each antigen, giving it double the valency than would be found in a traditional BsAb. The advantage to this format is that it only has one heavy and light chain, eliminating the production of unwanted species found with BsAbs having three or four different chains.

The molecules discussed so far contain Fc regions, which are useful for purification with protein A, extending half-life, or having effector function. Antibody fragments have also been explored as a bispecific format to simplify production and remove effector function. These formats typically include only the antigen-binding Fv domains that are tethered to each other *via* short linkers. One of the earlier formats is the diabody that has two chains with linkages V_H(a)–V_H(b) and V_L(a)–V_L(b) [24]. A recently approved BsAb fragment, the bispecific T-cell engager (BiTE™), blinatumomab (Blincyto®), retargets T-cells to kill tumor cells. Blinatumomab is a tandem scFv fragment binding both CD3 and CD19.

Alternative scaffolds have also been the focus of much attention in the bispecific molecule space. These scaffolds are generally well-behaved proteins that are engineered at multiple positions to recruit binding to an antigen. Most of these scaffolds also have the advantage of being modular in their architecture, allowing multiple single units to be linked together for the desired antigen-binding valency or extending binding specificity beyond two antigens. Due to these advantages, many different combinations can be easily screened for the best potency. Two examples of these scaffolds are the Affibody molecule [25], engineered from the Z-domain of protein A, and the DARPins [26], designed from the ankyrin repeat proteins.

With all the advantages that come with producing tethered bispecific molecules there are some potential downsides to consider. For many of these formats, peptide linker optimization can be necessary to obtain the optimal antigen-binding affinity [27] and reduced aggregation [28]. Additionally, when engineered linkers are used between antibody-binding domains or small antigen-binding domains are linked together in nonnatural configurations there is the possibility of immunogenicity and rapid proteolysis *in vivo* [28].

8.2.3 Bispecific Molecules with Engineered Mutations

The untethered human BsAb format is most desired therapeutically due to the native-like antibody architecture; however, it is the most complex to design and characterize. These molecules have engineered, ideally nonsolvent exposed mutations that drive preferential pairing between heavy chains, while different solutions are implemented to solve the light chain-pairing problem. The earliest designed heavy chain heterodimerization solution was the "knobs-into-holes" technology [29, 30], which revolutionized the generation of BsAb molecules.

This platform has differential engineering in each C_H3 domain, where one side has a large amino acid such as tryptophan to disfavor homodimerization (knob), while the other side has smaller compensating mutations (hole), which also disfavor homodimerization. These asymmetric heavy chain arms are robust in their pairing with various Fab arms. Alternative strategies to favor heavy chain heterodimerization have also been developed that use different mutations, electrostatic engineering, or exchanging structurally similar C_H3 strands of IgA and IgG to preferentially bind to each other but not to themselves [31–35]. While these bispecific engineering solutions were developed using the IgG1 isotype, some are also compatible with the IgG2 and IgG4 isotypes [33, 36].

When using mutations that drive heavy chain heterodimerization, the simplest solution to avoid the light chain-pairing problem is to use the same light chain (common light chain) [9, 37]. A common light chain can be identified by phage display or animals transgenic for a single light chain [38]. Due to the difficult engineering efforts that go into finding a common light chain, the most common strategy to produce a BsAb is to express the two component antibodies in separate cells, followed by *in vitro* assembly into the intact BsAb [33, 39–41]. However, strategies to express the BsAb in a single cell do exist. One solution is using domain swaps of the antibody constant or variable domains [42], and two molecules with this CrossMab format are currently in clinical development [1]. More recently, orthogonal Fab interfaces have been engineered [7, 43, 44] with mutations in both the V_H/V_L and C_H1/C_L interfaces. When these mutations are combined with a heavy chain heterodimerization strategy, they seem broadly applicable to allow efficient BsAb production from a single host cell [7, 43, 44].

8.2.4 Native Bispecific IgG with Dual Binding Behavior

One way to obviate the heavy or light chain-pairing problems altogether is to engineer the Fab domain to bind to two distinct antigens *via* two different binding sites on the antibody (paratopes). This approach produces a molecule that is a regular antibody with only a single heavy and light chain, however the antibody can now bind to two different antigens. There are two BsAb formats in this category, the dual-action Fab (DAF) [45, 46] and the DutaMab [47]. The DAF paratopes use the light and heavy chain complementarity determining regions (CDRs) and somewhat overlap, whereas the DutaMab uses diversity in only three CDRs for each antigen, creating two separate paratopes. For certain applications, a dual binding antibody could be very desirable as it allows for different types of antigen-binding valency, but when monospecificity is required this approach will not work. The benefit of these formats is producing and characterizing a molecule with the same properties as a MAb. However, there is a large amount of upfront engineering that is required to recruit a second antigen binding to a Fab domain, and not all molecules may be amenable to this process.

8.2.5 Bispecific Antibody Conjugates

In the early days before recombinant protein expression, antibodies or antibody fragments were chemically conjugated to create a BsAb. The first BsAbs produced were bispecific $F(ab')_2$ molecules that were monovalent for each antigen [48, 49]. These were made by digesting two different antibodies with pepsin followed by a reduction/oxidation procedure to conjugate into the bispecific $F(ab')_2$. However, these initial conjugation reactions had poor yields, with many undesirable side products. Thus, a more advanced conjugation strategy based upon cysteine-reactive crosslinking agents linking recombinantly produced full-length IgG or Fab′ was used to improve efficiency [50–52]. These bispecific molecules can be bivalent or monovalent for each antigen, respectively. Chemical conjugation to amine groups has also been used to crosslink two different IgG into a bispecific format with bivalency to each antigen [53]. This strategy has been applied to a number of different tumor targets [54–56] as well as cytomegalovirus infection [57]. Although site-specific conjugation does increase BsAb yield, this process is still not very efficient with much optimization required to decrease unwanted side products [58].

8.3 Alternatives to Bispecific Antibodies: Antibody Mixtures

Depending on the requirements of the biological system of interest, both antigen-binding activities of a BsAb molecule may be "obligatory" or "optional" to reside in the same molecule. Examples of "obligatory" BsAbs are the retargeting molecules that simultaneously engage cytotoxic T-cells and tumor cells, targeting molecules that use a transcytosis receptor to transport an antibody to the brain, and ligand mimetics that dimerize cellular receptors. "Optional" BsAbs include those that block or neutralize independent signaling pathways or those that can have multiple molecules bound to an antigen to "decorate" it. Only "optional" but not "obligatory" BsAbs can be formulated as mixtures, where two or multiple antibodies are combined, or coadministered [59].

There are some advantages that mixtures or coadministrations have over a BsAb. Besides both antibodies being bivalent and the benefits of avidity for antigen binding, the antibody ratios can be modulated. Additionally, coadministration allows each antibody to be dosed separately, providing more flexible timing in each dosing schedule. Antibody mixtures can also mimic the natural polyclonal response to decorate the target antigen and this may provide a solution to prevent drug resistance in oncology or infectious disease. Moreover, using mixtures to test different antibody combinations for additive and synergistic effects is preferred in a research setting due to the simpler production and characterization of MAbs compared to BsAbs. However, there are also some disadvantages to using antibody mixtures such as choosing the proper formulation and analyzing multiple antibodies in the same mixture, which introduces complexity into the manufacturing process. Some examples of antibody mixtures are the three-antibody cocktail to the Ebola virus called ZMAb [60, 61], the mixture of two anti-EGFR antibodies binding to distinct epitopes which is superior at inhibiting EGFR-dependent cancer cells in preclinical [62] and early clinical studies [63], and antibody mixtures to target viruses or botulinum toxin [64–66].

8.4 Characterization of the Bispecific Molecule

Access to and establishing quantitative analytics is crucial during the engineering of the initial bispecific format, as well as to ensure high product quality of the produced BsAb for research, development, and clinical use. For example, when engineering mutations that steer heterodimerization of the two heavy chains or orthogonal pairing of the light chains, quantitative assays are paramount to determine success. Understandably, once a bispecific format is selected careful characterization of the expressed or assembled bispecific product is of great importance. Partially assembled side products as well as homodimeric species need to be removed during purification and their presence should be assessed in the purified material. "Obligatory" BsAbs that rely on both arms of the BsAb for activity require high purity for correct interpretation of research data. While mispaired light and heavy chains are commonly binding-inert, side products such as homodimers can pose unwanted modes of action and safety liabilities [67, 68]. BsAbs should be characterized by chromatographic and mass spectrometric methods to ensure product quality. Additionally, binding and functional *in vitro* and *in vivo* assays need to confirm that the desired properties of the parental antibodies were successfully transferred to the BsAb. In the following sections, the most common assays will be described, and they are summarized in Table 8.2.

8.4.1 Characterization by Bioanalytical Methods

To assess the quality of the purified antibodies, denaturing capillary electrophoresis–sodium dodecyl sulfate (CE-SDS) or standard denaturing sodium dodecyl sulfate–polyacrylamide gel electrophoresis (SDS-PAGE) and size-exclusion chromatography (SEC) analysis are commonly performed. Nonreducing electrophoresis can give information on the integrity of the antibody inter- and intrachain disulfides, whereas reducing conditions can provide an estimate on relative chain distribution and purity of the material [36, 69]. SEC analysis can detect fragments and aggregates and in combination with a high-performance liquid chromatography system (size-exclusion high-performance liquid chromatography (SEC-HPLC)), improved resolution can be

Table 8.2 Techniques to purify and characterize bispecific antibodies.

Analytical technique	Purpose	Advantages	Limitations
Gel or capillary electrophoresis (SDS-PAGE or CE-SDS)	Verify integrity of interchain disulfide bonds Evaluate purity	High throughput (CE-SDS)	Limited mass resolution
Size-exclusion chromatography (SEC)	Preparative: remove fragments and aggregates Analytical: determine amount of fragments and aggregates	Universal buffer conditions Analytical: high throughput Higher resolution with SEC-HPLC	Not scalable to manufacturing
Reverse-phase high-performance liquid chromatography (RP-HPLC)	Determine mispaired species, fragments, and aggregates	High throughput and high resolution	Peak identity needs to be confirmed by other method Only analytical
Hydrophobic interaction chromatography (HIC)	Preparative: remove mispaired species, fragments, and aggregates Analytical: determine amount of mispaired species, fragments, and aggregates	Universal pH and salt gradient conditions	Elution times need to be determined empirically Species with similar hydrophobicity cannot be separated
Cation-exchange chromatography (CEX)	Preparative: remove mispaired species Analytical: determine amount of mispaired species	Elution times can be estimated from theoretical pI	Species with similar pI cannot be separated

(*Continued*)

Table 8.2 (Continued)

Analytical technique	Purpose	Advantages	Limitations
Differential scanning calorimetry/fluorimetry (DSC or DSF)	Thermal stability	Rapid formulation screening High throughput (DSF)	Individual domains, such as Fab or Fc may be required for straightforward data interpretation
HPLC/UHPLC-ESI-Q-TOF	Confirm expected mass of molecule Detect chemical stability under accelerated stress conditions	Increased resolution over HPLC separation (UHPLC)	
Orbitrap	MS analysis of complex mixtures	Increased resolution over ESI-Q-TOF (baseline separation of species close in mass)	
BV-ELISA, protein panel profiling, polyspecificity reagents	Risk mitigation to identify off-target binding and potential fast clearance	High throughput	Target-mediated clearance not predicted

achieved [70]. To improve resolution further, reverse-phase high-performance liquid chromatography (RP-HPLC) can be used to evaluate assembly efficiency as it allows the separation of homodimers from the assembled BsAb [70]. These techniques are often combined with the additional analytical tools discussed here.

For purification and characterization of BsAbs both cation-exchange chromatography (CEX) and hydrophobic interaction chromatography (HIC) have proven particularly useful and versatile [23, 33, 40, 41]. CEX has the advantage that elution of the bispecific as well as the mispaired species can be reasonably estimated based on the theoretical pI of the component molecules. In contrast, HIC has the advantage of using universal pH and salt gradient conditions but elution times cannot be predicted. Most laboratories characterize BsAbs by CEX and/or HIC in combination with MS methods, which will be described next.

8.4.2 Characterization by Mass Spectrometry Methods

8.4.2.1 General Considerations

To evaluate the molecular integrity and bispecific composition, the use of mass spectrometry (MS) for characterization is essential. There are a number of strategies that can be employed to reduce heterogeneity, making the bispecific sample more amenable to MS. Removing glycosylation prior to analysis, such as deglycosylation of N-linked glycans with PNGase-F, reduces heterogeneity of the BsAb [71]. For many basic engineering approaches the presence of the N-linked glycan in the antibody C_H2-domain is unimportant; thus mutations that abrogate glycosylation are a simple engineering strategy to improve product homogeneity [72]. Furthermore, incomplete processing of the C-terminal lysine residue in the antibody heavy chain is another source of product heterogeneity [73–75]. To simplify analytical characterization by MS, deletion of the terminal lysine residue in the protein construct can also be considered.

The difficulty in MS characterization of BsAbs scales with the number of chains. The simplest BsAb to characterize are the ones containing only one chain, for example, the BiTE™, as no pairing is required. BsAbs containing two chains that must pair, such as the DAF and DVD-IgG™, can be characterized like a regular monoclonal antibody. However, to create a BsAb with native-like architecture that has three or four different chains it is important to note that mispairing can occur during expression or *in vitro* assembly. The assembly of a BsAb from multiple chains requires that two heavy–light chain pairs with similar masses to be distinguished during the analytical process. To facilitate an easier analytical process, mutations steering chain pairing should be placed onto each antibody chain to achieve the maximum mass differences between all possible species.

8.4.2.2 Purity Analysis of the Final Bispecific Antibody

In addition to using MS analysis for the characterization of the produced BsAb, it is essential to validate the lack of side products during the BsAb engineering process [7, 33, 76]. Woods *et al.* demonstrated the use of an HPLC-MS-based assay using an advanced electrospray ionization and quadrupole time-of-flight (ESI-Q-TOF) mass spectrometer to characterize the purity of a heterodimeric antibody with a common light chain [77]. Relative quantification of homodimer and half-antibody in the samples was achieved by first preparing pure samples of the heterodimer and the separately expressed homodimers, and performing subsequent spiking experiments. One complication is that side products such as half-antibodies can have greater ionization than the intact antibody, which can lead to overrepresentation of the half-antibody and underrepresentation of the intact BsAb. Nevertheless, this heterodimer characterization assay allows purity assessment and is able to detect low levels (2%) of homodimers. Recently, multiple groups have used HPLC-ESI-Q-TOF MS to validate the success of engineered mutations in the Fab region that steer cognate heavy–light chain pairing in a single cell host [7, 43, 78].

Several bispecific strategies rely on the assembly of the intact BsAb from two separately expressed and purified parental homodimers [33, 41] or half-antibodies [40]. A redox reaction facilitates efficient formation of the covalent heterodimers. Under these assembly conditions, the hinge disulfides of the homodimer and the cyclized hinge disulfides of half-antibodies are reduced; however, the disulfide between the antibody light and heavy chain is also reduced, resulting in potential light chain exchange. Intact MS of the antibody cannot distinguish between the correctly paired and double-mispaired light chain species; therefore, manipulation of the BsAb prior to MS analysis must be performed. One possibility is the proteolytic digestion by LysC to release the individual Fab fragments for separate analysis [40]. Although this method was applied for assembled BsAbs, the same strategy could also be applied to unambiguously identify the success of BsAbs expressed in a single cell host.

MS has also been utilized to inform cell line development of BsAbs. During cell line development, high throughput and rapid feedback is necessary to optimize transfection strategies and clone selection. To analyze a BsAb with a common light chain, cell culture samples were subjected to a protein A capture step linked to inline MS [79]. By estimating the ratio of each species from the deconvoluted spectra, the relative amounts of homodimer, heterodimer, as well as partially assembled species can be estimated. As it provides more detail of the side products than other commonly used assays such as the enzyme-linked immunosorbent assay (ELISA), it may enable a targeted approach in optimizing transfection or clone selection.

8.4.2.3 Antibody Mixtures

In-depth characterization is necessary to gain approval of composite antibody mixtures for clinical development. Traditionally, CEX and ELISA have been used to characterize antibody mixtures, but the high resolution of native MS predisposes it for this application. It has been demonstrated that native MS using an ESI-TOF platform is able to qualitatively and semi-quantitatively analyze a complex mixture of up to 10 antibodies [71]. The relative abundance of each antibody could be quantified with an accuracy of 3%, and 9 out of 10 antibodies in the mixture were reliably identified with superior resolving power compared to CEX. The automated peak fitting of the native MS technology combined with minimal sample amounts and handling requirements make it a fast and robust tool. In addition to the characterization of antibody integrity, native MS was used to interrogate the dynamics of half-antibody formation for IgG4 isotype antibodies [80] by providing kinetic details about natural bispecific molecules formed by Fab arm exchange [81, 82]. Furthermore, native MS using ESI-TOF has been used to determine the stoichiometry of antigen–antibody complexes [83, 84]. While these studies have been limited to MAbs, they suggest that large antigen size can limit occupancy. As many BsAb have unique Fab geometry and valency, it seems intriguing to also apply this technology to bispecific molecules in order to elucidate the antigen occupancy.

8.4.2.4 Increasing Resolution

Often the masses of correctly paired and mispaired BsAb species are close and demand high-resolution analytics to reliably resolve differences. To increase resolution and separation compared to traditional RP-HPLC columns used in HPLC-MS applications, reverse-phase ultra-high-performance chromatography (UHPLC) was newly established and optimized for analyzing variants of intact antibodies [85]. While this technology was used for the characterization of a free thiol variant, the increased resolution has promise to streamline variant analysis of BsAbs as well. To also satisfy these high-resolution requirements, native Orbitrap MS has recently provided a powerful approach to the qualitative and semi-quantitative characterization of intact IgG [86, 87] and composite mixtures [88]. The high mass-resolving power allows baseline separation of antibodies close in mass. For example, Thompson *et al.* [88] were able to differentiate between two antibodies that differed by only 42.27 Da. This resolution enables not only superior identification of all products in a mixture, but it enables accurate relative quantification of desired and off-pathway species. In addition, the high resolution of the Orbitrap instrument allows differentiation of antibodies without enzymatic deglycosylation. This brings particular value to BsAbs that are glycosylated or glycoengineered to enable or enhance antibody effector function, respectively.

8.4.3 Characterization of Bispecific Antibodies by Binding Assays

For larger panels of BsAbs, it is easiest to screen binding using a bridging ELISA assay. Here, the first antigen is coated on the plate and after incubation with the BsAb, the second antigen is added and detected. To obtain binding kinetics of the BsAb, surface plasmon resonance (SPR) techniques can be used with a setup analogous to the bridging ELISA assay [7, 33, 42, 89]. For an "optional"

BsAb, however, it is often sufficient to verify the affinity to each antigen independently by SPR [36]. If fully binding-competent BsAb reference material is available, both ELISA- and SPR-based analysis can determine the relative binding competency of the sample BsAb [44, 90]. This is particularly useful in many engineering situations for ranking the success of BsAb mutations. After confirming the binding properties by *in vitro* techniques, cell-based assays to study antibody potency are then performed.

When characterizing the binding and potency of a BsAb, there are a number of things to consider. For BsAb formats that require monovalent target engagement, suitable parental antibodies need to be selected based on their monovalent affinity as there is often a disconnect between monovalent and bivalent binding affinity. On the other hand, the decrease in affinity due to the monovalent nature of some BsAb formats can also be beneficial. Trivalent BsAb molecules, for example, allow avidity to one target, while preventing receptor dimerization or avidity to the second target. This was leveraged in the case of anti-Aβ/transferrin receptor BsAbs to modulate cellular trafficking by facilitating transcytosis instead of lysosome sorting [12]. Two other features that can impact target engagement are the order of binding domains and the length of the linker between them. This has been studied in detail with several DVD-Ig™ molecules [27] and single-chain diabodies [91], and was shown to be antigen or format dependent [92].

8.4.4 Developability Assessment of the Bispecific Antibody

The complex composition of BsAbs from two parental antibodies raises the bar for developability. A number of assays to select favorable properties of parental antibodies have been established and reviewed in detail elsewhere [93, 94]. In summary, to enable developability of an antibody as a therapeutic, a number of criteria including expression, favorable physicochemical properties like thermal stability, aggregation, solubility, and viscosity, as well as chemical modifications, off-target binding, and *in vivo* stability need to be considered. Generally, both parental antibodies need to have favorable properties from a developability perspective as the combination into the BsAb rarely improves liabilities. Thus, during the discovery process and the selection of the two arms, it is necessary to first characterize both parental antibodies independently and then confirm developability with the intact bispecific molecule.

8.4.4.1 Expression

BsAbs have been successfully expressed in prokaryotic, yeast, mammalian, and cell-free systems [37, 39, 78, 95]. Ideally, multiple expression platforms should be compared to achieve optimal product quality and yield as each BsAb format has unique properties.

8.4.4.2 Physicochemical Properties

Since many BsAbs divert from their natural protein architecture or use mutations to achieve the correct chain pairing, it is recommended to determine the stability of the BsAb [31, 33, 96]. Many proteins with long serum half-life are characterized by high thermostability and decreased stability of proteins can impact serum half-life [97], expression [98], immunogenicity [99], and the propensity to aggregate [100]. Commonly used techniques to determine protein stability are differential scanning calorimetry (DSC) and differential scanning fluorimetry (DSF) [100–102]. Historically, DSC has been used to determine the thermostability of proteins. For antibodies, both DSC and DSF lead to comparable melting temperatures [103]. Since DSF is a high-throughput technique with minimal protein requirement, it is particularly attractive to guide development and formulation of the BsAb [103, 104].

About one-third of the BsAbs currently under clinical investigation are in autoimmune and inflammation indications [1]. These BsAbs are usually delivered by subcutaneous injection, which requires low viscosity and high solubility. Viscosity is most accurately determined with a rheometer to ensure that sample processing during purification and delivery by syringe is possible [105, 106]. Commonly used techniques to determine solubility include filtration and precipitation, as well as dialysis, evaporation, and chromatographic methods [107]. For the selection and engineering of highly soluble antibodies, cross-interaction chromatography was also successfully applied [108]. In this technique, a purified monoclonal or polyclonal IgG is chemically coupled to a chromatography resin. The sample antibody is then passed over the column and its retention time is inversely correlated to its solubility. Thus, antibodies prone to precipitation show delayed retention times and broader peaks. As many BsAb formats deviate from the traditional architecture of a monoclonal antibody or contain linkers, they can also challenge solubility. Viscosity and solubility are properties unique to the BsAb format and cannot be predicted; however, evaluation of different formats can help to mitigate developability issues [109].

8.4.4.3 Chemical Modifications

Chemical modifications can introduce product variability during storage and impact *in vitro* and *in vivo* binding behavior. These modifications include isomerization, deamidation, and succinimide formation of aspartate and asparagine residues, as well as oxidation of methionine and tryptophan residues. Chemical modifications can reduce binding affinity or lead to a complete loss of binding. "Optional" BsAbs can still have partial function and therapeutic efficacy with these modifications, although this is undesired. However, obligate BsAbs that rely on simultaneously engaging two cells or receptors require both binding partners to be fully functional.

Chemical modifications in antibodies are commonly analyzed after accelerated stress by MS identification and quantification of proteolytic fragments [93, 110].

This makes the chemical modifications time and labor intensive to identify and commonly analysis is limited to the three CDRs. The composition of most BsAbs with six CDRs introduces additional complexity to the analysis. As chemical modifications translate from the parental CDRs into the BsAb, it is often sufficient to only evaluate the parental antibodies.

Recently, *in silico* methods have been described that aid the analysis of a larger number of antibodies. Using predicted solvent accessibility and conformational flexibility, the degradation propensity of asparagine and aspartate residues as well as oxidation of methionine and tryptophan residues in the CDRs can be calculated [111, 112].

To study the impact of the identified chemical modification on the binding behavior, SPR- or cell-based assays can be performed. As demonstrated with an anti-Ang2/anti-VEGF BsAb, SPR-based binding assays could determine the fraction of fully binding competent BsAbs [90]. As the signal depends on simultaneous binding of both antigens, it can reliably determine the active concentration of BsAbs after accelerated stress conditions. However, such studies are limited to BsAbs with at least one high-affinity Fab, as this ensures good association of the BsAb to the immobilized antigen and accurate kinetic assessment of the second antigen binding.

8.4.4.4 Characterization of *In Vivo* Properties

In addition to having the desired *in vitro* characteristics, the BsAb needs to have suitable *in vivo* properties. Before evaluating the BsAb in pharmacokinetic assays to determine the *in vivo* half-life, it is often helpful to first determine its serum stability at physiological temperatures [89, 113]. This is of particular value when evaluating a new bispecific format as it is an easy and fast assay. After the molecule demonstrates suitable serum stability, pharmacokinetic studies are typically performed in rodent and cynomolgus monkeys to determine the *in vivo* half-life.

The antibody Fc region mediates the long pharmacokinetic half-life of many BsAbs. The immunoglobulin Fc binds to the neonatal Fc receptor, FcRn, providing for extended serum half-life [114]. Hence, bispecifics lacking the Fc region have a short serum half-life as they are not rescued after receptor-mediated endocytosis or pinocytosis. In addition, molecules of small size (below 70 kDa) are subject to rapid renal clearance [115, 116]. One possibility to overcome these limitations is to deliver the therapeutic by continued intravenous infusion. This has been successfully done with the tandem scFv blinatumomab (Blincyto®) that has a serum half-life of ~2 h [117]. Alternatively, the half-life can be extended by strategies used for other protein therapeutics [118]. Commonly exploited routes include PEGylation [118] and fusion to human serum albumin or albumin binding proteins [119–121]. Binding to human IgG can also extend pharmacokinetic half-life as shown for a diabody molecule that is an anti-carcinoembryonic antigen/anti-human IgG bispecific [122].

Similar to MAbs, off-target binding can account for reduced serum half-life of bispecific molecules. Several *in vitro* assays have been established for risk mitigation. Nonspecific binding to baculovirus particles can identify antibodies with increased risk for fast clearance [123]. This correlation allowed the implementation of a simple, but high-throughput ELISA assay to rapidly evaluate a large number of candidates. Rapid clearance can also be predicted by *in silico* analysis based on the charge profile and hydrophobicity [112]. Off-target binding can also be determined by the Protein Panel Profiling (*3P*) method that was recently developed. In this fully automated immunoassay, binding to a comprehensive set of test proteins is evaluated in a high-throughput manner [124]. Alternatively, counterselection against a polyspecificity reagent can be applied during the screening process [125]. These risk mitigation assays should be performed during lead generation and optimization to identify parental antibodies with minimal off-target binding risk and confirmed with the final BsAb.

A long serum half-life is not always advantageous. For example, in ocular indications it is desirable to minimize systemic exposure. Intravitreally injected antibody-based therapeutics have been used for the treatment of neovascular (wet) age-related macular degeneration and other neovascular diseases of the retina [126]. These antibodies either contain mutations in the FcRn binding site that reduce the half-life [127] or have a format that lacks the Fc altogether, such as a Fab or $F(ab')_2$ [52, 128]. In addition, many alternative scaffolds display a short systemic half-life due to their small size, and seem particularly suitable for these indications as well [129].

8.5 Conclusions

BsAbs are finally coming of age with two approvals and many more in clinical development. A broad number of bispecific formats have evolved accordingly, each with varying challenges in production and characterization. In addition, therapeutic modes of action for BsAbs often require highly pure material and even at an early research stage, careful characterization is necessary to interpret experimental results. Historically BsAbs were only characterized by bioanalytical methods; however, MS is now playing a leading role in ensuring product quality and purity due to its increased resolution. Having high-quality BsAb enables a reliable readout from *in vitro*, pharmacokinetic, and pharmacodynamic experiments. With many BsAbs being explored in exciting new applications, the development and refinement of analytical characterization tools is in the early stages.

References

1 Spiess, C, Zhai, Q, Carter, P J (2015) Alternative molecular formats and therapeutic applications for bispecific antibodies. *Mol Immunol*, 67 (2 Pt A), 95–106.

2 Marvin, J S, Zhu, Z (2005) Recombinant approaches to IgG-like bispecific antibodies. *Acta Pharmacol Sin*, 26 (6), 649–658.

3 Kontermann, R E, Brinkmann, U (2015) Bispecific antibodies. *Drug Discov Today*, 20 (7), 838–847.

4 Weiner, L M, Clark, J I, Davey, M, *et al.* (1995) Phase I trial of 2B1, a bispecific monoclonal antibody targeting c-erbB-2 and Fc gamma RIII. *Cancer Res*, 55 (20), 4586–4593.

5 Satta, A, Mezzanzanica, D, Turatti, F, *et al.* (2013) Redirection of T-cell effector functions for cancer therapy: bispecific antibodies and chimeric antigen receptors. *Future Oncol*, 9 (4), 527–539.

6 Joanne, O, Bent, K J (2013) ImmTACs: Novel bi-specific agents for targeted cancer therapy. *Oncoimmunology*, 2 (2), e22891.

7 Lewis, S M, Wu, X, Pustilnik, A, *et al.* (2014) Generation of bispecific IgG antibodies by structure-based design of an orthogonal Fab interface. *Nat Biotechnol*, 32 (2), 191–198.

8 Boersma, Y L, Chao, G, Steiner, D, *et al.* (2011) Bispecific Designed Ankyrin Repeat Proteins (DARPins) targeting epidermal growth factor receptor inhibit A431 cell proliferation and receptor recycling. *J Biol Chem*, 286 (48), 41273–41285.

9 Kitazawa, T, Igawa, T, Sampei, Z, *et al.* (2012) A bispecific antibody to factors IXa and X restores factor VIII hemostatic activity in a hemophilia A model. *Nat Med*, 18 (10), 1570–1574.

10 Kolumam, G, Chen, M Z, Tong, R, *et al.* (2015) Sustained brown fat stimulation and insulin sensitization by a humanized bispecific antibody agonist for fibroblast growth factor receptor 1/βKlotho complex. *EBioMedicine*, 2 (7), 730–743.

11 Yu, Y J, Zhang, Y, Kenrick, M, *et al.* (2011) Boosting brain uptake of a therapeutic antibody by reducing its affinity for a transcytosis target. *Sci Transl Med*, 3 (84), 84ra44.

12 Niewoehner, J, Bohrmann, B, Collin, L, *et al.* (2014) Increased brain penetration and potency of a therapeutic antibody using a monovalent molecular shuttle. *Neuron*, 81 (1), 49–60.

13 Farrington, G K, Caram-Salas, N, Haqqani, A S, *et al.* (2014) A novel platform for engineering blood-brain barrier-crossing bispecific biologics. *FASEB J*, 28 (11), 4764–4778.

14 Bhatnagar, P K, Das, D, Suresh, M R (2008) Sequential affinity purification of peroxidase tagged bispecific anti-SARS-CoV antibodies on phenylboronic acid agarose. *J Chromatogr B Analyt Technol Biomed Life Sci*, 863 (2), 235–241.

15 Sharkey, R M, Cardillo, T M, Rossi, E A, *et al.* (2005) Signal amplification in molecular imaging by pretargeting a multivalent, bispecific antibody. *Nat Med*, 11 (11), 1250–1255.

16 Kaushik, A, Schulze, D H, Bonilla, F A, *et al.* (1990) Stochastic pairing of heavy-chain and kappa light-chain variable gene families occurs in polyclonally activated B cells. *Proc Natl Acad Sci USA*, 87 (13), 4932–4936.

17 Suresh, M R, Cuello, A C, Milstein, C (1986) Bispecific monoclonal antibodies from hybrid hybridomas. *Meth Enzymol*, 121, 210–228.

18 Schaefer, W, Völger, H R, Lorenz, S, *et al.* (2016) Heavy and light chain pairing of bivalent quadroma and knobs-into-holes antibodies analyzed by UHR-ESI-QTOF mass spectrometry. *MAbs*, 8 (1), 49–55.

19 Suresh, M R, Cuello, A C, Milstein, C (1986) Advantages of bispecific hybridomas in one-step immunocytochemistry and immunoassays. *Proc Natl Acad Sci USA*, 83 (20), 7989–7993.

20 Lindhofer, H, Mocikat, R, Steipe, B, Thierfelder, S (1995) Preferential species-restricted heavy/light chain pairing in rat/mouse quadromas. Implications for a single-step purification of bispecific antibodies. *J Immunol*, 155 (1), 219–225.

21 Chelius, D, Ruf, P, Gruber, P, *et al.* (2010) Structural and functional characterization of the trifunctional antibody catumaxomab. *MAbs*, 2 (3), 309–319.

22 Smith, E J, Olson, K, Haber, L J, *et al.* (2015) A novel, native-format bispecific antibody triggering T-cell killing of B-cells is robustly active in mouse tumor models and cynomolgus monkeys. *Sci Rep*, 5, 17943.

23 Fischer, N, Elson, G, Magistrelli, G, *et al.* (2015) Exploiting light chains for the scalable generation and platform purification of native human bispecific IgG. *Nat Commun*, 6, 6113.

24 Holliger, P, Prospero, T, Winter, G (1993) "Diabodies": small bivalent and bispecific antibody fragments. *Proc Natl Acad Sci USA*, 90 (14), 6444–6448.

25 Löfblom, J, Feldwisch, J, Tolmachev, V, *et al.* (2010) Affibody molecules: engineered proteins for therapeutic, diagnostic and biotechnological applications. *FEBS Lett*, 584 (12), 2670–2680.

26 Stumpp, M T, Amstutz, P (2007) DARPins: a true alternative to antibodies. *Curr Opin Drug Discov Devel*, 10 (2), 153–159.

27 Jakob, C G, Edalji, R, Judge, R A, *et al.* (2013) Structure reveals function of the dual variable domain immunoglobulin (DVD-Ig™) molecule. *MAbs*, 5 (3), 358–363.

28 Whitlow, M, Bell, B A, Feng, S L, *et al.* (1993) An improved linker for single-chain Fv with reduced aggregation and enhanced proteolytic stability. *Protein Eng*, 6 (8), 989–995.

29 Ridgway, J B, Presta, L G, Carter, P (1996) "Knobs-into-holes" engineering of antibody CH3 domains for heavy chain heterodimerization. *Protein Eng*, 9 (7), 617–621.

30 Atwell, S, Ridgway, J B, Wells, J A, Carter, P (1997) Stable heterodimers from remodeling the domain interface of a homodimer using a phage display library. *J Mol Biol*, 270 (1), 26–35.

31 Kreudenstein, T S V, Escobar-Carbrera, E, Lario, P I, *et al.* (2013) Improving biophysical properties of a bispecific antibody scaffold to aid developability: quality by molecular design. *MAbs*, 5 (5), 646–654.

32 Labrijn, A F, Meesters, J I, de Goeij, B E C G, *et al.* (2013) Efficient generation of stable bispecific IgG1 by controlled Fab-arm exchange. *Proc Natl Acad Sci USA*, 110 (13), 5145–5150.

33 Strop, P, Ho, W-H, Boustany, L M, *et al.* (2012) Generating bispecific human IgG1 and IgG2 antibodies from any antibody pair. *J Mol Biol*, 420 (3), 204–219.

34 Gunasekaran, K, Pentony, M, Shen, M, *et al.* (2010) Enhancing antibody Fc heterodimer formation through electrostatic steering effects: applications to bispecific molecules and monovalent IgG. *J Biol Chem*, 285 (25), 19637–19646.

35 Davis, J H, Aperlo, C, Li, Y, *et al.* (2010) SEEDbodies: fusion proteins based on strand-exchange engineered domain (SEED) CH3 heterodimers in an Fc analogue platform for asymmetric binders or immunofusions and bispecific antibodies. *Protein Eng Des Sel*, 23 (4), 195–202.

36 Spiess, C, Bevers, J, Jackman, J, *et al.* (2013) Development of a human IgG4 bispecific antibody for dual targeting of interleukin-4 (IL-4) and interleukin-13 (IL-13) cytokines. *J Biol Chem*, 288 (37), 26583–26593.

37 Merchant, A M, Zhu, Z, Yuan, J Q, *et al.* (1998) An efficient route to human bispecific IgG. *Nat Biotechnol*, 16 (7), 677–681.

38 Dhimolea, E, Reichert, J M (2012) World Bispecific Antibody Summit, September 27–28, 2011, Boston, MA. *MAbs*, 4 (1), 4–13.

39 Jackman, Janet, Chen, Yongmei, Huang, Arthur, *et al.* (2010) Development of a two-part strategy to identify a therapeutic human bispecific antibody that inhibits IgE receptor signaling. *J Biol Chem*, 285 (27), 20850–20859.

40 Spiess, C, Merchant, M, Huang, A, *et al.* (2013) Bispecific antibodies with natural architecture produced by co-culture of bacteria expressing two distinct half-antibodies. *Nat Biotechnol*, 31 (8), 753–758.

41 Labrijn, A F, Meesters, J I, Priem, P, *et al.* (2014) Controlled Fab-arm exchange for the generation of stable bispecific IgG1. *Nat Protoc*, 9 (10), 2450–2463.

42 Schaefer, W, Regula, J T, Bähner, M, *et al.* (2011) Immunoglobulin domain crossover as a generic approach for the production of bispecific IgG antibodies. *Proc Natl Acad Sci USA*, 108 (27), 11187–11192.

43 Liu, Z, Leng, E C, Gunasekaran, K, *et al.* (2015) A novel antibody engineering strategy for making monovalent bispecific heterodimeric IgG antibodies by electrostatic steering mechanism. *J Biol Chem*, 290 (12), 7535–7562.

44 Mazor, Y, Oganesyan, V, Yang, C, *et al.* (2015) Improving target cell specificity using a novel monovalent bispecific IgG design. *MAbs*, 7 (2), 377–389.

45 Bostrom, J, Yu, S-F, Kan, D, *et al.* (2009) Variants of the antibody herceptin that interact with HER2 and VEGF at the antigen binding site. *Science*, 323 (5921), 1610–1614.

46 Schaefer, G, Haber, L, Crocker, L M, *et al.* (2011) A two-in-one antibody against HER3 and EGFR has superior inhibitory activity compared with monospecific antibodies. *Cancer Cell*, 20 (4), 472–486.

47 Beckmann, R (2014) Dual targeting. WO2012163520.

48 Mandy, W J, Rivers, M M, Nisonoff, A (1961) Recombination of univalent subunits derived from rabbit antibody. *J Biol Chem*, 236, 3221–3226.

49 Nisonoff, A, Rivers, M M (1961) Recombination of a mixture of univalent antibody fragments of different specificity. *Arch Biochem Biophys*, 93, 460–462.

50 Albrecht, H, Denardo, G L, Denardo, S J (2006) Monospecific bivalent scFv-SH: effects of linker length and location of an engineered cysteine on production, antigen binding activity and free SH accessibility. *J Immunol Methods*, 310 (1–2), 100–116.

51 Cao, Y, Axup, J Y, Ma, J S Y, *et al.* (2015) Multiformat T-cell-engaging bispecific antibodies targeting human breast cancers. *Angew Chem Int Ed Engl*, 54 (24), 7022–7027.

52 Gadkar, K, Pastuskovas, C V, Le Couter, J E, *et al.* Design and pharmaocokinetic characterization of novel antibody formats for ocular therapeutics. *Invest Ophthalmol Vis Sci*, 56 (9), 5390–5400.

53 Sen, M, Wankowski, D M, Garlie, N K, *et al.* (2001) Use of anti-CD3 x anti-HER2/neu bispecific antibody for redirecting cytotoxicity of activated T cells toward HER2/neu + tumors. *J Hematother Stem Cell Res*, 10 (2), 247–260.

54 Davol, P A, Smith, J A, Kouttab, N, *et al.* (2004) Anti-CD3 x anti-HER2 bispecific antibody effectively redirects armed T cells to inhibit tumor development and growth in hormone-refractory prostate cancer-bearing severe combined immunodeficient beige mice. *Clin Prostate Cancer*, 3 (2), 112–121.

55 Reusch, U, Sundaram, M, Davol, P A, *et al.* (2006) Anti-CD3 x anti-epidermal growth factor receptor (EGFR) bispecific antibody redirects T-cell cytolytic activity to EGFR-positive cancers in vitro and in an animal model. *Clin Cancer Res*, 12 (1), 183–190.

56 Yankelevich, M, Kondadasula, S V, Thakur, A, *et al.* (2012) Anti-CD3 × anti-GD2 bispecific antibody redirects T-cell cytolytic activity to neuroblastoma targets. *Pediatr. Blood Cancer*, 59 (7), 1198–1205.

57 Lum, L G, Ramesh, M, Thakur, A, *et al.* (2012) Targeting cytomegalovirus-infected cells using T cells armed with anti-CD3 × anti-CMV bispecific antibody. *Biol Blood Marrow Transplant*, 18 (7), 1012–1022.

58 Ellerman, D, Scheer, J M (2011) *Generation of bispecific antibodies by chemical conjugation*, Springer-Verlag Berlin/Heidelberg, pp. 47–63.

59 Robak, T (2013) The emerging therapeutic role of antibody mixtures. *Expert Opin Biol Ther*, 13 (7), 953–958.

60 Qiu, X, Wong, G, Audet, J, *et al.* (2014) Reversion of advanced Ebola virus disease in nonhuman primates with ZMapp. *Nature*, 514 (7520), 47–53.

61 Qiu, X, Audet, J, Wong, G, *et al.* (2013) Sustained protection against Ebola virus infection following treatment of infected nonhuman primates with ZMAb. *Sci Rep*, 3, 3365.

62 Koefoed, K, Steinaa, L, Søderberg, J N, *et al.* (2011) Rational identification of an optimal antibody mixture for targeting the epidermal growth factor receptor. *MAbs*, 3 (6), 584–595.

63 Dienstmann, R, Patnaik, A, Garcia-Carbonero, R, *et al.* (2015) Safety and activity of the first-in-class Sym004 Anti-EGFR antibody mixture in patients with refractory colorectal cancer. *Cancer Discov*, 5 (6), 598–609.

64 Tolstrup, A B, Frandsen, T P, Bregenholt, S (2006) Development of recombinant human polyclonal antibodies for the treatment of complex human diseases. *Expert Opin Biol Ther*, 6 (9), 905–912.

65 Nowakowski, A, Wang, C, Powers, D B, *et al.* (2002) Potent neutralization of botulinum neurotoxin by recombinant oligoclonal antibody. *Proc Natl Acad Sci USA*, 99 (17), 11346–11350.

66 Nayak, S U, Griffiss, J M, McKenzie, R, *et al.* (2014) Safety and pharmacokinetics of XOMA 3AB, a novel mixture of three monoclonal antibodies against botulinum toxin A. *Antimicrob Agents Chemother*, 58 (9), 5047–5053.

67 Riethmüller, G (2012) Symmetry breaking: bispecific antibodies, the beginnings, and 50 years on. *Cancer Immun*, 12, 12.

68 Couch, J A, Yu, Y J, Zhang, Y, *et al.* (2013) Addressing safety liabilities of TfR bispecific antibodies that cross the blood-brain barrier. *Sci Transl Med*, 5 (183), 183ra57–1–12.

69 Wu, C, Ying, H, Bose, S, *et al.* (2009) Molecular construction and optimization of anti-human IL-1alpha/beta dual variable domain immunoglobulin (DVD-Ig) molecules. *MAbs*, 1 (4), 339–347.

70 Williams, A J, Giese, G, Persson, J (2015) Improved assembly of bispecific antibodies from knob and hole half-antibodies. *Biotechnol Prog*, 31 (5), 1315–1322.

71 Rosati, S, Thompson, N J, Barendregt, A, *et al.* (2012) Qualitative and semiquantitative analysis of composite mixtures of antibodies by native mass spectrometry. *Anal Chem*, 84 (16), 7227–7232.

72 Sazinsky, S L, Ott, R G, Silver, N W, *et al.* (2008) Aglycosylated immunoglobulin G1 variants productively engage activating Fc receptors. *Proc Natl Acad Sci USA*, 105 (51), 20167–20172.

73 Harris, R J (1995) Processing of C-terminal lysine and arginine residues of proteins isolated from mammalian cell culture. *J Chromatogr A*, 705 (1), 129–134.

74 Liu, H, Gaza-Bulseco, G, Faldu, D, *et al.* (2008) Heterogeneity of monoclonal antibodies. *J Pharm Sci*, 97 (7), 2426–2447.
75 Dick, L W, Qiu, D, Mahon, D, *et al.* (2008) C-terminal lysine variants in fully human monoclonal antibodies: investigation of test methods and possible causes. *Biotechnol Bioeng*, 100 (6), 1132–1143.
76 Wranik, B J, Christensen, E L, Schaefer, G, *et al.* (2012) LUZ-Y, a novel platform for the mammalian cell production of full-length IgG-bispecific antibodies. *J Biol Chem*, 287 (52), 43331–43339.
77 Woods, R J, Xie, M, Kreudenstein, T S V, *et al.* (2013) LC-MS characterization and purity assessment of a prototype bispecific antibody. *MAbs*, 5 (5), 711–722.
78 Xu, Y, Lee, J, Tran, C, *et al.* (2015) Production of bispecific antibodies in "knobs-into-holes" using a cell-free expression system. *MAbs*, 7 (1), 231–242.
79 Prentice, K M, Wallace, A, Eakin, C M (2015) Inline protein A mass spectrometry for characterization of monoclonal antibodies. *Anal Chem*, 87 (4), 2023–2028.
80 Rose, R J, Labrijn, A F, den Bremer, van, E T J, *et al.* (2011) Quantitative analysis of the interaction strength and dynamics of human IgG4 half molecules by native mass spectrometry. *Structure*, 19 (9), 1274–1282.
81 der Neut Kolfschoten, van, M, Schuurman, J, Losen, M, *et al.* (2007) Anti-inflammatory activity of human IgG4 antibodies by dynamic Fab arm exchange. *Science*, 317 (5844), 1554–1557.
82 Labrijn, A F, Buijsse, A O, den Bremer, van, E T J, *et al.* (2009) Therapeutic IgG4 antibodies engage in Fab-arm exchange with endogenous human IgG4 in vivo. *Nat Biotechnol*, 27 (8), 767–771.
83 Oda, M, Uchiyama, S, Robinson, C V, *et al.* (2006) Regional and segmental flexibility of antibodies in interaction with antigens of different size. *FEBS J*, 273 (7), 1476–1487.
84 Oda, M, Uchiyama, S, Noda, M, *et al.* (2009) Effects of antibody affinity and antigen valence on molecular forms of immune complexes. *Mol Immunol*, 47 (2–3), 357–364.
85 Liu, H, Jeong, J, Kao, Y-H, Zhang, Y T (2015) Characterization of free thiol variants of an IgG1 by reversed phase ultra high pressure liquid chromatography coupled with mass spectrometry. *J Pharm Biomed Anal*, 109, 142–149.
86 Rosati, S, Rose, R J, Thompson, N J, *et al.* (2012) Exploring an orbitrap analyzer for the characterization of intact antibodies by native mass spectrometry. *Angew Chem Int Ed*, 51 (52), 12992–12996.
87 Rose, R J, Damoc, E, Denisov, E, *et al.* (2012) High-sensitivity Orbitrap mass analysis of intact macromolecular assemblies. *Nat Methods*, 9 (11), 1084–1086.
88 Thompson, N J, Hendriks, L J A, de Kruif, J, *et al.* (2014) Complex mixtures of antibodies generated from a single production qualitatively and quantitatively evaluated by native orbitrap mass spectrometry. *MAbs*, 6 (1), 197–203.

89 Johnson, S, Burke, S, Huang, L, *et al.* (2010) Effector cell recruitment with novel Fv-based dual-affinity re-targeting protein leads to potent tumor cytolysis and in vivo B-cell depletion. *J Mol Biol*, 399 (3), 436–449.
90 Gassner, C, Lipsmeier, F, Metzger, P, *et al.* (2015) Development and validation of a novel SPR-based assay principle for bispecific molecules. *J Pharm Biomed Anal*, 102, 144–149.
91 Le G, Fabrice, R, Uwe, L, Melvyn, K, Sergey M (2004) Effect of linker sequences between the antibody variable domains on the formation, stability and biological activity of a bispecific tandem diabody. *Protein Eng Des Sel*, 17 (4), 357–366.
92 Schanzer, J M, Wartha, K, Croasdale, R, *et al.* (2014) A novel glycoengineered bispecific antibody format for targeted inhibition of epidermal growth factor receptor (EGFR) and insulin-like growth factor receptor type I (IGF-1R) demonstrating unique molecular properties. *J Biol Chem*, 289 (27), 18693–18706.
93 Beck, A, Wagner-Rousset, E, Ayoub, D, *et al.* (2013) Characterization of therapeutic antibodies and related products. *Anal Chem*, 85 (2), 715–736.
94 Jarasch, A, Koll, H, Regula, J T, *et al.* (2015) Developability assessment during the selection of novel therapeutic antibodies. *J Pharm Sci*, 104 (6), 1885–1898.
95 Braren, I, Greunke, K, Umland, O, *et al.* (2007) Comparative expression of different antibody formats in mammalian cells and *Pichia pastoris*. *Biotechnol Appl Biochem*, 47 (4), 205.
96 Moore, G L, Bautista, C, Pong, E, *et al.* (2011) A novel bispecific antibody format enables simultaneous bivalent and monovalent co-engagement of distinct target antigens. *MAbs*, 3 (6), 546–557.
97 Sharma, B (2007) Immunogenicity of therapeutic proteins. Part 1: impact of product handling. *Biotechnol Adv*, 25 (3), 310–317.
98 Schaefer, J V, Plückthun, A (2012) Transfer of engineered biophysical properties between different antibody formats and expression systems. *Protein Eng Des Sel*, 25 (10), 485–506.
99 Ohkuri, T, Nagatomo, S, Oda, K, *et al.* (2010) A protein's conformational stability is an immunologically dominant factor: evidence that free-energy barriers for protein unfolding limit the immunogenicity of foreign proteins. *J Immunol*, 185 (7), 4199–4205.
100 Cheng, W, Joshi, S B, He, F, *et al.* (2012) Comparison of high-throughput biophysical methods to identify stabilizing excipients for a model IgG2 monoclonal antibody: conformational stability and kinetic aggregation measurements. *J Pharm Sci*, 101 (5), 1701–1720.
101 Niesen, F H, Berglund, H, Vedadi, M (2007) The use of differential scanning fluorimetry to detect ligand interactions that promote protein stability. *Nat Protoc*, 2 (9), 2212–2221.

102 Chaudhuri, R, Cheng, Y, Middaugh, C R, Volkin, D B (2014) High-throughput biophysical analysis of protein therapeutics to examine interrelationships between aggregate formation and conformational stability. *AAPS J*, 16 (1), 48–64.
103 Goldberg, D S, Bishop, S M, Shah, A U, Sathish, H A (2010) Formulation development of therapeutic monoclonal antibodies using high-throughput fluorescence and static light scattering techniques: role of conformational and colloidal stability. *J Pharm Sci*, 100 (4), 1306–1315.
104 Menzen, T, Friess, W (2013) High-throughput melting-temperature analysis of a monoclonal antibody by differential scanning fluorimetry in the presence of surfactants. *J Pharm Sci*, 102 (2), 415–428.
105 Connolly, B D, Petry, C, Yadav, S, *et al.* (2012) Weak interactions govern the viscosity of concentrated antibody solutions: high-throughput analysis using the diffusion interaction parameter. *Biophys J*, 103 (1), 69–78.
106 Jezek, J, Darton, N J, Derham, B K, *et al.* (2013) Biopharmaceutical formulations for pre-filled delivery devices. *Expert Opin Drug Deliv*, 10 (6), 811–828.
107 Shire, S J, Shahrokh, Z, Liu, J (2004) Challenges in the development of high protein concentration formulations. *J Pharm Sci*, 93 (6), 1390–1402.
108 Jacobs, S A, Wu, S-J, Feng, Y, *et al.* (2010) Cross-interaction chromatography: a rapid method to identify highly soluble monoclonal antibody candidates. *Pharm Res*, 27 (1), 65–71.
109 Durben, M, Schmiedel, D, Hofmann, M, *et al.* (2015) Characterization of a bispecific FLT3 X CD3 antibody in an improved, recombinant format for the treatment of leukemia. *Mol Ther*, 23 (4), 648–655.
110 Haberger, M, Bomans, K, Diepold, K, *et al.* (2014) Assessment of chemical modifications of sites in the CDRs of recombinant antibodies: susceptibility vs. functionality of critical quality attributes. *MAbs*, 6 (2), 327–339.
111 Sydow, J F, Lipsmeier, F, Larraillet, V, *et al.* (2014) Structure-based prediction of asparagine and aspartate degradation sites in antibody variable regions. *PLoS One*, 9 (6), e100736.
112 Sharma, V K, Patapoff, T W, Kabakoff, B, *et al.* (2014) In silico selection of therapeutic antibodies for development: viscosity, clearance, and chemical stability. *Proc Natl Acad Sci USA*, 111 (52), 18601–18606.
113 Silverman, J, Liu, Q, Lu, Q, *et al.* (2005) Multivalent avimer proteins evolved by exon shuffling of a family of human receptor domains. *Nat Biotechnol*, 23 (12), 1556–1561.
114 Roopenian, D C, Akilesh, S (2007) FcRn: the neonatal Fc receptor comes of age. *Nat Rev Immunol*, 7 (9), 715–725.
115 Knauf, M J, Bell, D P, Hirtzer, P, *et al.* (1988) Relationship of effective molecular size to systemic clearance in rats of recombinant interleukin-2 chemically modified with water-soluble polymers. *J Biol Chem*, 263 (29), 15064–15070.

116 Tryggvason, K, Wartiovaara, J (2005) How does the kidney filter plasma? *Physiology (Bethesda)*, 20, 96–101.

117 Oak, E, Bartlett, N L (2015) Blinatumomab for the treatment of B-cell lymphoma. *Expert Opin Investig Drugs*, 24 (5), 715–724.

118 Kontermann, R E (2011) Strategies for extended serum half-life of protein therapeutics. *Curr Opin Biotechnol*, 22 (6), 868–876.

119 Stork, R, Müller, D, Kontermann, R E (2007) A novel tri-functional antibody fusion protein with improved pharmacokinetic properties generated by fusing a bispecific single-chain diabody with an albumin-binding domain from streptococcal protein G. *Protein Eng Des Sel*, 20 (11), 569–576.

120 Müller, D, Karle, A, Meissburger, B, *et al.* (2007) Improved pharmacokinetics of recombinant bispecific antibody molecules by fusion to human serum albumin. *J Biol Chem*, 282 (17), 12650–12660.

121 Tijink, B M, Laeremans, T, Budde, M, *et al.* (2008) Improved tumor targeting of anti-epidermal growth factor receptor nanobodies through albumin binding: taking advantage of modular nanobody technology. *Mol Cancer Ther*, 7 (8), 2288–2297.

122 Holliger, P, Wing, M, Pound, J D, *et al.* (1997) Retargeting serum immunoglobulin with bispecific diabodies. *Nat Biotechnol*, 15 (7), 632–636.

123 Hotzel, I, Theil, F-P, Bernstein, L J, *et al.* (2012) A strategy for risk mitigation of antibodies with fast clearance. *MAbs*, 4 (6), 753–760.

124 Frese, K, Eisenmann, M, Ostendorp, R, *et al.* (2013) An automated immunoassay for early specificity profiling of antibodies. *MAbs*, 5 (2), 279–287.

125 Xu, Y, Roach, W, Sun, T, *et al.* (2013) Addressing polyspecificity of antibodies selected from an in vitro yeast presentation system: a FACS-based, high-throughput selection and analytical tool. *Protein Eng Des Sel*, 26 (10), 663–670.

126 Volz, C, Pauly, D (2015) Antibody therapies and their challenges in the treatment of age-related macular degeneration. *Eur J Pharm Biopharm*, 95, 158–172.

127 Olafsen, T (2012) Fc engineering: serum half-life modulation through FcRn binding. *Methods Mol Biol*, 907, 537–556.

128 Avery, R L, Castellarin, A A, Steinle, N C, *et al.* (2014) Systemic pharmacokinetics following intravitreal injections of ranibizumab, bevacizumab or aflibercept in patients with neovascular AMD. *Br J Ophthalmol*, 98 (12), 1636–1641.

129 Souied, E H, Devin, F, Mauget-Faÿsse, M, *et al.* (2014) Treatment of exudative age-related macular degeneration with a designed ankyrin repeat protein that binds vascular endothelial growth factor: a phase I/II study. *Am J Ophthalmol*, 158 (4), 724–732.e2.

9

Bio-Repository

Anne Baldwin[1,2,*], *Kurt Schroeder*[1,2,*], *Lovejit Singh*[1,2], *and Karen Billeci*[1]

[1] *Department of Proteomics & Biological Resources, Genentech Inc., South San Francisco, CA, USA*
[2] *Department of Biomolecular Engineering, Genentech Inc., South San Francisco, CA, USA*

* *These authors contributed equally to the preparation of this manuscript.*

Abbreviations

CRISPR	clustered regularly interspaced short palindromic repeats
DNA	deoxyribonucleic acid
Fc	crystallizable fragments
FLAG	polypeptide protein tag DYKDDDDK
His	polyhistidine tag
LIMS	laboratory information management systems
NCBI	National Center for Biotechnology Information
NHGRI	National Human Genome Research Institute
PLOS	Public Library of Science
RNA	ribonucleic acid
RNAi	RNA interference
shRNA	small hairpin RNA
siRNA	small interfering RNA

9.1 Introduction

It is human nature to preserve, document, and provide access to knowledge through the development of libraries and repositories. The oldest evidence of this was in the establishment of the great library of Alexandria which was

Analytical Characterization of Biotherapeutics, First Edition. Edited by Jennie R. Lill and Wendy Sandoval.

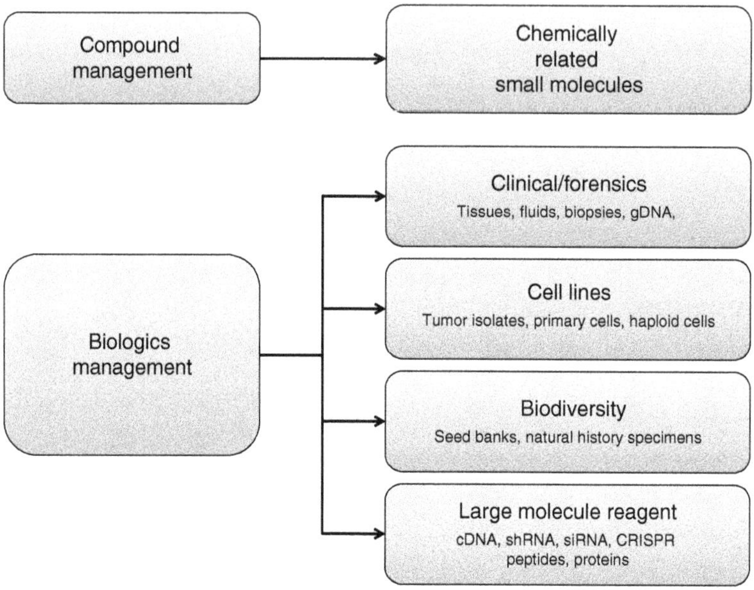

Figure 9.1 Repositories have been established to manage a diversity of sample types. Compound management facilities may contain millions of different structures but the management of the compounds is similar for each chemical type. Biologics repositories encompass a more diverse range of sample types that may require diverse and unique sample management practices [2, 3].

dedicated to the muses and goddesses of arts, and was one of the largest and important libraries of the ancient world [1]. In modern times, this endeavor has been extended to the establishment of various curations and activities including the development of biorepositories and compound management facilities within the biopharma industry. The purpose and types of materials these facilities manage are diverse (Figure 9.1). Compound management libraries established at most large pharmaceutical companies and research hospitals are seen as highly valued assets that are foundational to discovering new therapies to meet unmet medical needs. Compound management infrastructure consists of extensive informatics, automation, and business rules that enable high-throughput drug discovery programs [3]. Biorepositories have greater diversity and complexity due to the heterogeneity of the types of samples and the needs of the samples that are managed. Biorepositories may consist of clinical sample repositories used in clinical drug discovery programs [4] as with the Svalbard Global Seed Vault and are established to provide insurance against both incremental and catastrophic loss of crop diversity [5]. Access to these resources is intended to be rare so infrastructure is focused on documentation of detailed sample identity and disaster recovery [4]. In recent years, the establishment of a biorepository for the management of large molecules has become a priority. The primary and pragmatic drivers are increasing implementation of high-throughput proteomic and functional genomic research programs. In addition, as discussed in *Nature* [6] and the Public Library of Science (*PLOS*) [7] preclinical research is experiencing a "reproducibility crisis" that is associated with poor management of large molecule reagents such as antibodies, nucleotides, and proteins used in preclinical assays and characterization studies.

Centralized, high-throughput management of large molecules is becoming increasingly important in the drug discovery process. Integrated informatics (registration, inventory management, and request systems) and automation are necessary to allow for the request, archival, production, and supply of deoxyribonucleic acid (DNA), ribonucleic acid interference (RNAi), and proteins. These activities can then be utilized to support multiple areas of research, including pathway discovery, antibody discovery, and protein chemistry.

A typical workflow for the production of these reagents is shown in Figure 9.2. Briefly, nucleotide-based reagents such as DNA constructs, clustered regularly interspaced short palindromic repeats (CRISPR), small hairpin RNA (shRNA), or small interfering RNA (siRNA) are either cloned or synthesized. If the DNA construct is meant for bulk protein expression, the DNA may be scaled to produce sufficient quantities for transfection. The resulting protein may be produced in large quantities or batches. These lots may be further manipulated *via* conjugation, dilution, and so on. All of these activities may be performed in a high-throughput manner in multiple labs at a global scale. Therefore, there is a need to build infrastructure that can manage the location, status, and history of any given sample.

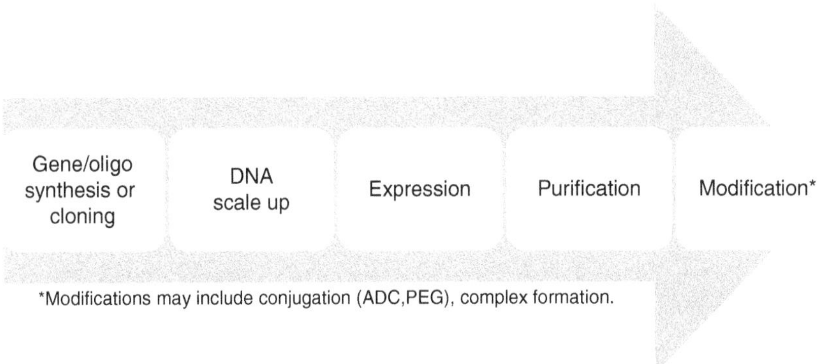

Figure 9.2 Production of large molecules though complex follows several basic steps from DNA synthesis through purification [8, 9].

Here we describe the establishment of a large molecule biorepository that encompasses the management of DNA constructs, proteins, and peptides. The informatics and automation infrastructure is modeled on what has been established for compound management libraries but modified to manage the diversity and nuanced annotation needs of large molecules. We also describe the establishment of automation and business practices more suitable to the mid-throughput requirement of biologics.

9.2 Large Molecule Repository Management

The development of business rules, informatics, and automation for the management of large molecule repositories can be modeled on elements form both compound management and biobanks. The primary goal for repository informatics is to catalogue unique entities and enable traceability. The implementation of automation has many drivers: throughput, precision, accuracy, and ergonomics. Development of best practices regarding logistics, sample storage conditions, and service level agreements (SLAs) regarding distribution timelines are dependent on the nature of a given reagent and research needs.

9.2.1 Informatics

Core informatic infrastructure consists of registration, process/inventory management, and search/request/archive functions. Informatic components can exist as separate software modules or may have overlapping functionality (Figure 9.3a). The development and implementation of a registration system is foundational to any repository management system to enable management of identity, ancestry, and traceability. Registration systems are well established for the management of identity of chemical entities. Registration for large molecules has been slow to develop for a number of possible reasons:

- The diversity of large molecules substance types has made registration complex.
- Sample management for large molecules has been siloed into multiple laboratories thus making a single registration system challenging
- The production and characterization of large molecules has generally been low throughput so the need for registration has been of lower priority

The initial implementation of registration may therefore cobble together multiple systems that enable sufficient traceability and sample management as a centralized facility builds infrastructure and as new biologic entities are added to the repertoire of sample management facility (Figure 9.3b).

As with compound management, identity should consist of three primary levels of identity: concept, batch, and container (Figure 9.4). The concept level

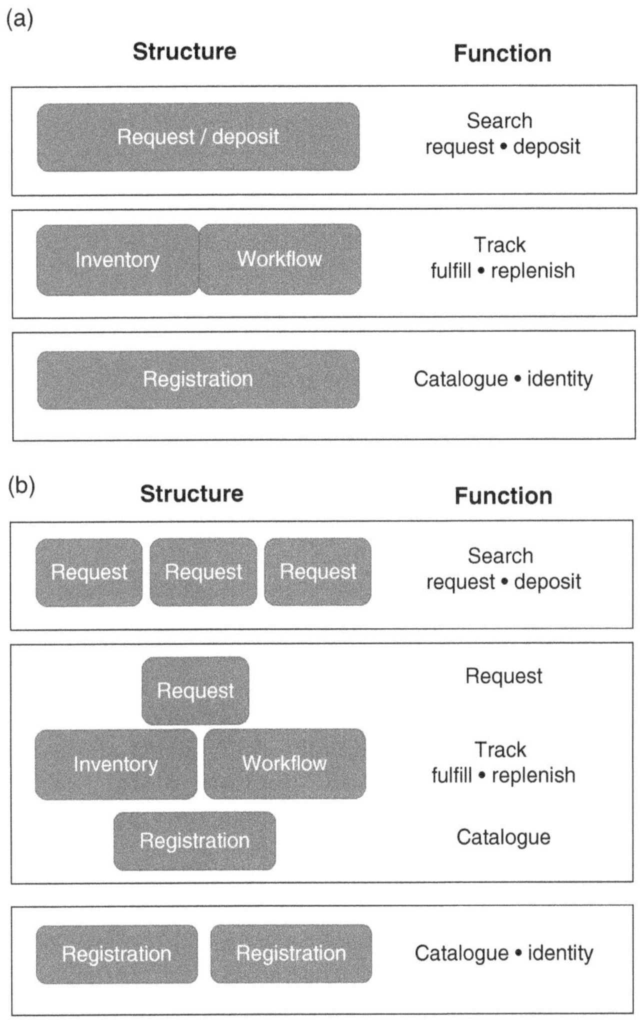

Figure 9.3 (a) Core infrastructure for sample management. Registration, inventory management, and request functionality are dispersed into modular systems that can be directly integrated *via* application programming interface (API) or manual processes. (b) In developing systems functions may overlap such that registration of different reagent types are in different registration systems; requests may be integrated into inventory management software.

definition is the most unique or defining characteristic of a given molecule. For small molecules the concept level is the structure of a chemical entity. In the case of large molecules, ideally the nucleotide or amino acid sequence can be used. It should, however, be noted that in some instances the sequence,

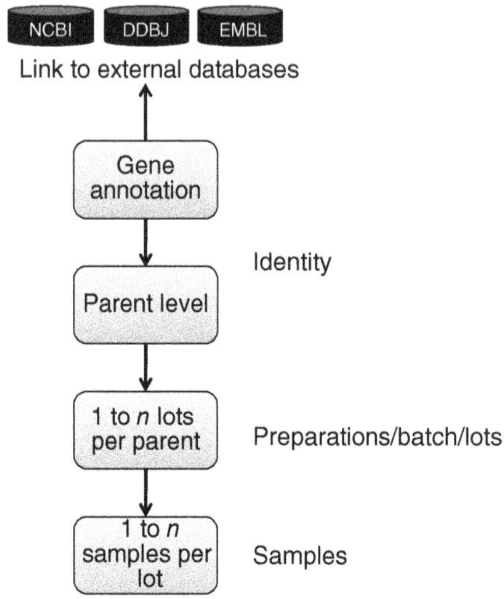

Figure 9.4 The hierarchy of registration for biologics is similar to that established in compound management: concept, batch/lot/sample.

specificity, or other defining characteristics may not be available for propriety or technical reasons [10, 11]. Additional attributes may include affinity purification tags (e.g., polyhistidine tag (His), polypeptide protein tag DYKDDDDK (FLAG), crystallizable fragment (Fc) tags), antigen or epitope specificity in the case of antibodies, and/or gene (entrez ID, symbol), to establish identity. Additional granularity in identity may define the components of a given substance such as the insert and vector of an expression construct or the building blocks of an antibody or complex protein. Identity of complex molecules may also include conjugation or other modifications. Modifications may include natural posttranslational modifications such as glycosylation or phosphorylation, or chemical or proteinaceous modifications such as a small molecule chemotherapeutic or albumin. However, it should be noted that natural posttranslational modifications may not be known at the time of initial registration but instead maybe documented at the lot/batch level *via* results management tools or lot characterization. The lot or batch level identity typically contains information such as source and production information. The final level is the container or aliquot of a given substance. The information maintained at this level of identity is concentration, volume, freeze/thaw cycle information, and so on. In general, sample level information is managed in an inventory management system, but may be staged in a registration when cataloguing the first aliquot of a given substance. Inventory management can be managed through laboratory information management systems (LIMS) and request systems and a number of commercial tools for curating samples by these means are available, for example, the Titian registration software Mosaic [12, 13]. Sample history and ancestry would be managed through links/

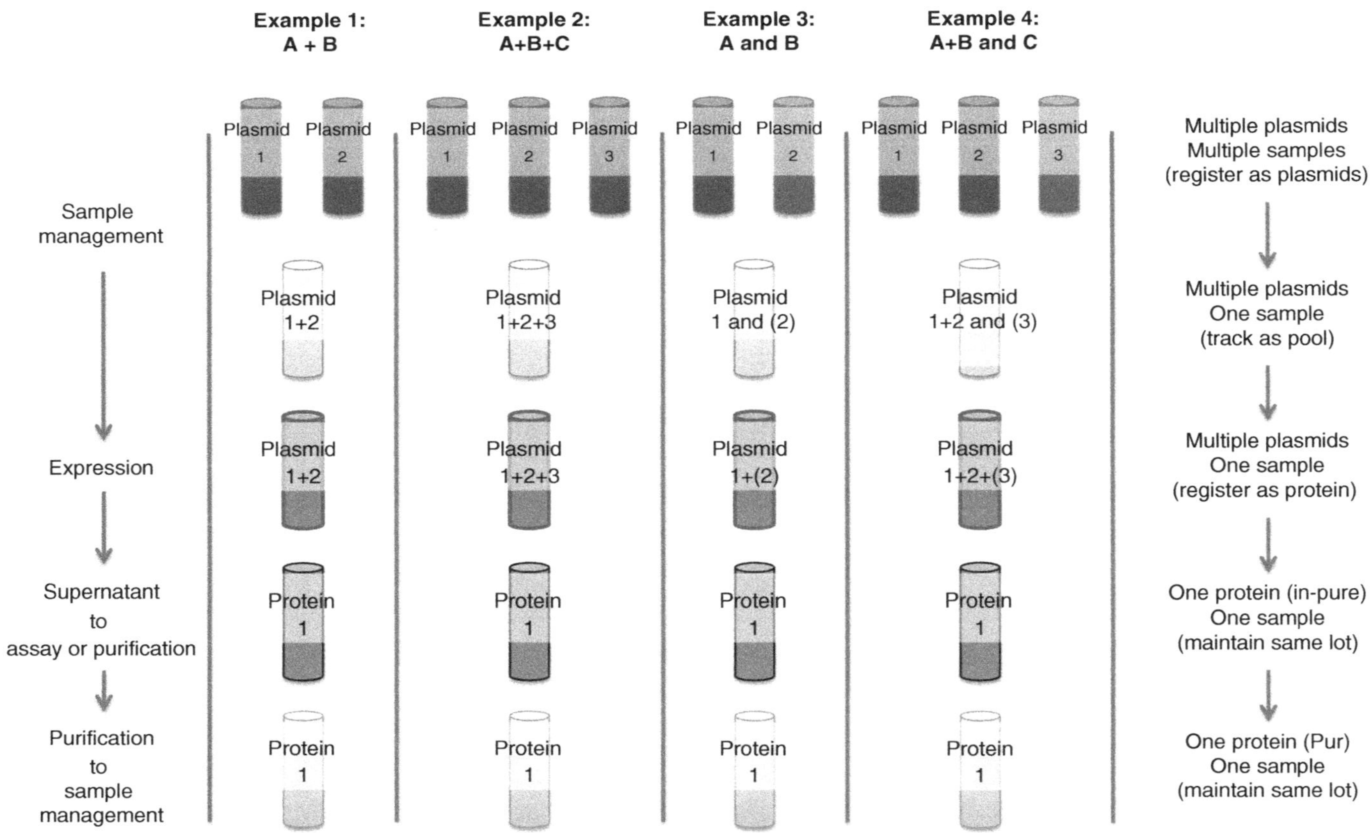

Figure 9.5 Ancestry or relatedness for biologics is more complex than observed for compounds. Here four examples are shown of how the ancestry of a protein may be tracked from DNA construct to protein.

integration with inventory/production management. Ancestry management for biologics can be more nuanced than seen for small molecules. Figure 9.5 shows a simple example of how the production of a protein might be tracked from the expression construct through transient transfection in a mammalian host to protein purification. There is added complexity when different DNA constructs can produce the same protein or if the same construct is used to make different proteins as with some antibodies. The primary function of inventory management is to track location, transfers *via* barcoded labware and other manipulations to allow precise monitoring of the physical inventory present in the laboratory and clinic.

9.2.2 Automation

There are two primary activities within a sample management facility that may be considered for automation: storage and sample manipulation. The primary considerations for implementing automation are to increase throughput, improve precision and accuracy, minimize unnecessary manipulation of samples which can lead to sample loss or sample contamination, and decrease ergonomic risk of injury due to repetitive manual processing. Improved throughput is the primary driver in typical compound management facilities that manage large collections of millions of compounds through drug discovery development and screening programs. The significant need for the development of technology that could rapidly cherry-pick, array, manipulate, and assay has been the major technology developmental catalyst in lab automation and storage.

In the biological space, the need for high-throughput sample management is increasing but the scale is unlikely to reach the throughput needs seen in the management of small molecules. The drive, therefore, for automation is to improve accuracy, precision, and traceability. There is also a significant desire to minimize the unnecessary manipulation of samples that may be sensitive to freeze thaw or other exposures.

9.2.2.1 Automated Refrigerated or Freezer Stores

The implementation of automated stores has been slow in the biologics space. The reason has been somewhat of a supply and demand problem. Store developers based on the demands from compound management groups focused on the development of automation at temperatures higher than required for the storage of biologics (−20°C or greater). The demand in the biologics space was small as biologics repositories were often decentralized. Therefore smaller, manual stores were considered advantageous as they are cheaper and easier to implement in individual labs, and do not require familiarity with automation. In addition, pick rates for biologics are typically small especially when compared to compounds management where literally millions of compounds can

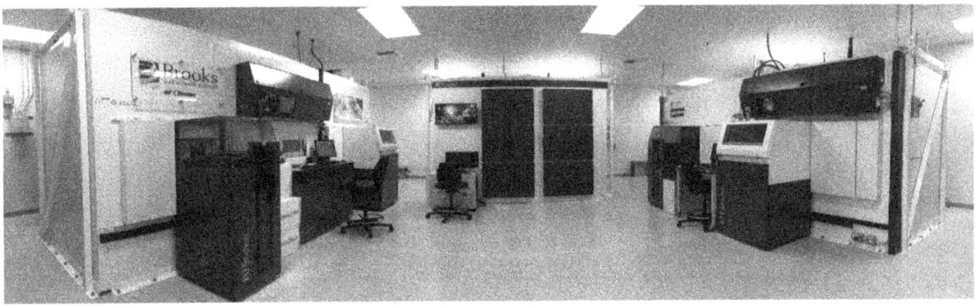

Store	Standardize Labware	Capacity	Substance
−80°C	REMP 300, CryoVials (matrix, FluidX) and fix well plates	1 million REMP	Transformed E.*coli*, cell lines, protein
−20°C	REMP 300 and fix well plates	1 million REMP or 10 000 plates	Purified plasmid, siRNA
4°C	4 and 20 ml vials	40–100K	Proteins

Figure 9.6 Implementation of automated stores for the management of large molecules.

be screened in a phenotypic or other high-throughput small molecule discovery campaign. Finally, unlike compound libraries that are frequently seen as assets to a drug discovery program, biologics are often seen as more transitory in the biologics discovery space. That at best there is only a need to archive the "good" reagents. The advantages for implementing automated stores, although expensive, especially if samples have different temperature requirement (Figure 9.6), for biologics are accuracy of sample placement and location tracking, improved temperature control, and decreased manipulation of the "innocent"—the inadvertent thawing of samples in racks or boxes containing the sample of interest [14–17]. The centralization of biologic sample management distributes the cost of implementation of such stores across multiple groups, and can enable improved quality of samples through standardized sample management distribution workstreams. Figure 9.6 shows an example implementation of automation stores for the management of biologics.

9.2.2.2 Lab Automation

The development of lab automation for large molecules is not dissimilar to that developed for compound management. There are, however, some exceptions. Pick rates are generally smaller; therefore highly integrated "tree-to-toothpick" automation platforms are not necessary. For large molecules the primary driver for automation implementation is accuracy, precision, and traceability. In the large molecule space the diversity of reagent types provides interesting challenges in developing automation platforms that can accommodate different liquid classes and densities. For example, plasmid-based reagents are often stored as transformed *Escherichia coli* in as much as 50% glycerol. The resulting solution is quite viscous and systems must be optimized to pipette accurately. Sample sterility and cross-contamination must also be considered when implementing lab automation. Equipping robots with enclosure, high efficiency particular air (HEPA) filters, or placement in biologics safety cabinets generally provide sufficient sterility for most downstream applications. In addition, some reagents must be isolated from one another to avoid cross-contamination. For example, proteins intended for animal studies must be processed in separate facilities from bacterial cultures to avoid endotoxin contamination. Finally, unlike compound management facilities in which the functionality of lab automation is focused on dilution and aliquoting, a facility responsible for the management of plasmid-based reagents may need to support DNA production from transformation to DNA purification [18].

9.3 Challenges and Future Perspectives for Working with Diverse Biological Reagent Types

The establishment of best practices for centralized biologics management is still a work in progress. There are many challenges in the development of informatics, automation, and business rules for large molecules. These challenges include the diversity of reagent types (e.g., cell to purified protein), rapidly evolving technologies (e.g., CRISPR [19], antibody–drug conjugates, bispecific antibodies), and workflows that are becoming increasingly global such that many of these reagents are sourced from multiple laboratories (e.g., through outsourcing efforts). Unlike traditional biobanks that frequently must comply with government regulation there is a significant effort toward established governance [3] for the management of patient samples. In compound management establishment of nomenclature for chemical entities was established nearly 100 years ago [20] and optimization of best practices for the management of compound libraries has been well established by compound management groups [21]. The establishment of best practices for centralized biologics management is still a work in progress. Biologic sample management groups, however, do not have to start from scratch. The establishment of public databases and consortiums such as the National Center for Biotechnology Information (NCBI) and National Human Genome Research Institute (NHGRI) amongst others may help to established standard nomenclature for gene-based reagents and proteins. Informatics and lab automation can be optimized from what has already been established for compound management. The establishment of business rules around sample stability/longevity (Figure 9.7) must be evidence-based [22–24] and dependent on regulatory consideration and project need.

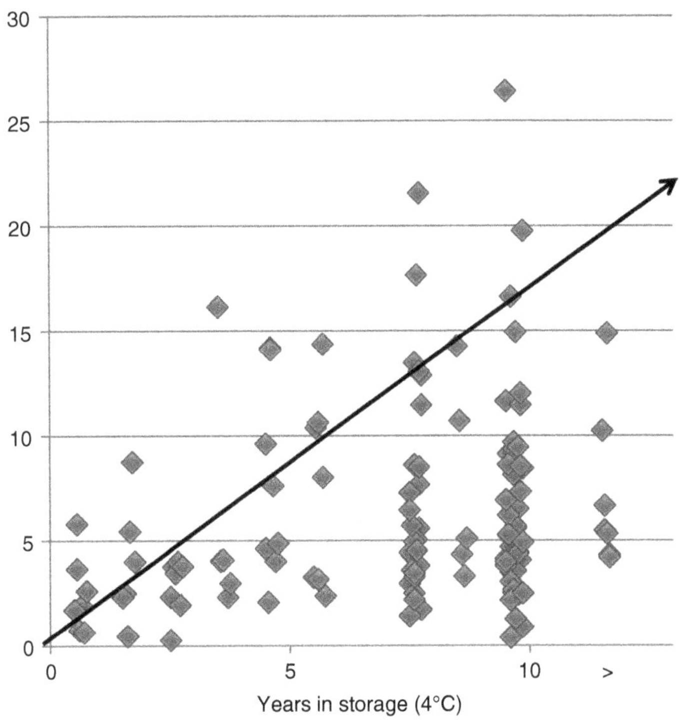

Figure 9.7 Stability of reagent antibodies stored in bulk at 4°C immediately after purification of the antibody preps contained <5% aggregation. Upon storage aggregation rates trended upward over time but were antibody dependent.

References

1 MacLeod, Roy, (2005) *The Library of Alexandria: Center of Learning in the Ancient World*, I.B. Tauris & Co Ltd, New York.

2 Wigglesworth, Mark and Wood, Terry, (2012) *Management of Chemical and Biological Samples for Screening Applications*, John Wiley & Sons, Ltd, Chichester, pp. 45–82.

3 Wigglesworth, Mark and Wood, Terry, (2012) *Management of Chemical and Biological Samples for Screening Applications*, John Wiley & Sons, Ltd, Chichester, pp. 243–274.

4 Campbell, Lori D., (2012) Best Practice for Repositories: Collection Storage, Retrieval and Distribution of Biological Materials for Research, ISBER, Third Edition.

5 https://www.croptrust.org/our-work/svalbard-global-seed-vault/ (accessed April 16, 2017).

6 Charles, Daniel, A Forever Seed Bank Takes Root in the Arctic, *Science* Jun 23, 2006, Vol. 312, Issue 5781, pp. 1730–1731.
7 http://www.nature.com/news/reproducibility-1.17552 (accessed April 30, 2016).
8 Freedman, Leonard P., Cockbur, Iain M., and Simcoe, Timothy S., The Economics of Reproducibility in Preclinical Research, *PLOS Biology* Jun 9, 2015, Vol. 13, Issue 6, e1002165.
9 Higgins, Steve J. and Hames, B. D., (1999) *Protein Expression: A Practical Approach*, Oxford University Press, Oxford.
10 http://adcreview.com/adc-university/adcs-101/antibody-drug-conjugates-adcs/ (accessed April 30, 2016).
11 Wigglesworth, Mark and Wood, Terry, (2012) *Management of Chemical and Biological Samples for Screening Applications*, John Wiley & Sons, Ltd, Chichester, pp. 23–58.
12 Vivien Marx, Find the right antibody for the job, *Nature Methods* 2013, Vol. 10, pp. 703–707.
13 http://www.ncbi.nlm.nih.gov (accessed April 30, 2016).
14 http://www.titian.co.uk/ (accessed April 30, 2016).
15 Mark Wigglesworth and Terry Wood, (2012) *Management of Chemical and Biological Samples for Screening Applications*, John Wiley & Sons, Ltd, Chichester, pp. 221–242.
16 Case Study: Brooks Automation, (2015) NIHR—National Biosample Centre Maximizes Flexibility, Accuracy, and Productivity with Centralized Biosample Management. http://www.brooks.com/applications-by-industry/life-science/library/case-study-uk-national-biosample-centre (accessed April 24, 2017).
17 http://www.biocor.umn.edu/library (accessed April 24, 2017).
18 Ian Pope, (2015) How to Plan Next Generation Biologics Repositories: Five Critical Element You Must Consider. Brooks Life Science Systems. https://www.laboratoryequipment.com/article/2017/02/how-plan-next-generation-biological-repositories (accessed April 24, 2017).
19 Billeci, Karen, Suh, Christopher, Di Ioia, Tina, *et al.* (2016) Implementation of a High-Throughput DNA Production Pipeline, *Journal of Laboratory Automation* Vol. 21, Issue 6, 765–778.
20 Doudna, Jennifer A. and Charpentier, Emmanuelle, (2014) The new frontier of genome engineering with CRISPR-Cas9, *Science* Vol. 346, Issue 6213, 1258096.
21 http://iupac.org/. (accessed April 30, 2016).
22 https://www.genome.gov/ (accessed April 30, 2016).
23 Cao, Enhong, Chen, Yahuei, Cui, Zhanfeng, and Foster, Peter R. Effect of freezing and thawing rates on denaturation of proteins in aqueous solutions. *Biotechnology and Bioengineering*, 2003, Vol. 82, Issue 6, pp. 684–690.
24 Anchordoquy, Thomas J. and Molina, Marion C., Preservation of DNA, *Cell Preservation Technology*, 2007, Vol. 5, Issue 4, pp. 180–188.

10

Characterization of Residual Host Cell Protein Impurities in Biotherapeutics

Denise Krawitz[1], Jason C. Rouse[2], Justin B. Sperry[2], Wendy Sandoval[3], and Martin Vanderlaan[1]

[1] *Department of Analytical Operations, Genentech Inc., South San Francisco, CA, USA*
[2] *Department of Analytical Research and Development, Pfizer, Cambridge, MA, USA*
[3] *Department of Microchemistry, Proteomics & Lipidomics, Genentech Inc., South San Francisco, CA, USA*

Abbreviations

1D SDS-PAGE	one-dimensional sodium dodecyl sulfate polyacrylamide gel
CFR	code of federal regulations
CHO	Chinese hamster ovary
CQAs	critical quality attributes
DIGE	2D difference gel electrophoresis
ELISA	enzyme linked immunosorbent assay
EMEA	European Medicines Agency
EUP	European pharmacopeia
HCPs	host cell proteins
ICH	International Council for Harmonisation of Technical Requirements for Registration of Pharmaceuticals for Human Use
LC-MS/MS	liquid chromatography–tandem mass spectrometry
MRM	multiple reaction monitoring
MS	mass spectrometry
USP	US pharmacopeia

Analytical Characterization of Biotherapeutics, First Edition. Edited by Jennie R. Lill and Wendy Sandoval.

10.1 Introduction

Recombinant protein therapeutics are produced in host cells that have been engineered to express a protein product at high levels, such as Chinese hamster ovary (CHO) cells, *Escherichia coli*, yeast, and plants. Purification processes are designed to separate the product from all cellular components, including proteins that are expressed by the host cells. Host cell proteins (HCPs) are a particularly complex impurity because they are a heterogeneous population of all proteins expressed by the host cell. CHO cells are a very common host cell for expression of biotherapeutics and contain greater than 24 000 genes [1]. Another common host, *E. coli*, contains ~4300 gene products [2]. Given posttranslational modifications, differential splicing, and proteolysis, HCPs constitute thousands of different proteins, making them an extraordinarily complex impurity to detect and quantify.

Regulatory agencies define HCPs as process-derived impurities (21 CFR 610.13, ICH Q6B, ICH Q11). Like other process-related impurities, HCP clearance needs to be monitored during purification of biotechnology products. However, the complexity of HCPs as an impurity sets it apart from other single analyte impurities, and there are unique challenges associated with HCP monitoring, quantification, and identification. While there are potentially thousands of proteins that need to be monitored, it is not clear how many of these proteins are expressed and which ones have a tendency to copurify with a product. Copurification of an HCP with a product may be driven by the HCP having similar biochemical properties as the product, and therefore having similar behavior during each purification step. More recent reports in the literature suggest that HCPs copurify with products because of a direct interaction with the product [3–5]. In each of these scenarios the HCP population that may copurify with the product is product and/or process specific.

Assays designed to monitor HCPs must be capable of casting a very wide net sampling hundreds to thousands of analytes simultaneously based on the starting population of HCPs that enter the recovery process. Throughout the biotechnology industry, different techniques such as the enzyme linked immunosorbent assay (ELISA), Western blotting, sodium dodecyl sulfate polyacrylamide gel (SDS-PAGE), and more recently, MS have been the workhorses for HCP monitoring. These methods and how they are used to monitor HCP impurities will be discussed in detail in this chapter.

10.2 HCP Measurement and Reporting

Manufacturers need to control and monitor HCPs in biotechnology products, but there is no guidance that stipulates acceptable limits. Such guidance would be difficult, if not impossible, to define with the analytical technologies available to us today. This is because HCPs are a uniquely complex impurity potentially consisting of 100s–1000s of proteins of varied compositions in protein products. At the end of cell culture, the starting population of HCPs that enter the recovery process is often relatively consistent across products produced by the same platform cell culture process. As purification proceeds through multiple chromatography steps the HCPs can vary both in the composition of proteins that are present and in the quantity of each of those proteins. The immunoassays report only one number for the aggregate HCP content. This number is obtained from HCP assays that are highly reagent (e.g., antibody preparation) and method (ELISA versus MS) specific, and assay standards vary from lab to lab. What this means is that the HCP value obtained for a sample tested in one assay/lab is likely going to differ from that same sample tested in a different assay. In 1997 the European Medicines Agency (EMEA) has stated that it would be problematic to standardize analytical methods for HCPs across the industry because the reagents used are "product and production system related." This makes setting industry-wide limits difficult, if not impossible.

Typically, HCP levels are expressed as a ratio relative to the product (ng HCP/mg product, or ppm). However, even though HCPs are referred to as a mass, the value obtained from immunoassays is not an actual mass of HCPs present in the samples. It is a measure of the relative immunoreactivity of the HCPs present in the sample relative to the immunoreactivity of the HCPs in the assay standard. Thus, when considering limits for any individual HCP that may be present in a final product, it is important to verify that the HCP ELISA accurately measures that HCP or to develop a quantitative assay specific for that HCP [6].

10.2.1 Antibodies to HCPs

Since immunoassays for HCPs depend critically on the antibodies used, it is important to show that the immunochemical reagents are suitable for their intended purpose. There is the expectation that host animals will have been extensively immunized with HCPs over an extended period producing a vigorous immune response, and that sufficient antisera will have been collected so that the same antibody reagents may be used for a long period of time, often the lifetime of the biopharmaceutical product.

"Coverage" is a term used to describe the fraction of the total HCP population for which there are antibodies in the anti-HCP antisera. While easy to define, coverage is often difficult to measure. One approach is to perform a 2D SDS PAGE gel separation of the HCP population in the blank run, used to immunize animals, and to immunoblot the proteins with the antisera. This approach provides a general image of whether the animals are generating a vigorous immune response with antibodies that recognize a majority of the

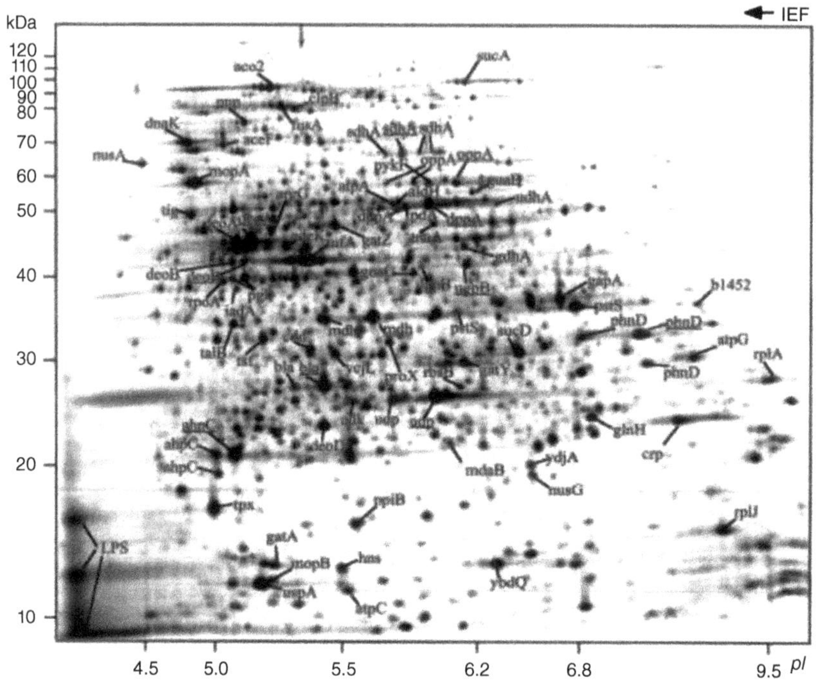

Figure 10.1 Two-dimensional gel electrophoresis profile of *E. coli* proteins at the end of anti-CD-18 blank run fermentation. Source: Kathleen *et al.* [7]. Reproduced with permission of John Wiley & Sons.

proteins present and include proteins of widely varying sizes and isoelectric points (pI) (Figure 10.1).

One must be cautious to not overinterpret 2D gel images, however, since the fraction of "positive" proteins is highly technique dependent and may be influenced by the amount of protein loaded on the gel, the dilution of the antisera, and the spot detection method. It is also worth noting that a single antibody that recognizes SDS-denatured proteins is sufficient for a "positive" in a gel, and that may or may not mean that there are at least two antibodies that will form a sandwich in an ELISA where the HCPs are present in solution in native configuration. Thus, recognition of a given protein in a gel does not mean that the same protein will give a positive single in an ELISA.

Another common question regarding anti-HCP antibodies is whether they must be produced specifically for each product, or whether the same antibodies can be used across a series of products produced by similar platform culture conditions. An extension of this question is whether commercially available anti-HCP reagents can be used. In both cases the issue is not how the antibodies were produced but whether the resulting antibody population is sufficiently broad to recognize a majority of the HCPs present at harvest, and whether the recovery process clears these proteins. It is not unusual to find that HCPs are present at a million ng/mg in harvest preparations, and that for therapeutic monoclonal antibodies (mAbs), for example, the HCP levels in the final product are in the low single digits of ng/mg. A million-fold clearance of HCPs as measured by an immunoassay is an indication of both the breadth of the HCP assay and the efficacy of the purification process.

It is unrealistic to expect that anti-HCP antibodies will recognize each individual HCP equally, or that all HCP proteins will be positive in the ELISA. However, these observations do not negate the utility of a broadly reactive anti-HCP ELISA. The HCP ELISA reflects the results of a sampling of the HCP population, and that can be a useful guide for purification process development. If the sampling is broad, chromatographic process steps that are optimized for reduction of HCPs detected in the ELISA will reduce the total HCP population as well. The speed, high throughput, low cost, ability to be QC validated, and broad reactivity of HCP immunoassays means that they are the workhorse assays for guiding process development at monitoring consistency in commercial product production. Nonetheless, orthogonal analytical methods are needed to ensure product purity because of incomplete coverage of the HCP ELISA [8].

10.2.2 Guidance on HCP Limits and Testing

Following purification, residual HCPs present in a drug substance (DS) have the potential to impact product quality, safety, and efficacy. HCP clearance during a purification process is also an important indicator that the process is performing as intended. Therefore, it is crucial that HCPs are closely monitored and controlled during manufacturing. However, it is inevitable that trace amounts of HCPs will be present in a biopharmaceutical DS.

Because they have the potential to impact product quality and patient safety, HCPs are classified as critical quality attributes (CQAs). However, there is no clear guidance from health authorities on the acceptable limits for HCPs in biotechnology products. ICH Q6B (Specifications for Biotechnology Products) and ICH Q11 (Development and Manufacture of Drug Substances) both address process-related impurities, including HCPs. These guides specify that HCP impurities must be monitored and reduced to acceptable levels to ensure product quality and patient safety. Additionally, 21 CFR 610.13 states that products be as free from impurities except when unavoidable.

It is typically expected that biotechnology products will contain less than 100 ng HCP/mg of product [9, 10]. Ultimately health authorities evaluate acceptable HCP levels for each product on a case-by-case basis. The use of a risk assessment, discussed later, is critical for setting HCP limits for every product. Acceptable levels of HCPs will depend on the phase of clinical development, with typical HCP specifications becoming lower as the product moves from Phase I to Phase III and commercial development. Acceptable levels will also depend on such things as the patient population, route of exposure, dose, and mode of action of the biopharmaceutical [11].

The US pharmacopeia (USP) has recently published a new chapter on Residual Host Cell Protein Measurement in Biopharmaceuticals (USP <1132>). This is a general chapter meant to provide stakeholders with best practices for preparation and characterization of HCP assay reagents, and development and validation of HCP measurement procedures. It was written with input from across the biopharmaceutical industry, US Food and Drug Administration (FDA), and USP. It allows some flexibility in the approach to HCP characterization, while creating consistency in analytical expectations and approaches across industry.

USP <1132> outlines an important and comprehensive overview on how to generate antibody preparations that recognize a wide variety of HCPs, as well as best practices in the development of HCP immunoassays. In addition, USP <1132> provides a robust discussion on setting HCP limits, and provides guidance on when assays for individual HCPs are required.

One important aspect of HCP ELISA method development discussed in USP <1132> is the establishment of dilution linearity for each sample type tested in the assay. When nonlinear sample dilution is observed, it must be investigated. Figure 10.2 depicts HCP ELISA results of a series of in-process purification samples. When a sample displays linear dilutions, the HCP content in ng/mg is independent of the dilution. However, when nonlinear sample dilution is observed, the HCP content in ng/mg is highly dependent on the sample dilution. This nonlinear dilution can be caused by antigen excess, a situation that occurs when the antigen being measured is present in excess of the available antibodies [6]. In this case, further HCP characterization is required to identify the offending HCP. USP <1132> also goes into some detail on supporting

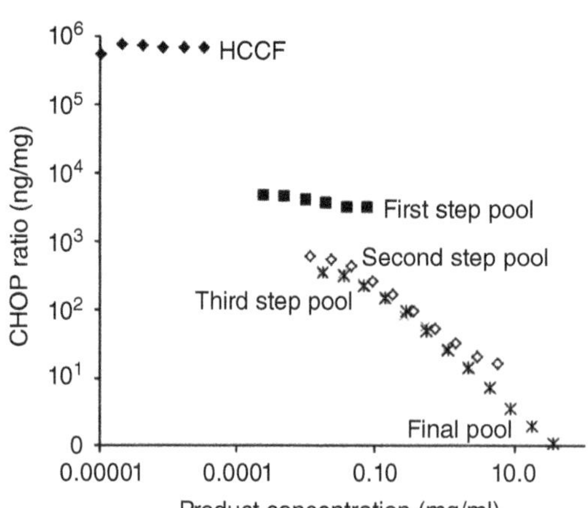

Figure 10.2 Nonlinear dilution of in-process pool samples for an MAb product in clinical development [6].

technologies that can be used for residual HCP detection, identification, characterization, and quantification. These orthogonal methods are critical for understanding product purity.

The European pharmacopeia (EUP) has also recently published a draft general chapter on HCP testing in biopharmaceuticals (EUP 2.6.34). This document was written with input from the biopharmaceutical industry, the Paul Erlich Institute, Swissmedic, and the European Directorate for the Quality of Medicines, and was published in 2016. This is a general chapter that provides guidance for the development and validation of HCP assays for biologics. It does not exclude the use of alternative approaches that are acceptable to relevant health authorities. However, general chapters become mandatory if they are referenced in a monograph. For this reason, there is minimal detail and instruction in EUP 2.6.34, and it serves as more of an overview for HCP testing. It is important to note that EUP 2.6.34 also indicates that nonlinear dilution behavior in an HCP ELISA must be investigated due to the risk of HCP impurities leading to antigen excess.

10.3 Methods to Characterize Host Cell Impurities

10.3.1 HCP-ELISA

Due to its ease, speed, and sensitivity a multianalyte ELISA is commonly used to monitor HCP in drug product (DP). The HCP-ELISA incorporates the use of polyclonal antibodies directed toward host cell impurities. Animals, typically goat or sheep, are challenged with either whole cell lysate or host cell culture fluid (HCCF) to generate antibodies. The resultant polyclonal antibodies contain a cornucopia of IgG against multiple HCPs.

In the HCP-ELISA assay, these capture antibodies are immobilized on a 96-well plate and the immunogenic HCP is sandwiched between the immobilized antibody and a detection antibody (Figure 10.3).

HCP antibodies rely on the immunogenicity of the particular protein in the animal used for immunization. Differences in reaction to a particular immunogen have a direct effect on the ELISA's ability to detect a particular HCP. For example, if a protein elicits a low response in the animal or resulting antibodies are of low affinity (or high), that impurity may not be adequately detected in the ELISA or conversely may be overrepresented. This and other limitations to the ELISA method for HCP detection have inspired the application of orthogonal methods such as SDS-PAGE and MS.

10.3.2 SDS-PAGE and Western Blots

SDS-PAGE based methods continue to be straightforward, fast, and accessible technologies which are used to assess HCP impurities in a sample. In order to detect impurities in final purification pools, SDS-PAGE gels can be overloaded

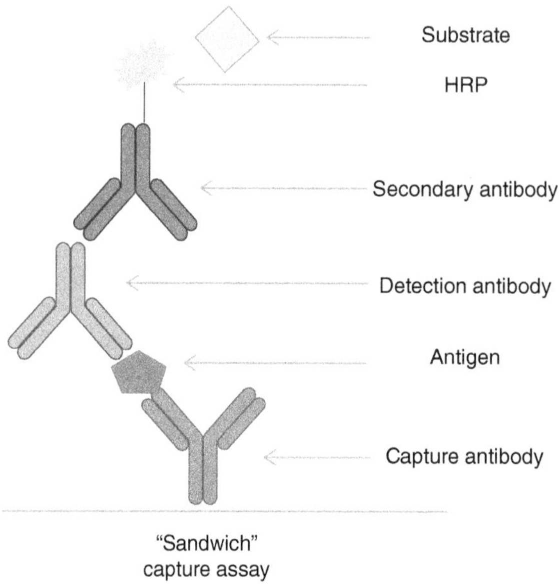

Figure 10.3 "Sandwich" capture ELISA assay.

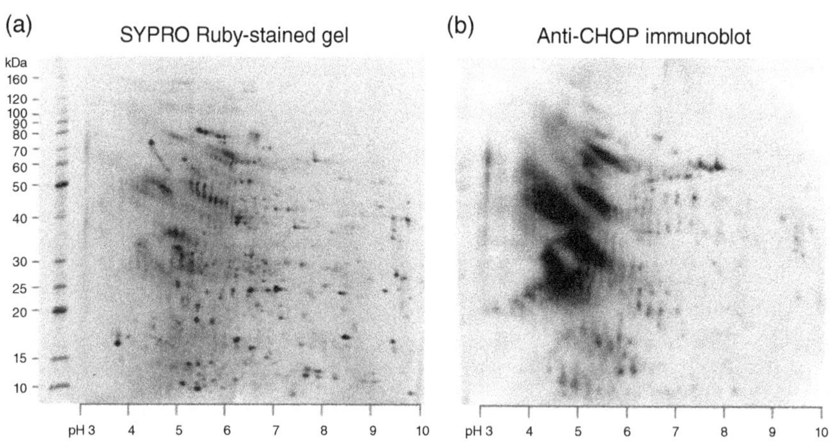

Figure 10.4 Two-dimensional gel electrophoresis-SDS-PAGE gel (a) with Sypro ruby stain and (b) Western blot detection using and anti-CHO protein (CHOP) antibody.

with 10–50 μg of product. This combined with a sensitive total protein stain, such as SyproRuby, silver stain, or another comparable technique, is a powerful orthogonal method to detect HCPs that are not reactive in the HCP ELISA, though sensitivity is limited to ~100 ng each HCP impurity/mg of product (Figure 10.4). Nonetheless, this is a relatively simple check of final purification pools to ensure the purification process is performed as expected. Furthermore, 1D SDS-PAGE is compatible with MS techniques that are very powerful for HCP characterization (described later).

In addition to stained gels, Western blot analysis using the polyclonal antibody mixture generated for the HCP-ELISA can demonstrate an abundance of HCPs in different downstream process pools and may also illuminate the differences between samples. One-dimensional Western blots are also very useful to further characterize HCPs in a sample when the HCP ELISA indicates HCPs are present. The Western blots can be used to differentiate between whether one HCP is present in great abundance or 10 HCP are present in low abundance.

In summary, SDS-PAGE is an older and less sophisticated method, but it remains a powerful tool that provides HCP characterization information and is easily executed in almost any laboratory.

10.3.3 MS Methods for HCP Analysis

MS-based proteomic methods are emerging as a standard approach for orthogonal residual HCP analysis in the biotechnology industry because individual HCPs can be detected, identified, and quantitated directly, complementing routine HCP-ELISA results and increasing confidence in the performance of both the assay and manufacturing process [8, 12]. The use of proteomic MS-based methodology and software for protein identification and quantitation is very well established and documented in the literature after nearly 20 years of research and development by various academic and industry groups worldwide. In contrast to HCP-ELISA, which relies on the presence of polyclonal antibodies for positive HCP detection, MS approaches have the distinct advantage of detecting a wider population of HCPs in a *de novo* manner, without prior knowledge, specialized polyclonal antibody reagents, or product- and process-specific method development. With this capability, and the continued availability of new, more capable MS technologies from multiple vendors and more intuitive specialized software, groups in the biopharmaceutical industry have been progressively adopting MS as an orthogonal approach to HCP analysis, changing the competitive landscape. In turn, this trend has swayed regulatory groups worldwide, who welcome HCP assessments with a more decisive analytical technique such as MS. It is envisioned that the use of MS as an orthogonal approach to HCP-ELISA is likely to become an expected exercise as part of normal process and product development. This section will briefly describe the use of MS as an orthogonal approach to HCP analysis in terms of current methods, recent applications, liquid chromatography–tandem mass spectrometry (LC-MS/MS) analysis of HCPs in practice, advantages and disadvantages, and future outlook.

The use of MS orthogonal to ELISA helps modernize HCP analysis relative to other orthogonal techniques used in the past (as discussed earlier). It is well known that some HCPs may copurify with the product, even with the use of affinity binding steps in the downstream purification process, especially if the molecular properties of both proteins are sufficiently similar. With respect to potential copurifying HCPs, the combination of ELISA and orthogonal MS clearly should provide greater assurance for higher product quality and safety, prevent delays in late-stage clinical studies, and promote timely regulatory approvals by minimizing regulatory scrutiny and risks during registration. As described earlier, quantitative HCP testing *via* traditional ELISA is still considered the "gold standard" approach in most biopharmaceutical process and product development laboratories. At present, HCP-ELISA remains the primary quantitative release test for monitoring total HCP content in each manufactured clinical and commercial product batch, and orthogonal methods such as MS are being used strategically to detect, identify, and quantitate individual HCPs, if present, at particular points during process development to provide confidence that the final, locked manufacturing purification process is clearing all HCPs above a predetermined threshold. Furthermore, an orthogonal method such as MS also helps confirm that the HCP-ELISA is providing comprehensive, process-, and product-specific HCP coverage, especially since hundreds of nonproduct proteins can be detected prior to protein A purification step for mAbs. MS can be used to verify coverage of anti-HCP antibodies since HCPs can be highly similar in molecular characteristics, abundance, and affinity to antibodies raised against them for immunoassays, which may limit the capability of ELISA to accurately represent the HCP population and content. Current MS approaches to identify and quantify individual HCP impurities will be discussed next.

10.3.3.1 Gel Electrophoresis and MALDI or nanoLC-MS/MS

One- and two-dimensional (1D/2D) SDS-PAGE and Western blotting is considered the gold standard of orthogonal approaches to HCP-ELISAs in the biopharmaceutical industry. Two-dimensional SDS-PAGE is used to assess coverage of polyclonal antibody reagent used in ELISA both qualitatively and quantitatively, and 1D or 2D SDS-PAGE is used to check product purity of fully purified DS with high sensitivity as well as evaluate/monitor HCP clearance after each downstream process step based on changing spot patterns. The 2D SDS-PAGE method can be time-consuming, tedious, prone to reproducibility error, and other gel-based assay formats have been developed and implemented such as 2D difference gel electrophoresis (DIGE) where different protein samples are labeled with separate fluorescent dyes prior to analysis, allowing direct quantitative comparison with minimal variability and higher sensitivity (0.2 ng), as compared with traditional silver staining (1 ng).

Identification of particular unknown spots from 1D/2D SDS PAGE or DIGE typically can be identified *via* matrix-assisted laser desorption/ionization time-of-flight (MALDI-TOF) MS using customary proteomic in-gel trypsin digestion following spot excising and staining dye removal. The trypsin peptides from each spot are extracted from the respective gel piece, desalted, mixed with alpha-cyano-4-hydroxycinnamic acid matrix, and spotted on the stainless steel MALDI target. Protein identification from gels is very established and involves submitting the mass spectral data to a protein database search engine such as Mascot (Matrix Science, Boston, MA); more specifically, for each gel spot, the accurately determined peptide molecular masses (also known as peptide mass fingerprinting) and the peptide fragmentation profiles (mass-intensity data containing partial amino acid sequence information) from the 10–20 most abundant tryptic peptides (following collisional activation and MALDI-TOF/TOF MS/MS) enable confident protein identification *via* robust scoring algorithms. The database search of the combined mass spectral dataset yields a list of proteins that are statistically ranked with the best match on top. The database contains known protein sequences that are digested *in silico* with trypsin, and the experimental mass information from peptides and their fragment ions is matched to all respective theoretical values using mass tolerances and other filters to limit or expand the number of proteins in the list. Nanoscale liquid chromatography–tandem mass spectrometry (nanoLC-MS/MS) can also be used for the analysis of in-gel trypsin digested gel spots; while the throughput may be slower due to the use of capillary HPLC systems, higher mass accuracies can be achieved resulting in more confident protein identifications.

10.3.3.2 Two Dimensional LC-MS/MS

Residual HCPs are several orders of magnitude lower in concentration than the protein therapeutic, which requires a method capable of at least 5 orders of magnitude of dynamic range for direct detection. Direct HCP detection by LC-MS/MS-based proteomic techniques in general can be hampered by overwhelming amount of therapeutic product present in the samples, especially since modern research-grade mass spectrometers are capable of 3–4 orders of magnitude detection routinely, with only the most sensitive, cutting-edge instruments achieving 5 orders of magnitude. Use of direct, traditional 1D LC-MS/MS with data-dependent acquisition, also known as label-free shotgun proteomics [13], is biased toward selection and fragmentation of the most abundant peptides [14], and for the case of residual HCP analysis, these intense signals would originate from the abundant therapeutic product rather than the desired low-level HCPs. Therefore, the most common MS-based strategy used to date for residual HCP analysis in biopharmaceutical products has been 2D LC-MS/MS [15], which is based on the proteomics approach, MudPIT, also known as multidimensional protein identification technology [16]. The performance characteristics of label-free quantitation for 2D LC-MS/MS was investigated using low-level spiked-in known protein "impurities," and the results indicated that these spiked "HCP impurities" were identifiable at 13 ppm (parts per million; ng HCP/mg mAb) and quantifiable to within twofold of their actual value [17].

Similar to traditional large-scale proteomics approaches like MudPIT, the addition of a second, upfront LC dimension prior to the standard shotgun proteomics workflow (i.e., trypsin proteolysis, online LC-MS/MS for peptide fragmentation, and protein database searching) increases the sensitivity for HCP detection in the presence of the therapeutic agent, with most recent 2D LC-MS/MS implementations reaching a full 5 orders of magnitude and previously unattainable detection/quantitation limits of 1 ppm, similar to ELISA, making for a robust, orthogonal method to HCP analysis [18].

The initial version of 2D LC-MS/MS for HCP analysis [15] involved a high pH reversed-phase (RP)-HPLC method in the first dimension coupled to a low-pH RP-HPLC method in the second dimension using capillary chromatography (0.3 mm i.d.). Proteolytic HCP peptides are loaded and retained on the RP-HPLC column at pH 10, where, then, the first fraction of peptides is step gradient-eluted with acetonitrile, diluted 10-fold with 0.1% trifluoroacetic acid to both reduce the pH and reduce the amount of organic solvent, such that all peptides can be effectively retained on a trap column. Finally, in the second dimension, the trapped peptides are flushed onto the low-pH RP-HPLC column for high-resolution peptide separation and online, high-resolution electrospray ionization (ESI)–quadrupole time-of-flight (QTOF) MS *via* data-independent acquisition (DIA, or LC/MS^E). In DIA, the instrument collision energy for collision-induced dissociation (CID) is alternated every 0.5 s between a low energy state (5 eV, normal operation) for accurate mass peptide analysis and a high energy state (ramped from 15 to 35 eV) for accurate mass peptide fragmentation, all without precursor ion selection, which prevents any missed high- or low-level peptide species, ultimately resulting in a more comprehensive HCP dataset. The initial peptide fractionation, as well as the chromatographic performance in the second dimension, was shown to be reproducible and stable during extended operation. As would be expected, by using the 5- and 10-step fractionation in the first dimension, more peptides from the coeluting HCPs emerge from the general peptide background increasing dynamic range 10- to 20-fold. Additionally, the high-pH fractionation overcame the situation commonly observed with strong cation-exchange chromatography where the same peptide would be split between multiple fractions. From a single injection, transfer of one high-pH fraction from the first dimension to the second dimension for HCP analysis takes 10–12 h to complete with replicates (four high-pH fractions will take four injections and approximately two days to complete). To demonstrate 2D LC-MS/MS for confident identification of low-abundance HCPs, five protein standards were spiked into the samples in a range of concentrations from 20 to 1000 ppm prior to trypsin digestion to assess dynamic range and quantify the HCPs, and HCPs as low as 50 ppm were routinely detected and identified with this initial approach for biotech labs. Routine quantitation of HCPs relies on the proven label-free relative quantification method.

The 2D LC-MS/MS approach can be used to more intelligently develop and refine manufacturing processes by monitoring individual HCP clearance through each downstream purification step [3], as well as monitoring batch-to-batch uniformity and comparing the purity of different batches when process or site changes are made. In a related application, the composition and quantity of individual residual HCPs as a result of changes to the manufacturing process were monitored by 2D LC-MS/MS, which permitted detailed tracking of one HCP that was not detected by ELISA, thereby minimizing impact and/or facilitating risk assessment of potential safety risks [17]. More recently, the 2D LC-MS/MS method was used to systematically investigate therapeutic mAb process parameters such as harvest time, protein A elution buffers, and subsequent downstream processing steps toward reducing HCPs in the final DS, providing a more rational assessment of potential safety risks by individual HCPs [19]. As mentioned earlier, the initial 2D LC-MS/MS method was recently modernized [18] to improve the HCP detection and quantitation limit to 1 ppm by use of a new RP-HPLC charge-surface-modified C18 stationary phase to prevent column saturation, peak tailing, and distortion commonly observed in HCP analysis under high sample loads; incorporation of ion mobility for separation of coeluting precursor ions prior to fragmentation; and correlation of collision energy with peptide ion mobility drift times to improve fragmentation efficiency. Last, a protein A enrichment step (for upfront depletion of the mAb product) was implemented prior to 2D LC-MS/MS to improve HCP detection sensitivity where in the flow-through fraction, 19 HCPs were detected as opposed to one HCP without Protein A enrichment [20].

10.3.3.3 Targeted MS Analysis

A targeted LC–multiple reaction monitoring (LC-MRM) assay can be used to monitor and accurately quantitate up to hundreds of HCPs with high specificity in a high-throughput, sensitive manner during process development, in conjunction with ELISA [15, 21]. Adopted from pharmacokinetic and biomarker discovery/validation studies, LC-MRM involves the use of a highly sensitive ESI–triple quadrupole mass spectrometer for analyzing targeted tryptic peptides in complex mixtures *via* 1D LC-MS/MS, where specific peptide precursor ions are selected and dissociated into fragment ions, and then multiple fragment ions for each precursor ion are sequentially selected for detection, and ultimately for quantitation. Typically, five replicate injections per sample are performed for more reproducible quantitation. In practice, the 1D and 2D LC-MS/MS methods discussed earlier are used to identify as many HCPs as possible in final bulk DS, as well as in intermediate in-process samples from early in the purification train, and then a targeted LC-MRM method is developed to routinely quantitate specific HCPs. Prior to LC-MRM assay development, target HCP peptide precursor ions and fragment ions are identified *in silico* for analysis using software tools. Then, the final HCP-specific LC-MRM assay can be used to robustly monitor specific HCPs as the process is tweaked/redeveloped using fast chromatography. LC-MRM appears especially useful for monitoring and accurately quantitating immunogenic HCPs at various process steps and in final DS. Linear quantitation (≥3 orders of magnitude) can be on a relative basis with respect to the therapeutic protein, or on an absolute basis in femtomole HCP/microliter with isotopically labeled ^{13}C or ^{15}N reference peptides. The LC-MRM quantitation results were found to compare well to 2D LC-MS/MS label-free (Hi3) results [15]. One downside is MS labs in process and product development organizations usually don't use ESI–triple quadrupole instruments for product/process characterization, which to date has limited the application of LC-MRM for targeted quantitation.

10.3.3.4 Ultrahigh-Resolution 1D LC-MS/MS

Contemporary ultrahigh-resolution (UHR) mass spectrometers offers significant improvements in critical performance parameters such as resolution, mass accuracy, sensitivity, dynamic range, and acquisition speed, resulting in higher-quality mass spectral data for trace level species and an increased rate of molecular interrogations that ultimately leads to a higher number of definitive protein identifications deeper in the proteome. Furthermore, the newest generation of UHR orbitrap mass spectrometers [22] paired with various proteomic workflows can reach unprecedented sensitivity levels and achieve even wider dynamic ranges in a reliable, quantitative manner, essentially finally complementing standard ELISA approaches for sensitive, direct HCP analysis *via* 1D LC-MS/MS with an extended gradient separation.

The application of 1D LC-MS/MS for the direct analysis of residual HCPs in mammalian- and bacterial-expressed biotherapeutic products has gained significant popularity in recent years due to ever-increasing instrument performance that is now standard on commercially available, top-of-the-line research-grade mass spectrometers. One-dimensional LC-MS/MS also boasts straightforward implementation, ease of use, greater robustness, and reasonable analysis times. In its simplest form, this shotgun proteomic approach involves the simultaneous analysis of residual HCPs and the particular protein therapeutic in high abundance, and employs customary reduced and alkylated trypsin proteolytic digestion, one-dimensional chromatographic separation by narrow-bore [23–25] or capillary [26, 27] column RP-HPLC, UHR hybrid linear ion trap (LIT) orbitrap MS/MS, traditional protein database search methods, and label-free relative abundance ranking—all to definitively detect, identify, and quantitate potential HCPs. One-dimensional LC-MS/MS with narrow-bore LC and label-free HCP quantitation has also been implemented on a research grade, UHR QTOF mass spectrometer using data-dependent acquisition that benefits from speed and high-resolution/accurate mass peptide fragmentation data for reduced false-positive hits, in contrast to lower-resolution

LIT peptide fragmentation [28]. Longer HPLC methods (180–200 min) with linear gradients (~160 min) are required with 1D LC-MS/MS approaches in order to create more separation between the abundant mAb proteolytic peptides, thereby increasing the chances for detection of potential HCPs. Additionally, single samples are analyzed in duplicate or triplicate to strengthen the identifications and relative quantitation (i.e., protein identifications are combined), and conducting blank injections between analyses is important to prevent peptide carry-over effects, although these blank injections can have shorter run times. Including sample preparation, run time, and automated data analysis, the final results (after data review) for a single sample analyzed in triplicate is available within 2–3 days.

Dedicated mass spectrometers typically are used for all LC-MS/MS proteomics work meaning that only desalted proteolytic peptide samples in formic acid–containing, high-purity LC mobile phases are administered to a particular instrument (using a divert valve for residual salts) as a way to maintain ultimate instrument sensitivity, which is crucial for residual HCP detection. Formic acid is added to the LC-MS/MS mobile phases to essentially enhance peptide ionization, minimizing any ion suppression effects/ion source fouling and maximizing overall MS sensitivity. Prior to HCP analysis, setup of the mass spectrometer involves cleaning the ion source region to ensure highest sensitivity and reproducible, sustained HCP method performance. With Thermo Scientific hybrid LIT orbitrap instruments, full-scan mass spectral data typically is acquired at 60k resolution or greater (at m/z 200) with a lock mass calibrant to both increase selectivity and enable accurate mass determinations of <3 ppm for potential trace-level HCP precursor ions, which is especially important in the presence of intense mAb signals and high chemical background. Then, CID is usually performed in the LIT device (if present) for the fastest, most sensitive *de novo* HCP sequencing using data-dependent MS/MS of the top 10–20 most abundant precursor ions with charge-state screening (2+/3+ precursor ions only). As an alternative approach to data-dependent acquisition with orbitrap instruments and DIA/LC/MS^E from Waters (described earlier), sequential window acquisition of theoretical spectra (SWATH) is a proprietary AB SCIEX DIA method for QTOF instruments that uses sequential 20–25 Da wide precursor ion windows between 400 and 1200 m/z for 1D (or 2D) LC-MS/MS analysis (i.e., multiplexed MS/MS data are deconvoluted prior to database searching), which has shown benefit for generating accurate mass, high-resolution composite peptide fragmentation data in HCP analyses without missing any precursor ions in a highly sensitive and reproducible manner, essentially enabling databases searches with narrower mass tolerances for increased specificity and fewer false positives [29].

Thermo Scientific Proteome Discoverer software, in combination with a protein database search program such as Mascot, is used to automatically interrogate the acquired mass spectral data for potential HCPs using mainly default parameters: precursor mass tolerance, 10 ppm; fragment mass tolerance, 0.8 Da (assuming LIT; acquisition of high-resolution/accurate mass peptide fragmentation data allows more selectivity and lower mass tolerance values); trypsin with one to two missed cleavages; carboxymethylation as a static modification; oxidation as a dynamic modification; target false discovery rate of 1% (ensures 99% confidence in a particular protein identification); and "search against decoy database" is enabled. Numerous postsearch filters can be applied to narrow the results to only high confidence identifications, as well as to specifically widen the search results to examine the positive hits near the detection limit. A customized protein database usually contains the therapeutic protein sequence of interest and the latest host cell proteome, as well as common contaminants and the proteolytic enzyme. The host cell–related proteomes are updated continually, which ideally results in more reliable protein identifications, but it also can result in new or missing hits upon reanalysis of old data. Additionally, searchable databases restricted to host organisms will limit the false positives as the protein search hits can be filtered specifically by the host species, that is, *Cricetulus griseus* for CHO cell–derived proteins. Single spectrum protein hits and small peptides <6 amino acids are usually discarded to further limit false-positive identifications. HCPs are considered identified if two or more unique peptides are found with high scores and the false discovery rate is ≤1%.

For quantitation, typically the label-free Hi3 method is used where the integrated peak areas of the top three most intense precursor ions for the therapeutic protein and each HCP are integrated and averaged. The relative level of each individual HCP is then compared to the therapeutic protein in terms of relative percent and parts per million levels (ng HCP/mg protein). Spiking studies with protein standards can be employed to determine the limits of quantitation and detection for a wide variety of possible HCPs. One-dimensional LC-MS/MS for HCP identification provides quantitation of individual HCPs between 1 and 50 ng/mg depending on the mass spectrometer and proteomics approach, although ~10 ng/mg is achievable in a reliable manner with the newer, cutting-edge orbitrap mass spectrometers. Protein standards may have different response factors in the mass spectrometer as compared to those for HCPs, confounding quantitation of specific HCPs. Following identification of HCPs, Thermo Scientific PinPoint software was used to identify, track, and quantify known HCPs during process development [26]. With this approach, the remaining tryptic peptides for a particular identified HCP that were not selected for fragmentation were extracted automatically from the 1D LC-MS/MS dataset by the software and the corresponding peak areas were integrated for more robust relative quantitation using the Hi3 method.

Many improvements to the 1D LC-MS/MS approach for detailed HCP analysis have been discussed recently. In one report [30], the active mAb therapeutic substance in a DP vial (40–100 mg) was depleted using protein A or protein L, and the HCP-containing flow-through was subjected to proteolysis and nanoLC-MS/MS. Nearly 700 HCPs were identified and quantified in these DP samples that otherwise had very low total HCP levels <10 ng/mg, thereby achieving six orders of magnitude detection and quantitation in a reliable manner. Moreover, the level of HCPs detected by this method would be similar to the exposure levels in patients. By depleting the abundant therapeutic mAb, the dynamic range of the LC column was improved, allowing more robust detection of the low-level HCPs. This 1D LC-MS/MS method employing affinity depletion overcomes the common situation when using direct 1D and 2D MS-based methods where usually no HCPs or few HCPs are detected, which can leave the analyst questioning if the method was appropriate for the particular sample and situation. Affinity enrichment of HCPs was also accomplished using ELISA polyclonal antibodies to detect 12 low-abundance bacterial proteins in the final bulk DS, improving process and product understanding [25]. As well, their quantitative results did not indicate significant changes to the HCP profile as a consequence of the affinity enrichment step. While HCP enrichment strategies, such as product removal *via* affinity chromatography/beads, carries the risk of losing HCPs bound to the product, if successful, enrichment methods for HCPs can result in a 10-fold increase in HCP detection sensitivity, but great care is needed to preserve the relative abundance information after enrichment. The use of exclusion and inclusion lists for more targeted HCP analysis of final purified DS can also increase the sensitivity of 1D LC-MS/MS [24]. Typically, the exclusion list represents the tryptic peptide masses of the abundant therapeutic protein and the inclusion list represents the tryptic peptide masses of the HCPs detected in the analysis of protein A–purified samples.

10.3.3.5 Top-down Proteomics

Emerging in the early 2000s, top-down proteomics involves accurate mass determinations of individual proteoforms and direct fragmentation of intact protein ions in the gas phase with high-resolution/UHR MS, allowing for identification of the protein and its posttranslational modifications [31]. Sometimes during early downstream process development or product characterization, one or more visible unknown peaks are observed in a particular chromatographic method/process step, requiring immediate analysis and direct identification of the resolved (enriched) components to remedy the situation. When contrasted with heightened characterization of final bulk DS, the complexity of the "proteoform mixture" may shroud possible low-level differences, especially pertaining to unexpected HCPs. In a representative study, two early-eluting unknown process-scale size-exclusion chromatography (SEC) peaks were found to contain 20 HCPs after proteolysis (Lys-C) of each collected fraction, accurate mass analysis by LC/MS peptide mapping, offline sequencing of isolated peptide peak fractions (*via* nano-electrospray ionization (nanoESI) MS/MS), and manual protein database searches [32]. Furthermore, as part of this same study, top-down proteomics, a new approach at the time, was conducted, where direct MS sequencing (without proteolysis) of the concentrated HCP mixture in the fractionated SEC peaks with nanoESI-QTOF MS/MS afforded partial amino acid "sequence tags" for efficient protein identifications, and elegantly provided immediate differentiation of full-length and truncated HCPs *via* accurate molecular mass data.

The enhanced performance attributes (sensitivity, resolution, and mass accuracy) of contemporary ESI-orbitrap [33] and MALDI-TOF [34] mass spectrometers, in combination with more reliable electron-transfer dissociation and in-source decay (ISD) techniques, respectively, continue to propel top-down MS approaches deeper into the proteome with progressively longer contiguous stretches of amino acid sequence data, resulting in the direct identification of intact proteins (or HCPs) at increasingly lower levels and overall accelerating analysis times. Furthermore, top-down proteomics on a chromatographic timescale has been demonstrated for the identification of smaller proteins <35 kDa [35], and is routine in many core MS laboratories using modern orbitrap and QTOF instruments for immediate identification of more abundant HCPs, thereby avoiding manual peak fractionation, vacuum centrifugation, reconstitution, proteolysis, and lengthy LC-MS/MS analyses, especially for low-level species where sample manipulation could impart further chemical modifications. However, significant challenges still remain for top-down proteomics in terms of improving dynamic range (i.e., complex mixtures can exhibit ion suppression effects in general, and large proteins undergo signal splitting from broad isotope and charge-state distributions) and increasing fragmentation efficiency for more contiguous sequence readouts, especially for large-scale studies involving proteins >35 kDa with and without posttranslational modifications [36].

10.4 Use of HCP-ELISA and Orthogonal 1D LC-MS/MS in Practice

In process and product development, the key focus areas for HCP testing include downstream process clearance studies and identifying particular HCPs that can affect final product quality attributes; however, cell line viability studies, cell culture condition comparisons, and ELISA reagent development work represent other applications of LC-MS/MS toward a better understanding of HCP profiles [23]. Knowing the HCP profile for a new developmental or commercial manufacturing process is advantageous if an issue should unexpectedly appear during a clinical trial, after a future process or site change, or in stability studies where protein degradation emerges. In several biotech laboratories, mAbs and recombinant therapeutic proteins are tested routinely for residual HCPs using 1D LC-MS/MS as the orthogonal technique to ELISA. The observed working sensitivity for detection and identification with 1D LC-MS/MS is 1–50 ng HCP per mg of product (1–50 ppm) using an ultrahigh-performance liquid chromatography (UHPLC) system with a 2.1 mm i.d. microbore C18 column coupled to a UHR Thermo Scientific ESI–hybrid LIT orbitrap mass spectrometer. The limit of detection appears to vary, which is expected given that the instrument performance will fluctuate over time and the general molecular properties of any HCP will influence detection, such as protein size, degree of posttranslational modifications, the number of proteolytic peptides with high response factors, and so on. Currently, the corresponding relative limit of quantitation is minimally 10 ng/mg with 1D LC-MS/MS using narrow-bore LC and the Thermo Scientific Orbitrap Elite™ based on commercial universal proteomics/dynamic range standards (Figure 10.5) [23]. It is very important to include a system suitability sample in the HCP sample set to ensure consistent method performance, as well as to develop a panel of three to five mAbs with known HCP profiles for training purposes and method cross-overs/exchange between analysts and secondary labs (i.e., commercial manufacturing, CROs/CMOs) as a means to build operational consistency.

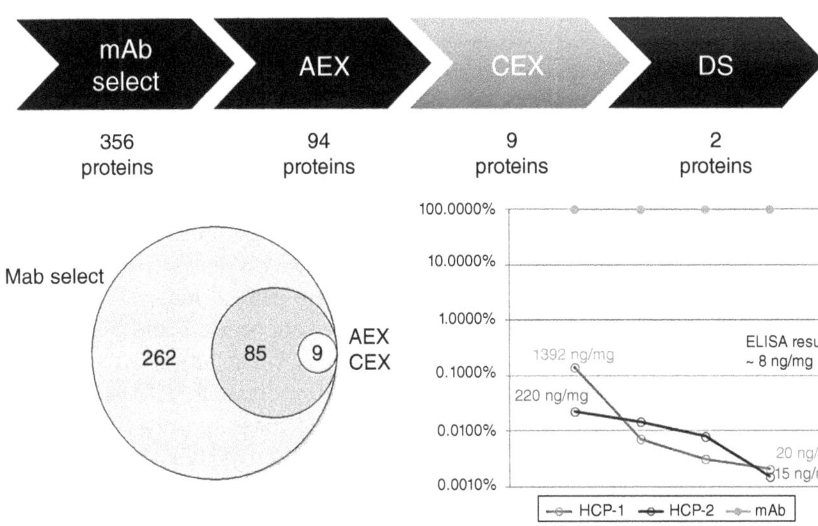

Figure 10.5 Assessment of HCP clearance during the purification of a mAb. HCPs are detected and quantitated for each process pool and the final drug substance.

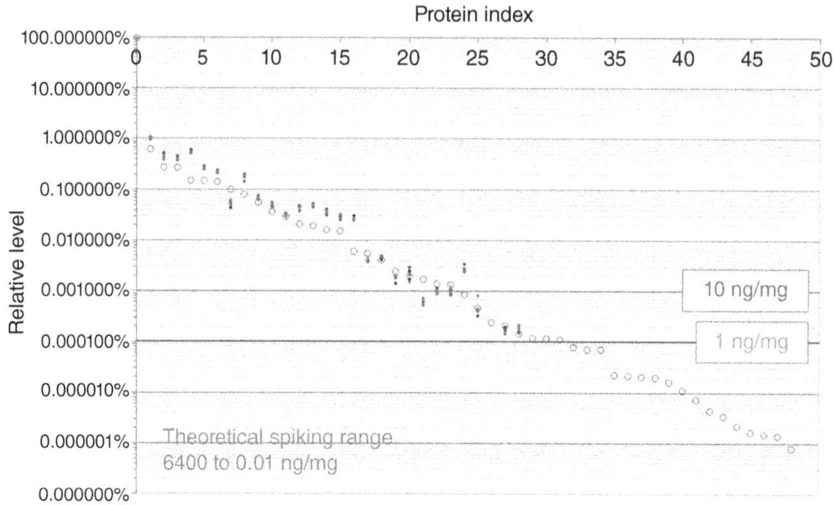

Figure 10.6 The detection of protein standards spiked in to mAb were definitiely achieved at 10 ng/mg and some at around 1 ng/mg.

The common industry specification limit for pooled and individual HCPs in therapeutic mAbs, as measured by ELISA or LC-MS/MS, respectively, is <100 ng/mg, meaning that any HCP signal greater than 100 ng/mg is considered out of specification. For mAbs produced by modern cell culture and purification equipment, practices, and processes (involving protein A affinity chromatography), detectable HCPs >10 ng/mg are not usually observed by 1D LC-MS/MS using orbitrap technology, consistent with HCP-ELISA results. HCPs have been found in final purified DS for some older, discontinued mAbs with antiquated upstream and downstream processing (Figure 10.6) [23]. LC-MS/MS analysis of the respective in-process purification pools can afford troubleshooting support to remove specific low-level HCPs in DS. For new developmental therapeutic programs, one HCP testing strategy that complements (and validates) ELISA results involves performing LC-MS/MS on final DS from a representative pilot-scale batch prior to manufacturing clinical material to ensure that the upstream and downstream processes are appropriate and performing properly. The residual HCP target amount for final DS ideally is <10 ng/mg by both ELISA and LC-MS/MS, while any pooled or individual HCP >100 ng/mg would be considered a regulatory risk and would likely require tweaking or redevelopment of the manufacturing process. Detection of HCP signal between 10 and 100 ng/mg falls within specification; however, with LC-MS/MS, specific HCPs will be identified and quantified, creating another decision point as to whether the HCP is known to be associated with any adverse events. Companies will need to decide how to report any intermediate residual HCPs between 10 and 100 ng/mg to regulatory agencies, and whether or not to remove them from the process. Most importantly, performing of LC-MS/MS in conjunction with HCP-ELISA early in process and product development ensures that no HCPs >100 ng/mg go undetected, which could affect clinical performance and safety.

The commercial manufacturing process is usually developed ahead of Phase III clinical trials, and, again, prior to the commercial process lock, it is prudent to test the near-final DS (nonclinical) produced by the representative upstream and downstream process conditions with LC-MS/MS for residual HCPs, as an orthogonal "validation" method to ELISA. This approach allows time for additional process development if an HCP is discovered before manufacture of clinical materials. Furthermore, prior to commercial lock, it is very powerful to use orthogonal LC-MS/MS alongside ELISA to decisively demonstrate clearance of residual HCPs through each step of the downstream manufacturing process by analysis of the in-process load and pool samples, where HCPs are identified, quantified, and individually tracked through the purification train. Sequential removal of each HCP at each process step adds further proof that the final purified bulk DS is HCP-free below 10 ng/mg target. Likewise, any detected, known offender HCPs can be monitored individually to further improve the safety profile of a particular product. In conclusion, application of MS-based orthogonal HCP testing alongside ELISA affords increased process and product understanding, and is projected to help toward the design of next generation upstream/downstream processes.

10.4.1 Pros and Cons of MS for Orthogonal HCP Analysis

As compared with ELISA, 1D and 2D LC-MS/MS are lower-throughput approaches, but they are especially definitive and quantitative with respect to HCP analysis as described earlier. While LC-MS/MS has been in existence for nearly 20 years and instrument manufacturers have made it easier than ever to carry out this hyphenated technique, current mass spectrometers still remain substantial capital investments relative to most other protein characterization technologies in terms of price, footprint, utilities, long-term upkeep, and service calls. Substantial proteomics MS expertise is still required to set up instruments and develop methods for HCP analysis (including effective system suitability checks), as well as maintain and overcome the finicky nature of mass spectrometer operation day to day, thereby ensuring that the final methods produce consistent and quantitative results week after week, especially in multiuser, core laboratories where instruments are shared and subjected to a multitude of samples and methods, causing performance variability and uncertainty. Last, absolute, reproducible HCP quantitation remains a challenge for MS-based approaches, as well as consistent detection/identification of HCPs between 1 and 50 ppm due to instrument and method fluctuations/limitations, although lots of progress has been made to date on both fronts.

10.4.2 Considerations and MS Evolution

Commercial HCP-ELISA kits have been shown to miss two-thirds of HCPs observed by MS, while an in-house HCP-ELISA, tailored to the upstream and downstream process, is better positioned to detect all residual HCPs in DS. However, a sensitive and selective orthogonal HCP analysis technique such as MS is viewed by both practitioners and regulators as a real necessity to cross-check the ELISA results during process development to make sure no HCPs >100 ng/mg are missed and ensure clinical (and commercial) product efficacy and safety. The latest mass spectrometers and proteomic methods approach the sensitivity of ELISA for residual HCP detection and MS offers the added analytical specificity to identify and quantify any residual HCPs greater than 1–50 ng/mg (at present depending on the instrumental setup and proteomic method). If an HCP is identified at high level or is viewed as a safety risk, MS-based targeted methods can assist real time in the process redevelopment effort by tracking specific HCPs between purification steps to ensure clearance. Supplementing HCP-ELISA with an orthogonal MS technique is clearly advantageous, and, at present, successful implementation tends to work well when trained mass spectrometrists and specialized laboratories exist to support the analyses. In contrast to other approaches, 1D LC-MS/MS has opened a practical, straightforward, and time-efficient proteomic approach to HCP analysis in the biopharmaceutical lab setting allowing both MS experts and nonexperts (after training) to readily assess HCP-ELISA results.

MS continues to evolve in terms of new capabilities, more intuitive software, and ease of use, and is expected to play an increasing role in HCP analysis, especially when it can be routinely performed in nonspecialized laboratories without its notorious finicky, overcomplicated nature. There is a current movement among vendors to develop simple, automated MS-based analyzer systems and "disposable" kits that will target regulated laboratories conducting routine testing where MS offers clear advantages over existing techniques and methodologies [37]. One can surely envision that these dedicated MS systems for the various characterization assays performed in biopharmaceutical laboratories such as HCP analysis will also become available in the near future. Strategies for multiattribute methodologies [38], MS-based release assays in QC laboratories [39], MS in manufacturing for product attribute control [40], and at-line real-time product release [41] concepts are also evolving currently to increase efficiency and reduce costs, and one can foresee HCP testing *via* MS becoming more integrated with batch release in the future [26].

10.5 Risk of HCPs Present in Products

Although substantial efforts are made to eliminate HCPs from biotherapeutics, most if not all biotechnology products contain at least trace levels of HCPs. These HCP impurities essentially become unintended components of the final

DP, and potentially pose risk to patients [11]. These include several patient safety risks, such as the development of antibodies to the HCPs. Potential safety issues in this situation would be cross-reactivity of the anti-HCP antibodies with endogenous proteins, an increase in infusion-related reactions, or anaphalaxis. Furthermore, depending on the HCP, the impurity could have a toxic effect on patients (e.g., endotoxin). HCPs also pose an efficacy risk for patients taking a biotherapeutic. The HCP may have an adjuvant effect and induce or increase the presence of antitherapeutic antibodies. Likewise, if the HCP impurity has activity in humans, such as a cytokine or enzyme, there may be an impact to patient safety or drug efficacy [42]. Select examples of these cases have been discussed in the literature and at scientific conferences [6] (i.e., Omnitrope scientific discussion), but biotechnology companies often do not disclose this information publically.

Beyond patient risks, there are also product quality risks associated with the presence of HCPs in DPs. Some HCPs, such as proteases or other enzymes that posttranslationally modify proteins, can substantially impact product quality. Recently there have been reports in the literature of visible particulate formation in mAb formulations, and an HCP impurity that enzymatically degrades polysorbates is thought to be the cause [43].

Typically when we think of HCPs as impurities, we consider a mixture of proteins that is essentially unknown. However, the HCPs that copurify with a product are often not a random HCP population, but rather tend to be specific HCPs that "hitchhike" with the product for a variety of reasons [4–6]. With dramatic improvements in various technologies over the past several years, it is now possible to identify HCP impurities that copurify with a product [15, 24, 44, 45]. Identification of the HCPs also changes how they are considered as impurities. Rather than being a low-level random mixture of various proteins, now we can consider specific proteins that will be administered with a therapeutic, and what risk that imparts to a patient and/or the product quality [11, 12, 42, 46].

10.6 Conclusions

HCP impurity clearance and monitoring is one component of an overall system that ensures consistent process performance and consistent product delivered to the patient. When processes are changed or manufacturing sites changed often there is more extensive characterization of in-process pools and final products than is normally performed during lot release testing, and this additional characterization should include HCP impurity levels. Quantitative sandwich immunoassays such as ELISAs with broadly reactive anti-HCP reagents are one component of the control system. Other components may include 1D and/or 2D gels stained with a sensitive stain such as silver or Sypro dyes. Similarly, anti-HCP immunoblots of the final product provide another means of demonstrating product purity. Other analytical tools that may also be part of the control system for other reasons, such as peptide maps and CE-SDS, may also be inspected for novel peaks that reflect unexpected HCP impurities. LC-MS/MS may also be used to look for HCP peptide fragments. Taken together these assays are a suite of orthogonal methods that provide assurance that HCP levels are low and consistent from batch to batch.

References

1 Xu, X., Nagarajan, H., Lewis, N. E., Pan, S., Cai, Z., Liu, X., *et al.* (2011) The genomic sequence of the Chinese hamster ovary (CHO)-K1 cell line. *Nat Biotechnol* 29(8), 735–742.
2 Blattner, F. R., Plunkett, G., Bloch, C. A., Perna, N. T., Burland, V., Riley, M., *et al.* (1997). The complete genome sequence of Escherichia coli K-12. *Science* 277, 1453–1462.
3 Zhang, Q., Goetze, A., Cui, H., Wylie, J., Trimble, S., Hewig, A., *et al.* (2014). Comprehensive tracking of host cell proteins during monoclonal antibody purifications using mass spectrometry. *mAbs*, 6(3), 659–670.
4 Sisodiya, V. N., Lequieu, J., Rodriguez, M., McDonald, P., & Lazzareschi, K. P. (2012). Studying host cell protein interactions with monoclonal antibodies using high throughput protein A chromatography. *Biotechnol J*, 7(10), 1233–1241.
5 Levy, M. J., Gucinski, A. C., & Boyne, M. T. (2015). Primary sequence confirmation of a protein therapeutic using top down MS/MS and MS3. *Anal Chem* 87(14), 6995–6999.
6 Vanderlaan, M., Sandoval, W., Liu, P., Nishihara, J., Tsui, G., Lin, M., *et al.* (2015). Hamster phospholipase B-like 2 (PLBL2): a host cell protein impurity in CHO-derived therapeutic monoclonal antibodies. *BioProcess Int* 13(4), 2–10.
7 Champion, K. M., Nishihara, J. C., Joly, J. C., & Arnott, D. P. (2001). Similarity of the *Escherichia coli* proteome upon completion of different biopharmaceutical fermentation processes. *Proteomics* 1, 1133–1148.
8 Zhu-Shimoni, J., Yu, C., Nishihara, J., Wong, R. M., Gunawan, F., Lin, M., *et al.* (2014). Host cell protein testing by ELISAs and the use of orthogonal methods. *Biotechnol Bioeng*, 111, 2367–2379.
9 Eaton, L. C. (1995). Host cell contaminant protein assay development for recombinant biopharmaceuticals. *J Chromatogr A* 705(1), 105–114.
10 Chamption, K., Madden, H., Dougherty, J., & Shacter, E. (2005). Defining your product profile and maintaining control over it, Part 2. *Bioprocess Int* 3(8), 52–57.
11 de Zafra, C. L., Quarmby, V., Francissen, K., Vanderlaan, M., & Zhu-Shimoni, J. (2015). Host cell proteins in biotechnology-derived products: a risk framework. *Biotechnol Bioeng* 112(11), 2284–2291.

12 Bracewell, D. G., Francis, R., & Smales, C. M. (2015). The future of host cell protein (HCP) identification during process development and manufacturing linked to a risk-based management for their control. *Biotechnol Bioeng* 112, 1727–1737.

13 Zhang, Q., Goetze A. M., Cui, H., Wylie, J., Tillotson, B., Hewig, A., *et al.* (2016). Characterization of the co-elution of host cell proteins with monoclonal antibodies during protein A purification. *Biotechnol Prog* 32(3), 708–717.

14 Michalski, A., Cox, J., & Mann, M. (2011). More than 100,000 detectable peptide species elute in single Shotgun proteomics runs but the majority is inaccessible to data-dependent LC-MS/MS. *J Proteome Res* 10, 1785–1793.

15 Doneanu, C., Xenopoulos, A., Fadgen, K., Murphy, J., Skilton, S., Prentice, H., *et al.* (2012). Analysis of host-cell proteins in biotherapeutic proteins by comprehensive online two-dimensional liquid chromatography/mass spectrometry. *mAbs* 4(1), 24–44.

16 Schirmer, E., Yates, J. I., & Gerace, L. (2003). MudPIT: a powerful proteomics tool for discovery. *Discov Med* 3(18), 38–39.

17 Schenauer, M., Flynn, G., & Goetze, A. (2012). Identification and quantification of host cell protein impurities in biotherapeutics using mass spectrometry. *Anal Biochem* 428(2), 150–157.

18 Doneanu, C., Anderson, M., Williams, B., Lauber, M., Chakraborty, A., & Chen, W. (2015). Enhanced detection of low-abundance host cell protein impurities in high-purity monoclonal antibodies down to 1 ppm using ion mobility mass spectrometry coupled with multidimensional liquid chromatography. *Anal Chem* 87(20), 10283–10291.

19 Farrell, A., Mittermayr, S., Morrissey, B., McLouchlin, N., Iglesias, N. N., Marison, I. W., *et al.* (2015). Quantitative host cell protein analysis using two dimensional data independent LC-MS^E. *Anal Chem* 87(18), 9186–9193.

20 Thompson, J., Chung, W., Zhu, M., Tie, L., Lu, Y., Aboulaich, N., *et al.* (2014). Improved detection of host cell proteins (HCPs) in a mammalian cell-derived antibody drug using liquid chromatography/mass spectrometry in conjunction with an HCP-enrichment strategy. *Rapid Commun Mass Spectrom* 28(8), 855–860.

21 Cott, K. V. (2015). Retrospective Analysis of a Host Cell Protein Perfect Storm: Identifying Immunogenic Proteins and Fixing the Problem. CMC Strategy Forum (p. 10). Washington, DC: CASSS.

22 Zubarev, R. A., & Makarov, A. (2013). Orbitrap mass spectrometry. *Anal Chem* 85(11), 5288–5296.

23 Sperry, J. (2014). Characterization of Host Cell Proteins in Biotherapeutics by Mass Spectrometry. BEBPA First Annual Workshop on Host Cell Protein Assays (p. 4). Dubrovnik, Croatia: Biopharmaceutical Emerging Best Practices Association.

24 Reisinger, V., Toll, H., Mayer, R., Visser, J., & Wolschin, F. (2014). A mass spectrometry-based approach to host cell protein identification and its application in a comparability exercise. *Anal Biochem* 463, 1–6.
25 Bomans, K., Lang, A., Roedl, V., Adolf, L., Kyriosoglou, K., Diepold, K., *et al.* (2013). Identification and monitoring of host cell proteins by mass spectrometry combined with high performance immunochemistry testing. *PLoS One* 8(11), e81639.
26 Rogers, R., Bell, A., Kowski, T., & Bailey, R. (2013). One-dimensional liquid chromatography analysis of host cell proteins in therapeutic antibodies by Orbitrap Velos. 10th Symposium on the Practical Applications of Mass Spectrometry in the Biotechnology Industry (p. 39). Boston, MA: CASSS.
27 Aboulaich, N., Chung, W., Thompson, J., Larkin, C., Robbins, D., & Zhu, M. (2014). A novel approach to monitor clearance of host cell proteins associated with monoclonal antibodies. *Biotechnol Prog* 30(5), 1114–1124.
28 Yuk, I., Nishihara, J., Walker, D. J., Huang, E., Gunawan, F., Subramanian, J., *et al.* (2015). More similar than different: host cell protein production using three null CHO cell lines. *Biotechnol Bioeng* 112(10), 2068–2083.
29 Selevsek, N., Chang, C., Gillet, L., Navarro, P., Bernhardt, O., Reiter, L., *et al.* (2015). Reproducible and consistent quantification of the *Saccharomyces cerevisiae* proteome by SWATH-mass spectrometry. *Mol Cell Proteomics* 14(3), 739–749.
30 Madsen, J. A., Farutin, V., Carbeau, T., Wudyka, S., Yin, Y., Smith, S., *et al.* (2015) Towards the complete characterization of host cell proteins in biotherapeutics via affinity depletions, LC-MS/MS and multivariant analysis. *MAbs* 7(6), 1128–1137.
31 Kelleher, N. (2004). Top-down proteomics. *Anal Chem* 76(11), 196 A–203 A.
32 Nemeth-Cawley, J. F., Tangarone, B. S., & Rouse, J. C. (2003). "Top Down" characterization is a complementary technique to peptide sequencing for identifying protein species in complex mixtures. *J Proteome Res* 2, 495–505.
33 Levy, N. E., Valente, K. N., Choe, L. H., Lee, K. H., & Lenhoff, A. M. (2014), Identification and characterization of host cell protein product-associated impurities in monoclonal antibody bioprocessing. *Biotechnol Bioeng* 111, 904–912.
34 Resemann, A., Jabs, W., Wiechmann, A., Wagner, E., Colas, O., Evers, W., *et al.* (2016). Full validation of therapeutic antibody sequences by middle-up mass measurements and middle-down protein sequencing. *mAbs* 8(2), 318–330.
35 Parks, B. A., Jiang, L., Thomas, P. M., Wenger, C. D., Roth, M. J., II, Boyne, M. T. 2nd, *et al.* (2007). Top-down proteomics on a chromatographic time scale using linear ion trap Fourier transform hybrid mass spectrometers. *Anal Chem* 79, 7984–7991.
36 Catherman, A. D., Skinner, O. S., & Kelleher, N. L. (2014). Top down proteomics: facts and perspectives. *Biochem Biophys Res Commun* 445, 683–693.

37 Vogeser, M. (2015, July). Mass spectrometry in the clinical laboratory—challenges for quality assurance. *Curr Trends Mass Spectrom* (Supplement to LCGC North America), 13(3), 14–19.

38 Rogers, R. S., Nightlinger, N. S., Livingston, B., Campbell, P., Bailey, R., & Balland, A. (2015). Development of a quantitative mass spectrometry multi-attribute method for characterization, quality control testing and disposition of biologics. *mAbs*, 7(5), 881–890.

39 Balland, A. (2013). MS in QC: A Single Multi-attribute Method for Quality Control and Release Testing of Biologics. 10th Symposium on the Practical Applications of Mass Spectrometry in the Biotechnology Industry (CASSS Mass Spec 2013) (p. 22). Boston, MA.

40 Zupke, C., Brady, L. J., Slade, P. G., Clark, P., Caspary, R. G., Livingston, B., *et al.* (2015). Real-time product attribute control to manufacture antibodies with defined n-linked glycan levels. *Biotechnol Prog* 31(5), 1433–1441.

41 Dasu, M., & Naresh, J. R. (2013). Real time release testing—a new quality paradigm for pharmaceutical development. *Int J Pharm Sci Rev Res* 19(2), 80–84.

42 Wang, X, Hunter, AK, & Mozier, N. M. (2009). Host cell proteins in biologics development: identification, quantitation, and risk assessment. *Biotechnol Bioeng* 103, 446–458.

43 Labrenz, S. (2014). Ester hydrolysis of polysorbate 80 in mAB drug product: evidence in support of the hypothesized risk after the observation of visible particulate in mAb formulations. *J Pharm Sci* 103(8), 2268–2277.

44 Silva, J. C., Denny, R., Dorschel, C., Gorenstein, M. V., Li, G. Z., Richardson, K., *et al.* (2006). Simultaneous qualitative and quantitative analysis of the *Escherichia coli* proteome: a sweet tale. *Mol Cell Proteomics* 5, 589–607.

45 Joucla, G., Le Sénéchal, C., Bégorre, M., Garbay, B., Santarelli, X., & Cabanne, C. (2013). Cation exchange versus multimodal cation exchange resins for antibody capture from CHO supernatants: identification of contaminating host cell proteins by mass spectrometry. *J Chromatogr B Analyt Technol Biomed Life Sci* 30, 942–943.

46 Bailey-Kellogg, C., Gutiérrez, A. H., Moise, L., Terry, F., Martin, W. D., & De Groot, A. S. (2014). CHOPPI: a web tool for the analysis of immunogenicity risk from host cell proteins in CHO-based protein production. *Biotechnol Bioeng* 111, 2170–2182.

11

Analytical Tools for Biologics Molecular Assessment

Wilson Phung[1], Wendy Sandoval[1], Robert F. Kelley[2], and Jennie R. Lill[1]

[1] *Department of Microchemistry, Proteomics & Lipidomics, Genentech Inc., South San Francisco, CA, USA*
[2] *Drug Delivery Department, Genentech Inc., South San Francisco, CA, USA*

Abbreviations

AAPH	2,2′-azobis(2-amidinopropane) dihydrochloride
AUC	area under the curve
CDR	complementarity determining regions
CHO	Chinese hamster ovary
DSC	differential scanning calorimetry
DTT	dithiothreitol
Eprex	erythropoietin
His	histidine
IEC	ion-exchange chromatography
ITC	isothermal titration calorimetry
Leu	leucines
mAbs	monoclonal antibodies
Met	methionine
MS	mass spectroscopy
Phe	phenylalanine
PTMs	post-translational modifications
SEC	size-exclusion chromatography
SP	sulfopropyl
TIC	total ion chromatogram
TPP	target product profile
Trp	tryptophan

Analytical Characterization of Biotherapeutics, First Edition. Edited by Jennie R. Lill and Wendy Sandoval.

11.1 Introduction to Molecular Assessment

The overall aim of molecular assessment is to identify from a panel of equipotent candidate molecules those that provide best ease of manufacture and formulation, shelf-life, and potential for maximum benefit to patients. Protein therapeutics consisting of natural amino acids can display chemical instability due to the nature of the side chains. Asn and Asp residues can undergo reaction, deamidation, and isomerization, respectively, that result in charge variation in the protein. Environmental conditions can result in oxidation of Trp, Met, Cys, and to a lesser extent Tyr and His. The susceptibility of these amino acids to degradation is influenced by the primary sequence and structure of the protein such that candidate molecules can differ in stability. These chemical changes are most bothersome when occurring in portions of the protein that contact the target in the complex and can thus have an impact on binding and biological activity. Proteins can also show physical instability, most notably aggregation that can alter potency. Chemical instability can exacerbate physical instability because protein degradants can have diminished physiochemical properties relative to the pristine materials. In addition to the effects on potency, instabilities can result in enhanced immunogenicity or unacceptable pharmacokinetic properties.

For a particular therapeutic target there may be few or no candidates that both meet potency requirements and have no chemical or physical liabilities. Thus, a secondary aim of molecular assessment is to generate early data on a clinical candidate that can inform technical teams on recommended measures to mitigate development risk. This may include selection of a formulation screen to maximize shelf-life, design of appropriate control systems to monitor degradants, and host selection to improve production.

In silico methods for performing molecular assessment are being developed (e.g., Sharma *et al.* [1]) but are yet to be fully implemented in the biopharma industry. Instead, biochemical methods are used to assess candidates, with typical molecular assessment activities outlined in Table 11.1. In order to generate measurable levels of degradants in a reasonable time period, accelerated stability tests are usually performed involving incubation of the protein containing formulation for an extended period at elevated temperature. In addition to performing "thermal stress" using the desired formulation buffer a physiological buffer may be employed to rank *in vivo* stability of candidates. Additional tests may be performed depending on the specific target candidate profile. For example, protein formulations intended for delivery through subcutaneous or intravitreal injection have volume limitations such that a high protein concentration may be required. High solubility and low viscosity are desirable in these formulations so that these properties are often investigated during molecular assessment.

Therapeutic antibodies and antibody derivatives comprise the majority of today's biotherapeutic reagents for the treatment of malignancies and other

Table 11.1 Typical molecular assessment activities.

Assessment	Test condition	Assays
Shelf-life stability	Thermal stress (>25°C) for ≥2 weeks in formulation buffer	Size exclusion chromatography (SEC) Ion exchange chromatography (IEC) Intact or reduced mass spectroscopy (MS)
In-use stability	Thermal stress (>25°C) for ≥2 weeks in physiological buffer	LC-MS/MS peptide mapping Target binding (e.g., ELISA or Biacore®)
Cell culture	Expression titer in host cell under standard conditions	ELISA or other
Target profile-specific tests	Solubility in physiological buffer	Turbidity by absorbance measurement Visual inspection for precipitate
	Viscosity of high concentration formulation	Cone-plate viscometry
	Thermal stress on high-concentration formulation (formulation and/or physiological buffer)	Size exclusion chromatography (SEC) Ion exchange chromatography (IEC) Intact or reduced mass spectroscopy (MS) LC-MS/MS peptide mapping Target binding (e.g., ELISA or Biacore)

maladies [2]. Monoclonal antibodies (mAbs) are mono-specific antibodies that bind to a specific antigen; typically 150 kDa in size antibodies consist of four polypeptide chains, two heavy and two light chains, which are joined together by disulfide bonds (Figure 11.1). For the purpose of discussing typical molecular assessment activities in this chapter we will focus our narrative mainly on mAbs although all the activities described herein can be translated to other protein-based therapeutics. An example is the Fc fusion protein class of biotherapeutic; notable examples are entaracept and aflibercept, where a receptor ectodomain is produced as a fusion protein with the Fc portion of a human IgG1 antibody. Soluble receptors can be antagonists of ligands that require binding to a membrane-bound receptor to produce a biological effect. Fusion to the Fc can enhance production, facilitate purification using well-known Fc binding resins, and improve half-life of the molecule by access to the FcRn recycling pathway [4] that promotes long half-life of IgG proteins. Candidate Fc fusion proteins can and should be subjected to similar molecular assessment procedures to determine if degradants affect activity. A chemical liability that Fc fusion proteins share with IgG antibodies is susceptibility of Fc methionine (Met) residues to oxidation that impairs FcRn binding. Extensive

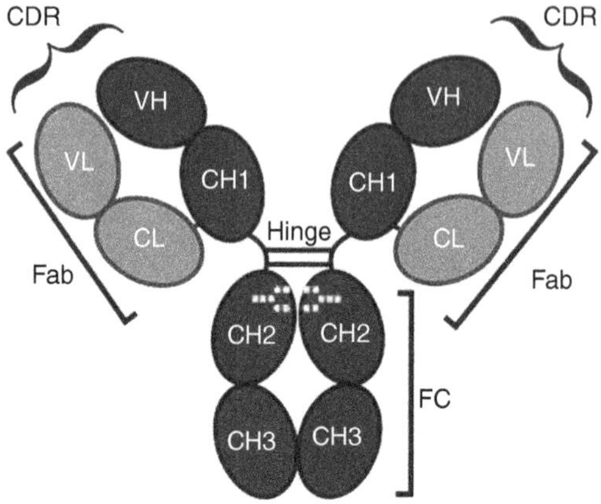

Figure 11.1 Three main regions can be distinguished in the IgG antibody structure: the Fc region, involved in the dimerization between the two heavy chains (black), and two Fab regions, which result from the interaction between one light chain (gray) and one heavy chain coupled by disulfide bridges. The dots denote glycan moieties. The two Fabs are connected with the Fc *via* the hinge region, in which interchain disulfide bridges between the two heavy chains occur. Glycosylation on a conserved site in the C_H2 domain is the dominant PTM on IgGs and is required for full effector function. The variable domains (VL, variable light; VH, variable heavy) determine antibody specificity and contain the antigen binding site (CDRs, complementarity determining regions). Source: Rosati *et al.* [3]. Reproduced with permission of Nature Publishing.

oxidation can result in unacceptable pharmacokinetic properties and should thus be avoided in clinical candidates [5].

The Fc fragment of a mAb is where the constant region binds to cell surface immunoglobulin (Ig) receptors. The Fab fragment of a mAb, specifically the complementarity determining regions (CDRs), is where the variable region binds to a specific antigen. There are various antibody isotypes in nature but the human IgG1 subclass is often chosen for therapeutic mAbs because IgG1 antibodies can elicit immune effector functions such as antibody-dependent cellular cytotoxicity, antibody-dependent cellular phagocytosis, and complement-dependent cytotoxicity. Most mAbs for therapeutic development are typically derived from a murine source [6]. To reduce potential immunogenicity of a murine antibody when administered to humans, chimeric mAbs are engineered comprising the variable regions from murine antibodies and the constant region from a human antibody source integrated. Immunogenicity is further decreased when humanized mAbs are engineered to have only the CDRs, and a few framework residues critical for CDR conformation, from murine moieties with the remainder of the framework being a human-derived sequence [7]. In addition, various other antibody isotypes and subclasses can be employed, and amino acids can be exchanged between different isotypes (i.e., “cross-isotypes”), which can be exploited to generate various novel therapeutic platforms [8].

The stability of therapeutic mAbs is important for any pharmaceutical drug product and it is imperative to understand the mechanisms of how degradation pathways occur in order to better engineer antibodies and increase their stability. Through molecular assessment, therapeutic mAbs are analyzed for chemical liability hotspots, specifically within the CDRs of an antibody, which are crucial regions for antigen binding activity. In addition aggregation, proteolysis, and other liabilities are also monitored to ensure that the antibodies or other biotherapeutics being generated are of optimum stability and potency for storage and delivery to patients. This chapter will focus on the utilization of analytical tools from a biopharma perspective to screen and assess therapeutic mAbs to minimize such liabilities before a lead molecule moves into manufacturing on a large scale and resultant clinical trials [9].

11.2 Molecular Assessment

During the molecular assessment process several parameters are investigated to determine the stability of an antibody or biotherapeutic reagent and many of these parameters are implemented in various iterations, depending on the type of molecule being assessed and indeed the importance of these tests as determined by the company carrying them out. Many physical and chemical modifications can occur during the development or manufacturing of an antibody. For example, chemical modifications such as deamidation, isomerization, and oxidation can occur during different production stages, including synthesis, purification, formulation, manufacturing, and storage. Therefore, screening for molecules that are more resistant to these reactions than others is critical from a pharmaceutical standpoint and lead candidates may be selected or further engineered to produce sequences that minimize these chemical instabilities from occurring. In this process, therapeutic mAbs undergo various assays, stability tests, and screenings to determine if they meet an optimal physicochemical profile.

In the preliminary phase of assessing a molecule's candidacy for manufacturing on a large scale a cell culture expression assay is performed. The antibodies may be expressed from a variety of sources including HEK293 cells, Chinese hamster ovary (CHO) cells, or *Escherichia coli*. Each antibody can be generated from multiple variants and clones in different isotypes. The fermentation titer (g/l) of the protein can be determined using enzyme-linked immunosorbent assay (ELISA)-based methods and judged if economically acceptable. This may include an assessment of large-scale purification development costs. Variants with acceptable expression levels are then subjected to further testing.

After protein expression, antibodies are purified with protein A resin, typically followed by sulfopropyl (SP) cation exchange chromatography. Stability testing is performed which includes thermal and oxidation stress tests and other types of testing for protein modification, both chemical and posttranslational modification. For thermal stress, antibodies are stressed under two conditions and this serves to create accelerated degradation conditions to mimic long-term formulation stability.

The first is under neutral pH conditions that approximate physiological conditions, using phosphate buffered saline (PBS), pH 7.4 at 40°C, for 2 and 4 weeks, to mainly probe for asparagine deamidation, stability, and solubility. The second is under formulation conditions, using a pH 5.5 formulation at 40°C for 2 and 4 weeks, to mainly probe for aspartic acid isomerization, stability, and solubility. Assessment of sensitivity to oxidation stress also involves two conditions. The first is AAPH (2,2′-azobis(2-amidinopropane) dihydrochloride) oxidation stress, where antibodies are incubated in pH 5.5 buffer with and without the addition of AAPH, at 40°C overnight to probe for Met and tryptophan (Trp) oxidation. Secondarily, light oxidation stress is applied, where antibodies are exposed to ultraviolet (UV) light to probe for Trp oxidation events. Each of these tests is discussed in more detail here.

In parallel or sometimes prior to the tests mentioned earlier, target product profile (TPP) testing is performed which includes high-concentration thermal stability, viscosity, solubility, and intravenous, subcutaneous, and ocular delivery analyses. TPP enables selection of protein therapeutic candidates with suitable physicochemical properties and manufacturability profiles. These molecule profile documents, which describe the critical attributes and expected performance of a specific potential product, are used during development to align teams, functional areas, and management on desired product features as a molecule moves from late stage research through to final product approval. An overview of a typical strategy for molecular assessment is depicted in Figure 11.2.

11.3 Biotherapeutic Stability

Biotherapeutic degradation can be separated into chemical and physical degradation. Major chemical degradation routes include deamidation, isomerization, oxidation, hydrolysis, and proteolysis as mentioned earlier. These can precipitate changes in hydrogen bonding and structure, leading to physical degradation. Physical degradation includes changes in the secondary, tertiary, and quaternary structures, which can result in protein misfolding, denaturation, fragmentation, and aggregation.

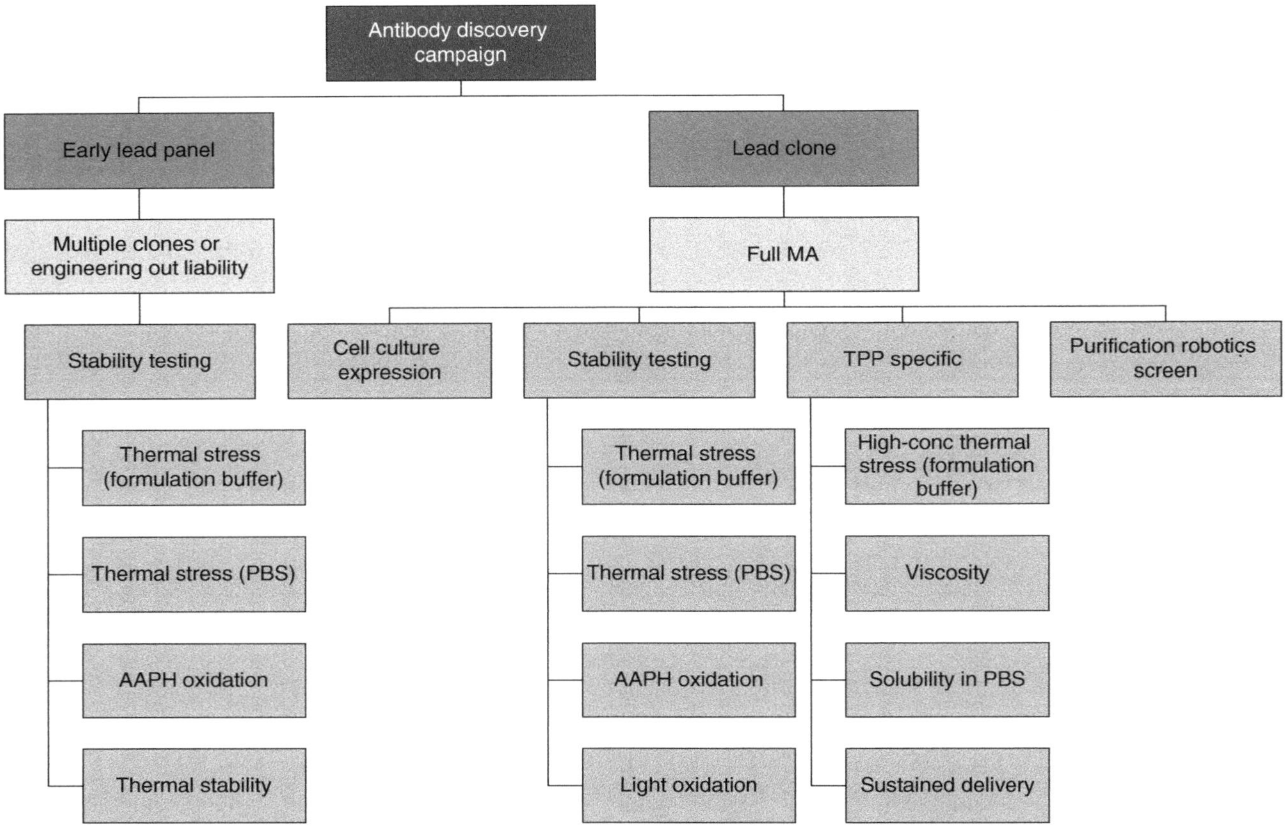

Figure 11.2 A typical molecular assessment workflow encompassing many of the techniques outlined in this chapter are displayed. These techniques can be interchanged with other assays, depending on the biomolecule being tested, the properties the company performing the tests is most interested in, and based upon previous testing used on other bioversions of the molecule of interest. The authors thank Christoph Spiess, Karthik Rajagopal, and Paul McDonald for their help in preparing this figure.

11.3.1 Deamidation and Isomerization of Asparagine

Deamidation is the process whereby the nitrogen on the asparagine amino acid side chain undergoes a nucleophilic attack to become a succinimide intermediate, which then spontaneously hydrolyzes to a mixture of aspartic acid and isoaspartic acid [3]. Deamidation and isomerization result in a mass change of +0.984 Da, the mass difference between –OH and $-NH_2$. Since aspartic acid and isoaspartic acid have the same mass, they can be distinguished by liquid chromatography–mass spectrometry (LC-MS) analysis separation due to their structural differences, but not by mass alone as the masses are isobaric. Factors influencing deamidation and isomerization include high pH and temperatures, which can accelerate their reaction. The residue following asparagine in the primary sequence can influence the rate of deamidation with bulky residues slowing the rate and faster rates observed with glycine or serine in the second position [10].

Aspartic acid residues in proteins can undergo direct isomerization, and like deamidation involves a succinimide intermediate. Unlike deamidation, isomerization is favored by lower pH conditions. At pH values of 4–7 the succinimide intermediate may be stable and can be detected as a basic peak on ion exchange chromatography (IEC). Hydrolysis of the succinimide intermediate, favored at higher pH, will produce a mixture of aspartic acid and isoaspartic acid. As for deamidation, the residue in the second position can affect the rate of isomerization with bulky residues slowing the reaction whereas glycine or serine in the second position results in a faster rate.

11.3.2 Oxidation

Oxidation of key residues within the CDRs of antibodies or antigen binding domains of other engineered biotherapeutic moieties can alter or eliminate the capacity of that molecule to interact with its antigen of interest. This may lead to a lower efficacy of the molecule *in vivo*, through either decreased potency or increased elimination, and an increased propensity to be less stable in general. Many factors can lead to oxidation of Met, Trp, and more rarely histidine (His) including those induced using thermal stress or enzymatic processing as well as exposure to light or reactive oxygen species [11]. To induce oxidation, and thereby assess the risk of oxidation of key antigen or effector recognition elements of a biotherapeutic, potent oxidizers can be employed. AAPH is a common example of such an oxidizer although others exist, and this is commonly employed to induce oxidation to replicate events that would happen over time with the biotherapeutic coming into contact with various metals, exposure to light, and so on. If potential "oxidative hotspots" are observed during an AAPH test, these sites can be assessed for reengineering using more stable amino acids and the resultant biotherapeutic tested to ensure no losses in binding to the antigen [12].

Oxidation typically occurs on the amino acid side chains of cysteine, Met, Trp, His, and tyrosine, in that decreasing order [7]. Because the thiol group in cysteine is usually not free in pharmaceutical antibodies and His and tyrosine oxidize at a much slower rate, Met and Trp are of a greater interest when analyzing therapeutic mAbs for oxidation. Met oxidation results in the formation of methionine sulfoxide (+16 Da) and, under extreme conditions, such as hydrogen peroxide, performic acid, or UV irradiation, it oxidizes to become methionine sulfone (+32 Da). Trp oxidation can result in the formation of many different metabolites, including kynurenine (+4 Da), hydroxytryptophan (+16 Da), hydroxykynurenine (+20 Da), *N*-formylkynurenine (+32 Da), and hydroxy-*N*-formylkynurenine (+48 Da). Oxidation can be influenced by factors including buffer, pH, and excipients [7]. There are several examples in the literature where oxidation of Trp in the CDR of an antibody has a devastating effect on its ability to bind to the desired antigen.

One such example was when a single Trp residue was demonstrated to be responsible for loss of binding and biological activity for MEDI-493, a humanized mAb against respiratory syncytial virus (RSV), upon UV light irradiation. Irradiation with UV light resulted in spectral changes typical of Trp photoproducts and in a progressive loss of MEDI-493 binding and biological activity as measured by ELISA, Biacore®, and a variety of cell-based assays. Mass spectrometric characterization of peptides generated from the UV-irradiated MEDI-493 confirmed that most Met and a few Trp residues were oxidized to varying degrees upon exposure to UV light. However, the only Trp showing a significant level of oxidation was Trp-105, residing within a CDR of the heavy chain, and representing the most solvent-exposed indole moiety in MEDI-493. By generating a second mAb (MEDI-524) with Trp-105 replaced with phenylalanine (Phe) a similar pattern of Met oxidation was observed but without loss of binding and biological activity following irradiation [13]. Another example where oxidation has caused problems with a molecule's integrity was for recombinant human granulocyte colony-stimulating factor (G-CSF), a hematopoietic growth factor where oxidation in Met residues 122 and 127 results in a loss in receptor binding and therefore biological activity. Oxidation of two other Mets, residues 1 and 138, exacerbated this problem by inducing an unstable protein conformation with further decrease in biological activity [14].

Due to observations such as those discussed earlier, many design strategies are employed to introduce oxidation-resistant forms of the therapeutic moieties. In the case of the unstable oxidized METs on G-CSF, leucines (Leu) were engineered in place of Met127 and 138, which translated to greater stability than the native form of the molecule and also resulted in retaining optimum *in vitro* biological activity [15]. Strategies such as these are now commonplace and a very important part in the engineering of stable biotherapeutics.

Overall, acceptable levels of deamidation, isomerization, and other chemical stability events may vary, depending on factors such as the impact the modification has on functionality, the product's desired shelf-life, the *in vivo* half-life, and a number of other parameters. The appropriate actions to address such potential "chemical hotspot" liabilities may include reengineering of the antibody to remove the hotspot, adjustment of the formulation (pH, ionic strength, addition of excipients, etc.) to minimize degradation during storage, or a combination of both.

11.4 Physical Degradation

Long-term stability is an important aspect for biopharmaceutical products because storage conditions may affect the quality and potency of the drug over time. Thus, different formulation conditions need to be tested to find the ones that ensure optimum drug function and structural integrity over time. One of the biggest concerns in assessing the stability and integrity of biotherapeutics is the appearance of sample aggregation. These can be either visible (>100 μm) or subvisible (<100 μm) particles. Aggregation or degradation can lead to loss of potency (as there is less active drug available), or a potential risk of immunogenicity as the aggregates are seen as "foreign" to the patient being treated with the biotherapeutic. Several case studies have been described whereby aggregation was problematic including one extreme example where the poor conformational stability of recombinant human erythropoietin (Eprex) was associated with the occurrence of antibody-mediated red cell aplasia [16]. Also, in another example, thermal stability analysis by differential scanning calorimetry (DSC) was employed for the biophysical characterization of different HIV vaccine candidates based on mimetics of the gp41 protein. A first generation of vaccine construct demonstrated a propensity to aggregate at physiological pH; therefore a second generation version was designed. DSC analysis of the constructs from the second generation showed improved stability, and additional isothermal titration calorimetry (ITC) experiments characterizing the interaction with a known ligand confirmed a native-like structure [17]. Chapter 5 provides a detailed overview of size exclusion chromatography (SEC), SEC-multiangle light scattering (MALS), DSC, ITC, and other techniques typically employed for assessing aggregation.

For larger particle sizes, that is, increased aggregation, problems are further exacerbated as these larger particulates could lead to capillary clogging, pulmonary dysfunction, needle clogging during injection into the patient in addition to induction of immunogenicity [18]. Sharma and Kelley published an example on solubility issues associated with a Fab molecule during early development showcasing the transition of a clear solution at pH5 to a cloudy gel at pH 7 [19]. This figure is reproduced here in Figure 11.3.

In addition to problems related to aggregation, proteolysis can also be a challenging event whereby the protein is cleaved by proteases or chemicals during

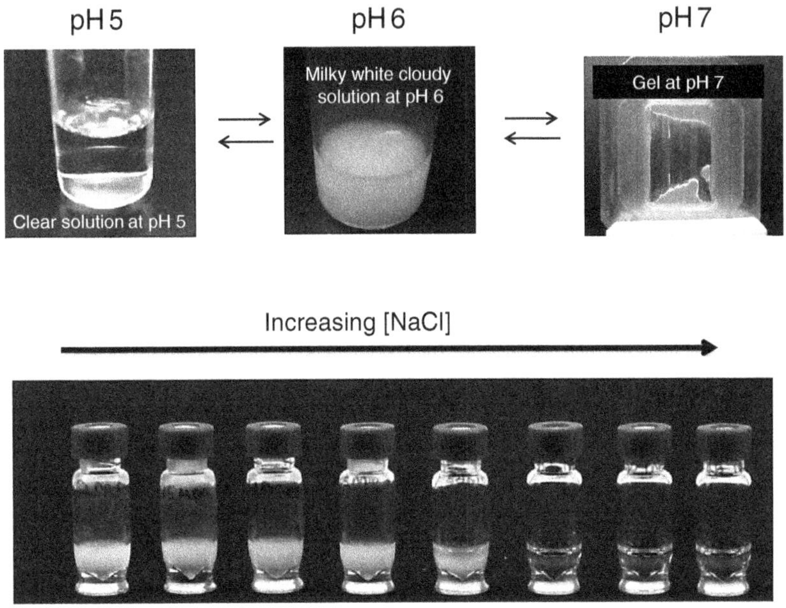

Figure 11.3 An example of a Fab molecule with solubility issues observed during early development. The molecule solubility is pH dependent where it transitions from clear solution (pH 5) to turbid solution (pH 6) to a gel (pH 7) (top panel). The molecule's solubility also showed salt dependence exhibiting higher solubility in solutions with high NaCl concentration (bottom pane). Source: Sharma and Kelley [19]. Reproduced with permission of Springer.

manufacturing or storage of a biotherapeutic or during administration of the drug to a patient. There are many predictive tools based on molecular modeling that allow assessment of the propensity of a molecular sequence or structure to be prone to proteolytic cleavage [20]. If proteolysis sites are predicted these can be engineered to be protease resistant, or other strategies such as tethering regions of a biotherapeutic together to enhance structural integrity can be performed. This can be achieved by introducing additional disulfide bonds, or by covalently crosslinking hydrocarbons. The number of stress points on a molecule will dictate the need for such engineering approaches [21].

Various approaches can therefore be undertaken to improve the physical stability of a biotherapeutic leading to increased efficacy and longer shelf-life.

11.5 Yield and Structural Stability

The yield and reproducibility of yield from the manufacturing process is another parameter that is taken into consideration upon lead selection. If yields are poor for one sequence over another, the one producing a higher yield may be more desirable moving forward, all other parameters such as stability and specificity being equal [22]. The other main concern in manufacturing a biotherapeutic is its structural stability during production, manufacturing, and upon patient administration. To assess structural integrity a battery of tools and techniques such as nuclear magnetic resonance (NMR), X-ray crystallography, ITC, and transversed relaxation optimized spectroscopy are employed to carefully assess the attributes related to stability (please refer to Chapters 2, 4, 5 and 6 in this book).

11.6 Posttranslational Modifications

Many modifications, both co- and posttranslational as well as synthetic derivatization, affect the stability and specificity of a biotherapeutic protein. Chapter 3 in this book, for example, discusses glycosylation, the *in vitro* and *in vivo* challenges of this posttranslational modification (PTM) in terms of biotherapeutics efficacy and immunogenicity and the problems associated with the presence of glycosylation sites at nonframework regions, that is, in the CDRs of an antibody. If an antibody with glycosylation in the CDRs is observed, it is often standard practice to reengineer the antibody to eliminate this site and potential source of heterogeneity [23]. When designing antibodies through phage display rather than *via* immunization of an animal, libraries can be designed to minimize or obliterate glycosylation sites occurring in the CDRs.

In addition to oxidation and glycosylation, antibodies in particular are susceptible to two key modifications. The first is the conversion of an *N*-terminal glutamate or glutamine residue to pyroglutamate, a modification that can easily be detected by mass spectrometry using both top-down and bottom-up approaches. Like many other PTMs, pyroglutamate formation at the N-terminus of an antibody increases the level of heterogeneity, always of concern in the manufacturing process, but may have the potential to decrease biological activity as well as alter pharmacokinetics and antigenicity [24]. The second is the potential influence on stability, pharmacokinetics, homogeneity of the product due to antibody charge variants whereby subtle differences in the C-terminal lysine of antibodies in various cell culture conditions are often observed during routine manufacturing or process changes. These variants can pose a challenge when demonstrating product comparability. Several solutions have been designed to overcome this cause of heterogeneity including engineering tactics with C-terminal deletions of either lysine or lysine and glycine mutants. Analytical and physiological characterization of these mAb mutants showed that through engineering one can obtain greatly reduced levels of basic variants without decreasing antibody biologic activity, structural stability, pharmacokinetics, or subcutaneous bioavailability in rats [25].

11.7 Analytical Techniques

Upon stability testing stress treatments, the following analytical techniques are used in molecular assessment to identify and characterize an antibody's physicochemical profile with respect to functional, chemical, and physical stability. These include the BIAcore binding assay that is employed to analyze the binding activity of an antibody sample, differential scanning fluorimetry to characterize the thermal stability of a biotherapeutic moiety, pH ion exchange chromatography (pH IEC) to characterize the charge heterogeneity of a biotherapeutic. Here, pH IEC is used to separate molecules based on charge. Control and stressed samples in the pH 5.5–10.8 range are separated on a shallow gradient at 40°C. The pH chromatographic profile is analyzed for percentage of acidic, main, and basic species.

To characterize the size distribution of an antibody sample, SEC is employed to separate molecules based on size. This allows detection of any dimers, trimers, aggregates, as well as mAb fragmentation products. Many of these techniques are covered in Chapter 10.

In addition to the analytical techniques mentioned earlier, intact molecular weight analysis by mass spectrometry is also employed. Mass spectrometry is a very versatile tool for verifying the intact mass of a given biotherapeutic molecule and to observe any discernible PTMs. Antibody samples are subjected to LC-MS analysis. If samples are glycosylated, PNGase F can be added first to remove N-linked glycosylation. Samples are then acidified and separated on a high-performance liquid chromatography (HPLC) polymeric reverse-phase column. After separation, ionized proteins are analyzed on a quadrupole time of flight (Q-TOF) or other high mass accuracy mass spectrometer. Many mass spectrometry vendors offer software for the qualitative and quantitative assessment of a protein. For example, when using Agilent's Qualitative Analysis software an average mass spectrum is generated from the eluted mass peak in the total ion chromatogram (TIC) of the acquired mass spec data file. Each envelope of m/z peaks in the average mass spectrum is then deconvoluted into a single mass peak, which is compared to the expected theoretical mass.

Peptide map analysis is to map and quantify sites for PTMs, specifically deamidation, isomerization, and oxidation in the CDRs; antibody samples are subjected to liquid chromatography–tandem mass spectrometry (LC-MS/MS) analysis. Samples are first treated with dithiothreitol (DTT) and guanidine HCl to reduce the disulfide bonds and unfold the protein. The free cysteine thiol groups are then capped with the alkylating reagent iodoacetamide. The reaction is quenched with additional DTT and buffer exchanged into digestion buffer. Depending on the peptide sizes in the CDRs, multiple digestion enzymes, including trypsin, Lys-C, and Asp-N, may be used in order to obtain peptide sequence coverage. Digested samples are acidified and separated on a HPLC reverse-phase column packed with CSH C18 resin. Peptides are finally analyzed on a Thermo Q-Exactive mass spectrometer with collision-induced dissociation (CID) fragmentation. The acquired mass spectrometry data is converted to a general, readable format using online software, and the file is searched against a putative protein sequence database using the Mascot algorithm. The search includes variable modifications of interest and fixed carbamidomethylation on cysteine. Peptides are identified by MS/MS fragment ions. After PTM sites of interest are identified, they are quantified *via* the label-free method. Peptide precursor masses of native and modified peptides are extracted from chromatograms in the mass spec data file. The area under the curve (AUC) of each extracted mass peak is used to calculate the percent of modification for each peptide of interest. As mentioned earlier, deamidated peptides have a mass addition of 1 Da, whereas oxidized peptides have various mass additions, typically 16 and 32 Da. Because aspartic acid and isoaspartic acid have the same mass, extracted peaks are distinguished and quantified by chromatographic separation, where the isomerized peptide is typically seen eluting at a slightly later retention time following the native peptide. Figure 11.4 shows a typical mass spectrometric workflow for characterizing and quantitating the CDRs of an antibody. This is performed mainly for monitoring deamidation and oxidation events in these critical antigen binding domains of an antibody.

11.8 Summary

Therapeutic mAbs have a very high molecular complexity. Any small change to mAb structure can have significant effects on its activity and potency. Such changes are contributed by the inherent chemical and physical instabilities of the molecule. Some molecules are more prone to these liabilities than others. Molecular assessment functions to identify such molecules to enable strategic decisions and resource planning early in the drug discovery process. The utilization of all the aforementioned analytical techniques, assays, and tests greatly helps characterize and assess these potential therapeutic mAbs. Testing on earlier versions of an antibody including the murine preclinical model can not only provide the opportunity to engineer out the liabilities but it can be used to possibly predict stability in the final humanized sequence. Furthermore, early liability characterization enables project teams to choose the strongest therapeutic candidate to move into clinical trials, which will enhance the likelihood of its success.

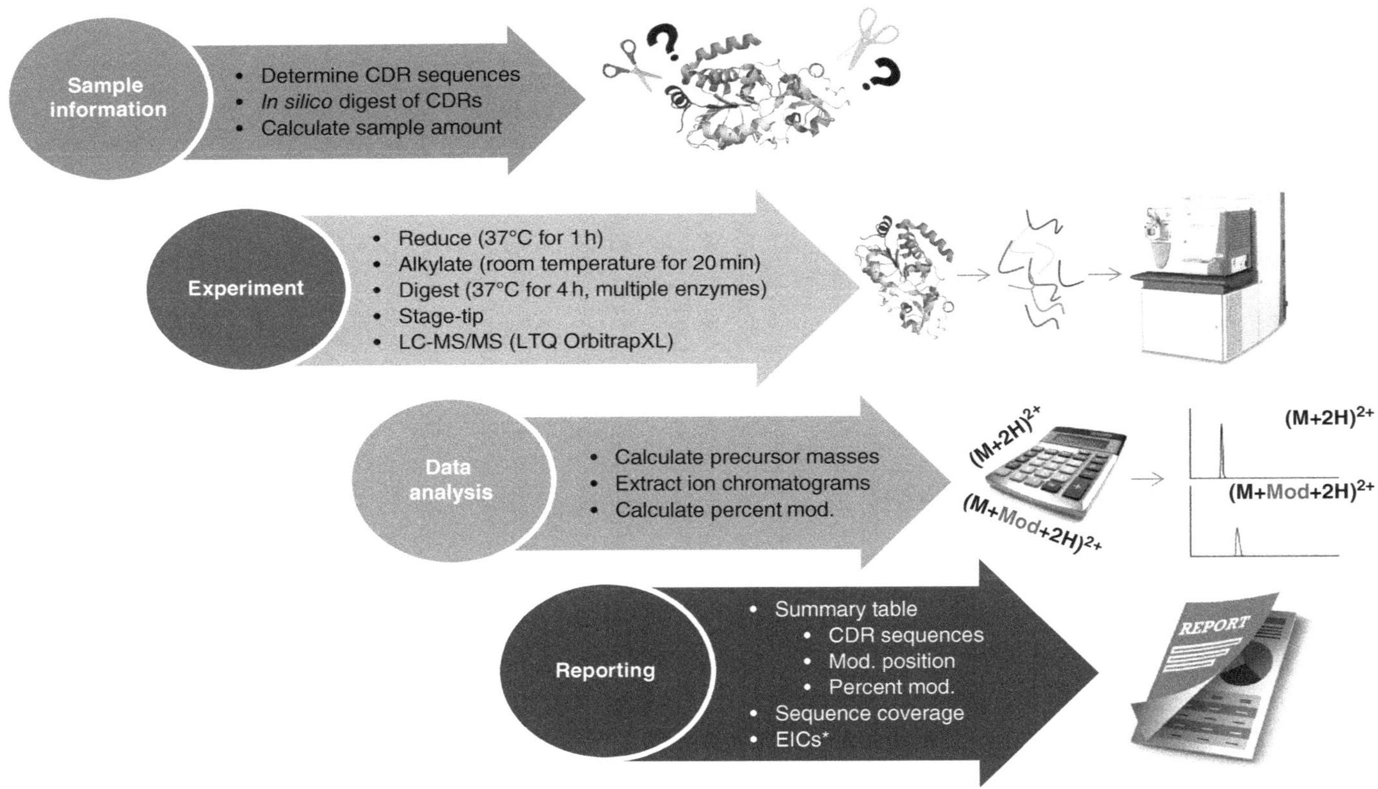

Figure 11.4 *In silico* digestions of the antibody or biotherapeutic of interest are performed. Samples are reduced and alkylated and then digested with a panel of proteolytic enzymes. Samples are then cleaned up using a reverse-phase resin (such as a stage-tip) and analyzed using LC-MS/MS. Extracted ion chromatograms corresponding to the masses of predicted peptides, with or without oxidation or deamidation (or both), are selected and quantified by measuring the area under the curve. A summary of these observations is produced allowing researchers to identify "hotspots" of modified residues after a battery of molecular assessment testing has been performed.

References

1 Sharma VK, *et al.* (2014) Biological sciences. *Proceedings of the national academy of sciences of the United States of America* 111(52):18601–18606.
2 Walsh G (2014) Biopharmaceutical benchmarks 2014. *Nature biotechnology* 32(10):992–1000.
3 Rosati S, Yang Y, Barendregt A, & Heck AJ (2014) Detailed mass analysis of structural heterogeneity in monoclonal antibodies using native mass spectrometry. *Nature protocols* 9(4):967–976.
4 Roopenian DC, Akilesh S (2007) FcRn: the neonatal Fc receptor comes of age. *Nature reviews immunology* 7:715.
5 Stracke J, *et al.* (2014) A novel approach to investigate the effect of methionine oxidation on pharmacokinetic properties of therapeutic antibodies. *mAbs* 6(5):1229–1242.
6 Imai K & Takaoka A (2006) Comparing antibody and small-molecule therapies for cancer. *Nature reviews. Cancer* 6(9):714–727.
7 Carter PJ (2011) Introduction to current and future protein therapeutics: a protein engineering perspective. *Experimental cell research* 317(9):1261–1269.
8 Brezski RJ & Georgiou G (2016) Immunoglobulin isotype knowledge and application to Fc engineering. *Current opinion in immunology* 40:62–69.
9 Furman JL, Chiu M, & Hunter MJ (2015) Early engineering approaches to improve peptide developability and manufacturability. *The AAPS journal* 17(1):111–120.
10 Rivers J, McDonald L, Edwards IJ, & Beynon RJ (2008) Asparagine deamidation and the role of higher order protein structure. *Journal of proteome research* 7(3):921–927.
11 Torosantucci R, Schoneich C, & Jiskoot W (2014) Oxidation of therapeutic proteins and peptides: structural and biological consequences. *Pharmaceutical research* 31(3):541–553.
12 Ji JA, Zhang B, Cheng W, & Wang YJ (2009) Methionine, tryptophan, and histidine oxidation in a model protein, PTH: mechanisms and stabilization. *Journal of pharmaceutical sciences* 98(12):4485–4500.
13 Wei Z, *et al.* (2007) Identification of a single tryptophan residue as critical for binding activity in a humanized monoclonal antibody against respiratory syncytial virus. *Analytical chemistry* 79(7):2797–2805.
14 Raso SW, *et al.* (2005) Aggregation of granulocyte-colony stimulating factor in vitro involves a conformationally altered monomeric state. *Protein science: a publication of the Protein Society* 14(9):2246–2257.
15 Lu HS, *et al.* (1999) Chemical modification and site-directed mutagenesis of methionine residues in recombinant human granulocyte colony-stimulating factor: effect on stability and biological activity. *Archives of biochemistry and biophysics* 362(1):1–11.

16 Schellekens H (2009) Biosimilar therapeutics-what do we need to consider? *NDT plus* 2(Suppl_1):i27–i36.
17 Crespillo S, *et al.* (2014) Single-chain protein mimetics of the N-terminal heptad-repeat region of gp41 with potential as anti-HIV-1 drugs. *Proceedings of the national academy of sciences of the United States of America* 111(51):18207–18212.
18 Doessegger L, *et al.* (2012) The potential clinical relevance of visible particles in parenteral drugs. *Journal of pharmaceutical sciences* 101(8):2635–2644.
19 Sharma VK, Kelley RF (2015) Molecular assessment of monoclonal antibody-based therapeutics enabling lead selection for clinical development. *Biobetters, AAPS advances* 19:153–180.
20 Song J, *et al.* (2011) Bioinformatic approaches for predicting substrates of proteases. *Journal of bioinformatics and computational biology* 9(1):149–178.
21 Bird GH, *et al.* (2010) Hydrocarbon double-stapling remedies the proteolytic instability of a lengthy peptide therapeutic. *Proceedings of the national academy of sciences of the United States of America* 107(32):14093–14098.
22 Jager V, *et al.* (2013) High level transient production of recombinant antibodies and antibody fusion proteins in HEK293 cells. *BMC biotechnology* 13:52.
23 Li B, *et al.* (2014) In vitro affinity maturation of a natural human antibody overcomes a barrier to in vivo affinity maturation. *mAbs* 6(2):437–445.
24 Chelius D, *et al.* (2006) Formation of pyroglutamic acid from N-terminal glutamic acid in immunoglobulin gamma antibodies. *Analytical chemistry* 78(7):2370–2376.
25 Jiang G, *et al.* (2016) Evaluation of heavy-chain C-terminal deletion on product quality and pharmacokinetics of monoclonal antibodies. *Journal of pharmaceutical sciences* 105(7):2066–2072.

12

Glycan Characterization: Determining the Structure, Distribution, and Localization of Glycoprotein Glycans

John B. Briggs

Department of Protein Analytical Chemistry, Genentech Inc., South San Francisco, CA, USA

Abbreviations

2-AA	2-aminobenzoic acid
2-AB	2-aminobenzamide
2-AET	2-aminoethanethiol
APTS	8-aminopyrene-1,3,6-trisulfonic acid
CID	collision-induced dissociation
DMB	1,2-diamino-4,5-methylenedioxybenzene
ESI-MS	electrospray ionization–mass spectrometry
ETD	electron-transfer dissociation
Fuc	fucose
Gal	galactose
GalNAc	*N*-acetylgalactosamine
GC-MS	gas chromatography–mass spectrometry
Glc	glucose
GlcNAc	*N*-acetylglucosamine
HILIC	hydrophilic interaction liquid chromatography
HPAEC	high-performance anion-exchange chromatography
HPAEC-PAD	high-performance anion-exchange chromatography with pulsed amperometric detection
mAb	monoclonal antibody
MALDI-TOF MS	matrix-assisted laser desorption/ionization time-of-flight mass spectrometry
Man	mannose
MS	mass spectrometry
MS^2	tandem mass spectrometry

Analytical Characterization of Biotherapeutics, First Edition. Edited by Jennie R. Lill and Wendy Sandoval.

m/*z*	mass-to-charge ratio
$NaBH_3CN$	sodium cyanoborohydride
NeuAc	*N*-acetylneuraminic acid
NeuGc	*N*-glycolylneuraminic acid
NMR	nuclear magnetic resonance
OPD	*o*-phenylenediamine
PNGase A	peptide-*N*-glycosidase A
PNGase F	peptide *N*-glycosidase F
TFA	trifluoroacetic acid

12.1 Introduction

Glycosylation is the most common co- and/or posttranslational modification of proteins, with studies estimating that greater than 50% of all proteins are glycosylated [1]. Protein glycosylation has been demonstrated to have a profound effect on both the structure and function of proteins, with studies indicating that glycans can affect the tertiary structure [2, 3], the solubility [4], the stability [3, 5], the immunogenicity [6–8], the *in vivo* clearance [9–11], and the biological activity [5, 12–14] of the protein. In order to produce biopharmaceutical products with consistent glycosylation and thereby ensure consistent product safety and efficacy, an efficient and reproducible manufacturing process as well as an analytical system that is capable of effectively monitoring glycan profiles must be in place. With a concern for consistent product quality and the emergence of biosimilars, regulatory agencies are paying particular attention to molecular attributes that are known to fluctuate in manufacturing processes, such as protein glycosylation [15].

Glycosylation is by far the most complex protein modification, with variation in the degree of occupancy at a potential glycosylation site within a protein (macroheterogeneity), as well as variation in the types of glycans observed at a particular glycosylation site (microheterogeneity). An example of the extent of complexity imparted by glycan microheterogeneity is provided by the diverse group of oligosaccharides observed in recombinant tissue plasminogen activator (rtPA). Although rtPA contains only 3 N-linked glycosylation sites, 18 different rtPA oligosaccharides have been isolated and characterized [16]. Fully defining the structure of a glycan requires knowing not only the monosaccharide composition and the sequence of the monosaccharides, but it requires determining the position of attachment of the monosaccharides to each other (glycosidic linkage), as well as whether the attachment is formed in an axial or equatorial orientation with respect to the hemiacetal or C-1 carbon of the sugar (anomericity). With respect to anomericity, if the glycosidic linkage is in an axial orientation it is referred to as an alpha (α) anomeric linkage; if the linkage is in an equatorial orientation it is referred to as a beta (β) anomeric linkage. Examples of α- and β-anomeric linkages are provided in Figure 12.1.

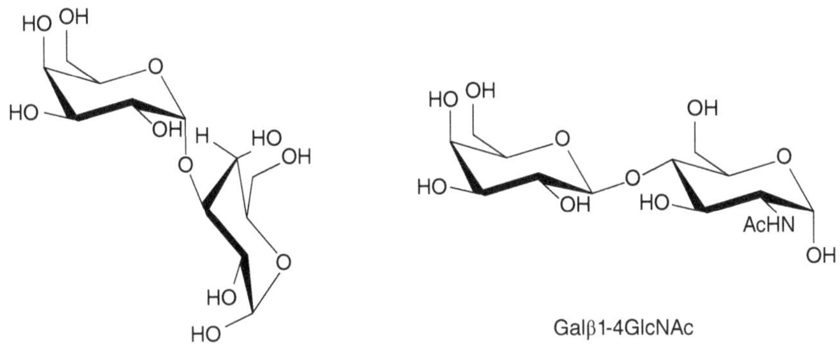

Galβ1-4GlcNAc

Galα1-3Gal

Figure 12.1 Examples of α- and β-anomeric linkages. Provided are the Galα1-3Gal and Galβ1-4GlcNAc glycan sequences. GlcNAc,*N*-acetylglucosamine; Gal, galactose.

Protein glycosylation is generally divided into two categories: N-linked glycosylation and O-linked glycosylation. This categorization is based upon the type of atom involved in the linkage between the glycan and the protein, with N-linked glycans bound to proteins *via* an asparagine nitrogen atom and O-linked glycans bound to proteins *via* either a serine or a threonine oxygen atom. The two types of protein glycosylation differ not only in the linkages that are involved in attaching the glycan to the protein, but also in their biosynthesis and overall structure.

N-linked glycosylation is a co- and posttranslational process that is initiated in the endoplasmic reticulum by the transfer of an oligosaccharide containing three glucose (Glc) residues, nine mannose (Man) residues, and two *N*-acetylglucosamine (GlcNAc) residues from a dolichol pyrophosphate-linked oligosaccharide to an asparagine residue on the protein. The transfer primarily takes place at sites in the protein that contain the consensus sequence asparagine–X–serine/threonine [NX(S/T)], where X can be any amino acid except proline. As the glycoprotein traverses through the endoplasmic reticulum and the Golgi apparatus, it encounters a series of glycosidases and glycosyltransferases that alter the structure of the glycan moiety and generally produce a heterogeneous mixture of glycan structures at the glycosylation site due to differential cleaving or addition of the monosaccharides by the various enzymes [17]. The N-linked glycan structures produced are classified into three groups; high Man glycans, hybrid glycans, and complex glycans (Figure 12.2). All N-linked glycans contain a common trimannosyl-chitobiose core structure consisting of three Man residues and two GlcNac residues ($Man_3GlcNAc_2$). The classification of glycans into one of the three groups is based upon the monosaccharides that extend beyond this core structure. In high Man glycans, the only monosaccharides that are observed in the antennae that extend beyond

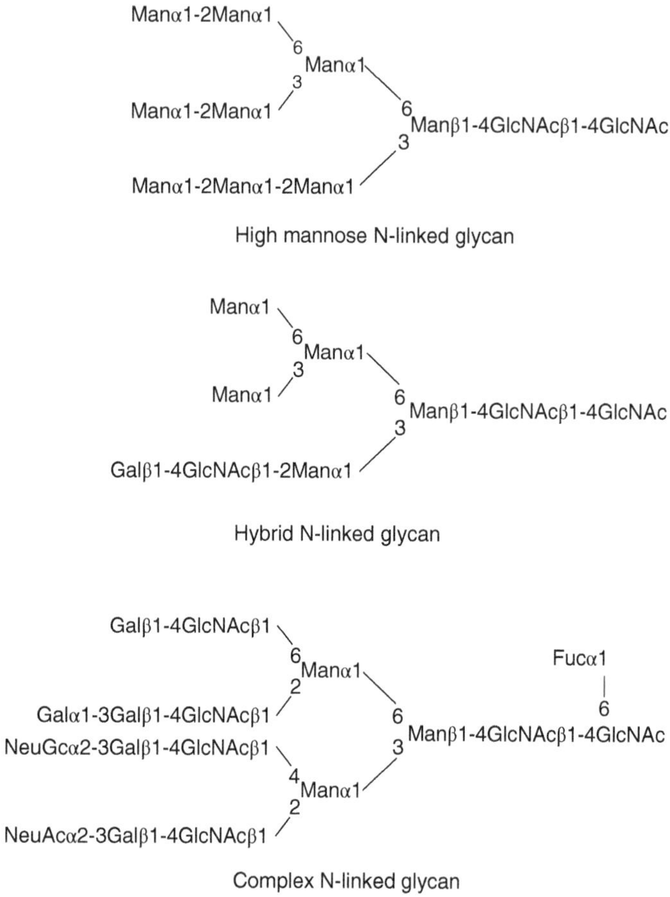

Figure 12.2 Examples of high-mannose, hybrid, and complex N-linked glycan structures. Man, mannose; GlcNAc, *N*-acetylglucosamine; NeuAc, *N*-acetylneuraminic acid; NeuGc, *N*-glycolylneuraminic acid; Gal, galactose; Fuc, fucose.

the glycan core structure are Man residues. In complex glycans, none of the antennae that extend beyond the glycan core contain Man residues, but are instead composed of a variety of monosaccharide constituents that are attached to the trimannosyl core *via* a GlcNAc residue. Common monosaccharide constituents of complex glycan antennae include *N*-acetylglucosamine (GlcNAc), galactose (Gal), *N*-acetylneuraminic acid (NeuAc), and *N*-glycolylneuraminic acid (NeuGc). In hybrid glycans, the antenna that extends from the α1,6-linked Man in the trimannosyl core contains only Man residues, while all other antennae extending from the α1,3-linked Man in the trimannosyl core are complex-type antennae that are attached to the core *via* a GlcNAc residue.

Another common constituent of complex N-linked glycans is fucose (Fuc), which in mammalian glycoproteins is most commonly found attached in an α1,6-linkage to the GlcNAc residue that links the glycan to the asparagine residue of the protein. This GlcNAc is referred to as the reducing terminal GlcNAc residue.

The order and linkage of the monosaccharides observed in N-linked glycan antennae are dictated by the specificity of the enzymes that form the glycans, so if the enzymatic repertoire of a particular host cell is known, predictable glycan structural motifs are produced by the cell. In high Man and hybrid glycans, all Man residues are attached to each other in specific, invariant sequences and linkages. In hybrid and complex glycans, one or more antennae are attached to the glycan core structure through a GlcNAc residue. This GlcNAc is attached to the trimannosyl core *via* a β-anomeric linkage and can be attached to one of the Man residues at the 2-, 3-, 4-, or 6-hydroxyl position depending on the Man residue involved in the linkage. Beyond this initial attachment to the trimannosyl core, there are predictable complex-type structural motifs that are associated with mammalian glycans. The most common motif is one in which a Gal residue is attached in a β1,4-linkage to the GlcNAc residue, which in turn can be capped by a sialic acid (either NeuAc or NeuGc) in either an α2,3- or α2,6-linkage. Example of a fully extended, complex-type antennae could include NeuAcα2-3Galβ1-4GlcNAcβ1-xMan or NeuGcα2-3Galβ1-4GlcNAcβ1-xMan. A variation on this motif involves incorporation of an extension of an antenna by the insertion of a Galβ1-4GlcNAc sequence between the terminal sialic acid residue and the Gal residue. This insertion is commonly referred to as an *N*-acetyllactosamine or LacNAc repeat.

In contrast to N-linked glycosylation, O-linked glycosylation is entirely a posttranslational modification that takes place in the Golgi apparatus and typically involves the step-wise addition of monosaccharides to serine or threonine residues within the protein. Unlike N-linked glycosylation, there is no common glycan core structure, or even a common monosaccharide-protein linkage, with α-GalNAc, β-GlcNAc, α-Man, α-Fuc, β-Xyl, β-Glc, and β-Gal glycan-protein linkages observed. The α-GalNAc glycan linkage forms the basis of the most common type of O-linked, termed mucin-type O-linked, glycosylation. There are eight different mucin-type core structures that differ from one another by the carbohydrate structures that extend beyond the reducing terminal GalNAc that attaches the glycan to the protein (Figure 12.3). The absence of a common core structure or structural motif and the similarity of the monosaccharide composition amongst the various core structures can make the deduction of O-linked glycan structure more challenging than that of the N-linked glycan structure.

The choice of cell line used in production can dramatically affect the types of oligosaccharides expressed on recombinant proteins [18]. Most biotherapeutics that rely on glycosylation for their mode of action are produced in

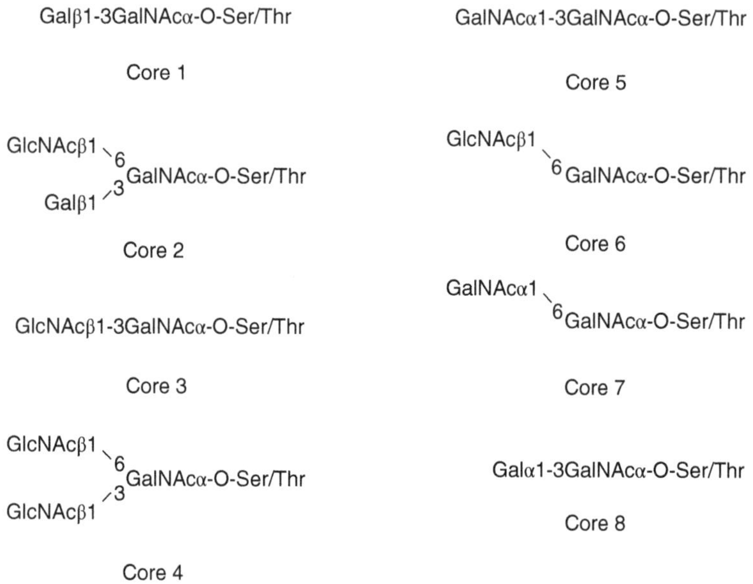

Figure 12.3 Mucin-type O-linked glycan core structures. Ser, serine; Thr, threonine; Gal, galactose; GalNAc, *N*-acetylgalactosamine; GlcNAc, *N*-acetylglucosamine.

mammalian cell lines, predominantly Chinese hamster ovary (CHO) cells, baby hamster kidney (BHK) cells, SP2/0 (murine myeloma cells), and NS0 cells (murine myeloma cells). These cell lines have become the most commonly used cell lines largely because of their ability to produce glycoproteins with glycosylation profiles that resemble human glycoproteins. Although the glycans produced by these cell lines have structures that are, for the most part, similar to those produced by human cells, these murine and hamster cell lines can produce glycan epitopes that are not expressed on human glycoproteins. One structural moiety that can be observed in murine and hamster glycans that is not observed in human glycans is the sialic acid NeuGc. While the genes that are responsible for the production of NeuGc are present in human cells, NeuGc is not expressed on human glycans due to a conserved inactivating mutation in one of the genes encoding for a key enzyme responsible for the synthesis of NeuGc. Most human sera contain anti-NeuGc antibodies, which are believed to be the result of an immune response to the incorporation of trace amounts of NeuGc into human tissues following dietary exposure to foods of animal origin [19]. Another structural difference between human glycans and hamster and murine glycans is that both murine and hamster cell lines can produce glycans containing the Galα1-3Gal epitope. This epitope is most commonly observed as an extension of complex-type N-linked glycan antennae as a Galα1-3Galβ1-4GlcNAc sequence. Antibodies to the Galα1-3Gal epitope are naturally occurring antibodies that are present in relatively high titer in all humans despite the absence of any known prior exposure [20]. The presence of these antibodies against glycan epitopes in normal human sera can pose significant safety risks for biotherapeutic glycoproteins. An example of the extent of risk introduced by the presence of these antigens is provided by the postmarket experience with cetuximab, a chimeric mouse–human IgG1 antibody specific for human epidermal growth factor receptor that is produced using the SP2/0 murine myeloma cell line. A high incidence of hypersensitivity reactions was observed in patients following the administration of cetuximab. Characterization of cetuximab glycans indicated that the antibody contained glycans that expressed both the NeuGc and the Galα1-3Gal epitopes [21]. An analysis of the serum samples from patients treated with cetuximab linked the observed anaphylaxis to the presence of IgE antibodies that were specific for the Galα1-3Gal epitope [6].

Cell culture conditions can also have a profound effect on protein glycosylation. Cell culture media composition, including monosaccharide levels, amino acid composition, and trace metal composition, can affect the glycan profile, as well as media additives such as butyrate and nucleotide sugar precursors. Cell culture process parameters like bioreactor pH and temperature, dissolved oxygen, ammonia, and cell viability can also affect protein glycosylation. Even the configuration of the bioreactor can significantly affect glycosylation. The complex interplay between the various cell culture parameters can make effective control of protein glycosylation quite challenging. Since the response to a change in a particular cell culture condition can vary in a cell- or protein-specific manner, the optimal cell culture conditions are generally determined on a case-by-case basis [15, 22]. A reliable evaluation of cell culture process control relies on the use of sensitive analytical tools capable of accurately assessing the types and relative quantities of glycans present on the glycoprotein therapeutic.

With the inherent complexity in characterizing oligosaccharides, it is not surprising that multiple analytical techniques are typically required for the characterization of the complex mixture of glycans observed in glycoproteins. Among the common analytical techniques used in the characterization of protein glycosylation are nuclear magnetic resonance (NMR), mass spectrometry (MS), gas chromatography (GC), liquid chromatography (LC), and capillary electrophoresis (CE). The use of some of these analytical methods can be somewhat challenging due to the physical characteristics of the glycans, which have no inherent chromophores and can have substantially different charge characteristics. This chapter will focus on some of the more commonly used methods for assessing glycan structure and distribution, with N-linked glycosylation being the primary focus, although some aspects of O-linked glycan analysis will also be reviewed.

12.2 Glycan Labeling

Since glycans have no inherent chromophores or fluorophores and neutral glycans have no charge, it is often advantageous to label the glycans in order to facilitate their analysis. The absence of a chromophore or fluorophore precludes the use of either the UV or fluorescence detection used in typical chromatographic or electrophoretic analyses. The lack of a charged group on neutral glycans results in the inability to use electrophoretic separations as a means of characterization. The differential charge on neutral and negatively charged glycans can also affect the ability to effectively characterize glycan mixtures by MS, since mass spectrometric acquisition is performed in either a positive or a negative ion mode, which can differentially affect the ionization of charged and neutral glycans. These limitations can be overcome by labeling the glycans with a compound containing the desired characteristics, whether that includes UV absorption or fluorescence, or a negatively charged functionality. The particular label used in glycan analysis will depend on the nature of the analyte, as well as the mode of analysis, with the most common property used in selection of the label being the ability to impart fluorescence to the labeled glycan. By far the most common means of glycan labeling is through reductive amination (Figure 12.4). In labeling by reductive amination, the carbonyl group on the acyclic aldehyde form of the reducing sugar reacts with the amine group of the selected fluorophore to form a Schiff base. The resulting imine is then reduced, typically by treatment with sodium cyanoborohydride ($NaBH_3CN$), to produce a stable fluorescently labeled glycan.

One very important factor in the reductive amination of glycans is the reaction pH. An acidic environment promotes the formation of the sugar aldehyde functionality and protonates the imine intermediate to form an iminium ion, both of which are advantageous to the reductive amination reaction. However, if the pH is too low, the amine group on the label can be protonated and the $NaBH_3CN$ reducing agent can undergo acid-catalyzed degradation, both of which are detrimental to the reductive amination process. An additional consideration is the effect that low pH can have on the structure of the glycans, since sialic acids are acid-labile and can be eliminated from the glycan if the pH is too low. Potential side reactions can also be influenced by the pH of the reaction mixture. One side reaction that can affect the efficiency of the glycan labeling is the reduction of the reducing sugar aldehyde group to form a sugar alditol, which is incapable of being labeled by reductive amination.

In order to optimize the reductive amination reaction it is beneficial to understand the effects of pH on the various chemical processes. At pH 3–4 the reduction of aldehydes in the presence of $NaBH_3CN$ is efficient, whereas reduction of the iminium ion intermediate is rapid at pH 6–7, and reduction of aldehydes is negligible in this pH range [23]. The propensity for protonation of the amine group of the label is dependent on the basicity of the amine,

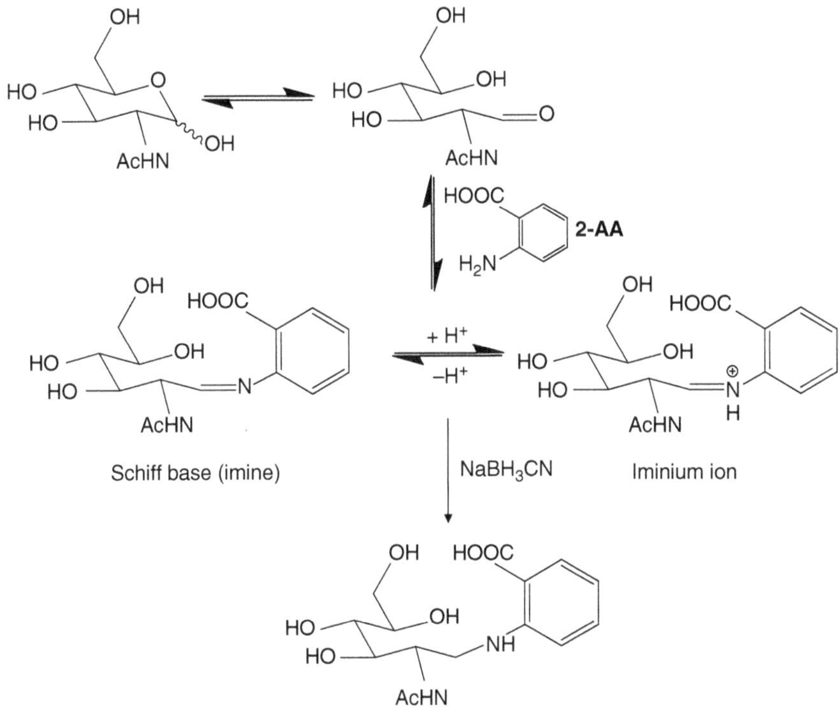

Figure 12.4 Reductive amination. Shown is the labeling of *N*-acetylglucosamine with 2-aminobenzoic acid (2-AA) by reductive amination.

so different labels might have different optimal pH ranges. Generally speaking, the reductive amination of glycans is optimal when the pH of the reaction mixture is kept between 6 and 8.

Other important factors in labeling by reductive amination are the concentrations of the label and the reducing agent. Reductive amination is typically carried out using a large molar excess of both the reducing agent and the label relative to the glycan. The molar excess of the reducing agent increases the labeling efficiency by irreversibly preventing the conversion of the glycosylimine back into the free glycan and label, thereby shifting the equilibrium of the first reaction toward the formation of more glycosylimine. The excess label increases labeling efficiency not only by pushing the equilibrium toward the formation of the iminium ion intermediate, but also by preventing the labeled glycan from undergoing further reaction with another carbonyl-containing compound, like a glycan in the aldehyde configuration. The secondary reaction is only of concern if the amine group of the label is a primary or secondary amine, since the reductive amination reaction requires that the amine group involved in the reaction contain a proton.

The medium in which the reaction is conducted is also important in reductive amination. Labeling is commonly conducted using dimethyl sulfoxide or tetrahydrofuran solvent systems containing either acetic acid or citric acid. Salts can interfere with the efficiency of the labeling reaction, so it is best to use labile buffer solutions that can be easily evaporated in the upstream glycan release and processing steps. The presence of water also tends to decrease labeling yields, so it is best to conduct the labeling reaction in water-free solvents.

A large number of glycan labels have been cited in the literature [24, 25], with some of the more common ones described elsewhere in the chapter. The reaction conditions used with each label vary widely, with temperatures ranging anywhere from 37 to 90°C and incubation times varying between 15 min and 16 h. Alternative reducing agents have also been proposed [26, 27]. One alternative reducing agent that has drawn particular interest is 2-picoline-borane. Unlike other reducing agents, effective reduction using 2-picoline-borane can be accomplished using water as the solvent. As a result, the substitution of 2-picoline-borane for $NaBH_3CN$ not only avoids the potential generation of hydrogen cyanide, which can result from the use of $NaBH_3CN$, but it eliminates the need to conduct the reductive reaction in toxic organic solvents. One limitation to the use of 2-picoline-borane is that the efficiency of the reduction is dependent on the nature of the label, with the use of poorly water-soluble labels resulting in good labeling efficiency and the use of highly water-soluble labels resulting in low labeling efficiency. This difference in labeling efficiency is attributed to increased hydrophobic association between the 2-picoline-borane reducing agent and more hydrophobic labels in an aqueous environment [28]. When adopting the use of any label it is advisable to evaluate the reaction conditions for both their effect on labeled glycan yield and glycan integrity.

12.3 Compositional Analysis

A good first step in characterizing glycoprotein glycosylation is to determine the types of glycans that are present. One means of assessing the potential types of glycans present is to determine the glycoprotein monosaccharide composition. As discussed previously, N- and O-linked glycans have characteristic structures and the presence or absence of the characteristic structural elements can be used to determine whether a particular type of glycan is present in the sample. For example, all mucin-type glycans are linked to the protein through a GalNAc residue, whereas GalNAc is not commonly found in N-linked glycans. Therefore, if monosaccharide analysis is performed and no GalNAc residues are observed, the presence of mucin-type O-linked glycans can be ruled out. In a similar manner the absence of Man residues in a monosaccharide analysis can infer the absence of N-linked glycans, since all N-linked glycans contain Man residues, whereas most mammalian O-linked glycans do not contain Man residues. Monosaccharide analysis can also be used to screen for the presence of uncommon types of glycosylation, like O-fucosylation. In attempting to assess the types of glycans present using monosaccharide analysis, it can be useful to perform the analysis with and without prior treatment with peptide *N*-glycosidase F (PNGase F) to remove N-linked glycans. The release of the N-linked glycans prior to monosaccharide analysis makes it possible to determine the potential core structures of any O-linked glycans that might be present in the glycoprotein.

Two types of analysis are required for full monosaccharide characterization: sialic acid analysis and neutral sugar analysis. In both types of analysis glycans are subjected to acid hydrolysis in order to break the glycosidic bonds linking the monosaccharides, thereby releasing the constituent monosaccharides. The need for separate sialic acid and neutral sugar analyses is based upon the relative instability of the glycan constituents to acid hydrolysis. In particular, sialic acid residues are more sensitive to acid hydrolysis than the other monosaccharides commonly observed in N- and O-linked glycans, and the acidic conditions required for the release of the neutral sugars can result in the destruction of sialic acids.

12.3.1 Neutral Sugar Analysis

In releasing neutral sugars for quantification, the glycoprotein is typically incubated in 2 M trifluoroacetic acid (TFA) for 2–6 h at 100°C. These hydrolysis conditions are designed not only to release the individual monosaccharide components, but to de-N-acetylate amino sugars like GlcNAc and GalNAc. The complete de-N-acetylation of the amino sugars is important not only because it makes the integration of a single peak for each amino sugar possible, but also because the native GlcNAc and GalNAc sugars can coelute with other monosaccharides in some chromatographic analyses. For accurate quantification of neutral sugars, a second hydrolysis is recommended. Studies have shown that GlcNAc tends to be more resistant to acid hydrolysis than other sugars, which leads to underrecovery of GlcNAc using the standard 2 M TFA hydrolysis conditions. To overcome this underrecovery, a second hydrolysis can be performed in which the glycoprotein is incubated in 4 M TFA for 5–6 h at 100°C. Unfortunately, the 4 M TFA hydrolysis cannot be used for quantification of all monosaccharides due to losses of Gal and Man under the harsher hydrolysis conditions [29]. It is important to subject the monosaccharide standards used to generate standard curves to the same hydrolysis conditions used in the release of the monosaccharides to compensate for any monosaccharide destruction introduced by the acid hydrolysis. Once the neutral sugars have been released from the glycoprotein, the samples are evaporated to dryness and then reconstituted with purified water. The resultant monosaccharide mixture can then either be directly analyzed by high-performance anion-exchange chromatography with pulsed amperometric detection (HPAEC-PAD) or be derivatized with a fluorophore and analyzed either chromatographically or by CE.

In HPAEC-PAD analysis, monosaccharides are separated by anion-exchange chromatography using hydroxide-based mobile phases with a pH of 12 or higher. The high pH of the mobile phase ionizes monosaccharide hydroxyl groups, which results in the ability to separate the sugars by anion-exchange chromatography. The separation of the individual sugars is largely based upon slight differences in the overall pK_as of the sugars [30]. Detection of the monosaccharides as they elute from the anion-exchange column is accomplished by pulsed amperometric detection. In pulsed amperometric detection, chromatograms are generated by a current that results from electrocatalytic oxidation of sugars at a gold electrode [31]. One potential issue with the HPAEC-PAD analysis is that amino acids present in the hydrolysates, in particular lysine and arginine, can elute within the retention time window of the monosaccharides and decrease the detector response to the monosaccharides that elute after them. To eliminate this interference, an Amino Trap guard column should be used. The introduction of the guard column results in a shift in the retention times of the amino acids relative to the monosaccharides and this retention time shift eliminates detector response inhibition [32]. The major advantage of the HPAEC-PAD method over other methods that require monosaccharide labeling is that no prior derivatization or cleanup of the monosaccharide mixtures is necessary. However, specially designed chromatographic systems that are able to withstand the high pH of the hydroxide-based mobile phases are required.

If HPAEC-PAD is not the method chosen for monosaccharide analysis, then most other analytical methods require that the hydrolyzed sugars be derivatized with a fluorescent tag prior to chromatographic or electrophoretic analysis. The fluorescent label not only enables the sensitive detection of the monosaccharides, but it also influences the modes of separation that can be used to quantify the sugars. The most common fluorescent label used in the chromatographic analysis of neutral sugars is anthranilic acid or 2-aminobenzoic acid (2-AA). The widespread use of the 2-AA label can be attributed to its greater molar fluorescent intensity relative to the other commonly used carbohydrate labels, as well as the fact that, unlike most other labels, neither removal of excess label nor re-N-acetylation of amino sugars after hydrolysis is required prior to analysis [24]. Additionally, there are commercially available kits that provide the materials necessary for 2-AA labeling that simplify the sample preparation.

As is typical of most carbohydrate labeling, the 2-AA label is attached to the monosaccharides *via* reductive amination. The labeling reaction takes place in the presence of molar excesses of both the 2-AA label and $NaBH_3CN$ in a methanol–acetate–borate solvent system, with incubation at 80°C for 30–60 min. One consideration in labeling by reductive amination is that carbohydrates can undergo a base-catalyzed rearrangement at the C-2 position to produce a structural isomer of the original sugar. This rearrangement, called epimerization, can lead to the production of a number of different labeled species for a given glycan that are separable either chromatographically or electrophoretically. In labeling monosaccharides with 2-AA, the hexosamine products formed in the hydrolysis of the amino sugars GalNAc and GlcNAc can undergo epimerization. In the chromatographic analysis of 2-AA-labeled monosaccharides derived from N-linked glycans, this has led to the observation of both glucosamine and mannosamine epimers, even though N-linked glycans contain no mannosamine or *N*-acetylmannosamine residues. The use of a methanol–acetate–borate solvent system in the derivatization has been shown to minimize the formation of epimers [33]. Once the monosaccharides have been labeled with 2-AA, they can be analyzed by either CE [34] or reversed-phase chromatography [35].

12.3.2 Sialic Acid Analysis

Sialic acids can be released from glycans either enzymatically or by acid hydrolysis. *Arthrobacter ureafaciens* sialidase is the enzyme most commonly used in the enzymatic release and quantification of sialic acids. It has broad specificity and is able to release α2,3-, α2,6-, α2,8-, and α2,9-linked sialic acids. One concern with the use of sialidases in quantitative methods is that they can have issues with specificity. Some commercially available sialidases tend to have lower activity toward *N*-glycolylneuraminic acid (NeuGc) than toward NeuAc, and they also tend to have lower activity toward O-acetylated forms of sialic acids than toward non-acetylated forms of sialic acid [36–38]. Since these specificity differences can affect the overall quantification, acid hydrolysis is most often the method of choice in the quantitative release of sialic acids.

In releasing sialic acids by acid hydrolysis, it is important to control the hydrolysis conditions, because if the hydrolysis conditions used are too harsh, the sialic acid can be either outright destroyed or some of the more labile structural elements of the sialic acids can be eliminated. Of particular concern are the O-acetyl groups. These functional groups cannot only be lost during acid hydrolysis, but also they can migrate from one hydroxyl position to another. Studies have shown that when the pH is either below 3 or above 6 migration of sialic acid O-acetyl groups can occur [38]. Incubation of glycoprotein samples in 2 M acetic acid at 80°C for 1–3 h has been demonstrated to provide quantitative release of sialic acids while minimizing de-O-acetylation and migration of O-acetyl groups [38, 39]. If sialic acid O-acetylation is not a concern, incubation in 0.1 M TFA at 80°C for 1–2 h has also been used in the quantitative release of sialic acids. Once the sialic acids have been released from the glycoprotein, either they are typically directly analyzed by HPAEC-PAD or they are derivatized with a fluorophore and then analyzed by reversed-phase chromatography. As in the analysis of neutral sugars, it is preferable to subject the sialic acid standards used to generate standard curves to the same hydrolysis conditions used in the release of the sialic acids to compensate for any sialic acid destruction introduced by the acid hydrolysis.

In contrast to the analysis of neutral sugars by HPAEC-PAD in which the monosaccharides are separated isocratically on anion-exchange columns using hydroxide-based mobile phases, the elution of the various sialic acid forms from the anion-exchange column requires the implementation of a salt gradient—most commonly a sodium acetate gradient. The overall mobile phase composition depends on the structure of the sialic acids in the glycoprotein sample. If the sialic acids are known to contain no O-acetyl groups, then the most commonly used mobile phase is hydroxide-based with sodium acetate salt dissolved into the eluting solvent. Typical mobile phase components are 100 mM sodium hydroxide (solvent A) and 100 mM sodium hydroxide containing 1 M sodium acetate (solvent B). If the sialic acids contain O-acetyl groups, then the hydroxide-based mobile phase cannot be used, since the basic environment can lead to de-O-acetylation. In cases where the conservation of O-acetylation is required, a sodium acetate/acetic acid-based mobile phase can be used to elute the sialic acids. Since PAD detection is based on electrocatalytic oxidation of sugars at a gold electrode and sugars are more readily oxidized under alkaline conditions, a post-column addition of sodium hydroxide is incorporated to enable effective detection of the sialic acids [40].

There are two fluorescent compounds that are commonly used to label sialic acids in preparation for quantifying them by reversed-phase chromatography: one is 1,2-diamino-4,5-methylenedioxybenzene (DMB) and the other is *o*-phenylenediamine (OPD). The reversed-phase methods are widely used because, unlike the HPAEC-PAD method, no special chromatographic instrumentation is required for the analysis.

The hydrolytic release and labeling of sialic acids with OPD is typically performed in a single step by incubating the glycoprotein sample in 0.25 M sodium bisulfate containing 10 mg/mL OPD for 60–70 min at 80°C. As illustrated in Figure 12.5, the reaction of OPD with sialic acids results in the formation of quinoxaline derivatives which are separable by reversed-phase chromatography and can be monitored by fluorescence using an excitation wavelength of 230 nm and an emission wavelength of 425 nm. The widespread use of the OPD method can be attributed to the ease and speed of the combined hydrolysis and derivatization. However, use of the standard hydrolysis and derivatization conditions leads to substantial de-O-acetylation of sialic acids. An alternative

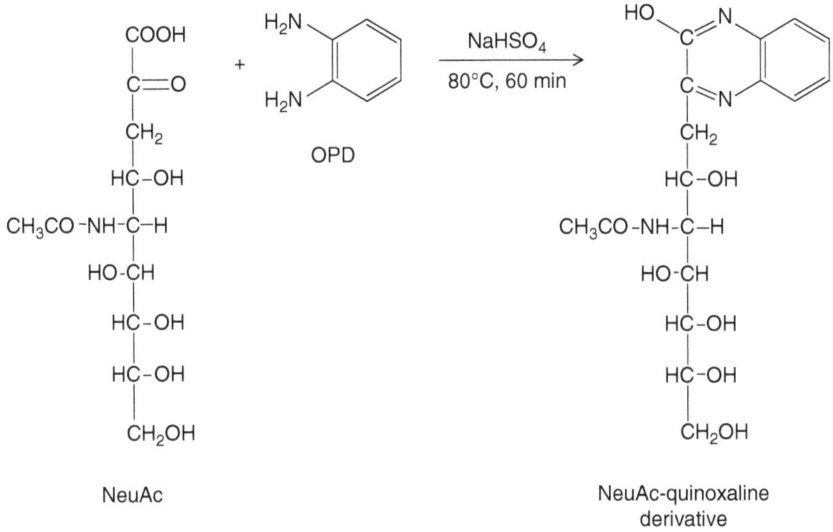

Figure 12.5 Derivatization of *N*-acetylneuraminic acid (NeuAc) with *o*-phenylenediamine (OPD).

method in which hydrolysis is carried out by incubation at 80°C for 20 min, followed by derivatization with the OPD reagent at 50°C for 2 h, has been shown to reduce but not eliminate de-O-acetylation [41].

In labeling with DMB, released sialic acids are typically incubated in a labeling solution containing 7 mM DMB, 0.75 M β-mercaptoethanol, and 18 mM sodium hydrosulfite in 1.4 M acetic acid for 2.5 h at 50°C in the dark [39]. The result of the reaction between DMB and sialic acids is the formation of quinoxalinone derivatives, which are highly fluorescent and are separable by reversed-phase chromatography. An excitation wavelength of 373 nm and an emission wavelength of 448 nm are typically used to monitor the separation of DMB-labeled sialic acids. The DMB/sialic acid reaction is illustrated in Figure 12.6. When used in conjunction with the acetic acid hydrolysis method, this labeling method has been demonstrated to result in no significant sialic acid de-O-acetylation or O-acetyl migration [39]. One of the reasons for the widespread use of the DMB labeling method is that there are commercially available kits that simplify the derivatization process. In addition, the method has been demonstrated to provide excellent separation of a wide variety of sialic acids and, unlike the other two widely used methods, the chromatography is amenable to sialic acid identification by liquid chromatography–mass spectrometry (LC-MS) with direct infusion of the column eluate into the mass spectrometer [42].

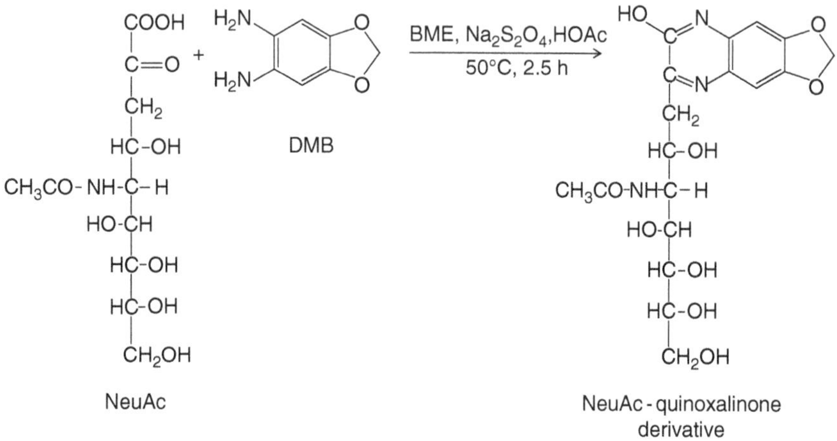

Figure 12.6 Derivatization of *N*-acetylneuraminic acid (NeuAc) with 1,2-diamino-4,5-methylenedioxybenzene (DMB). BME, β-mercaptoethanol; HOAc, acetic acid; $Na_2S_2O_4$, sodium hydrosulfite.

12.4 Glycan Release

12.4.1 Release of N-linked Glycans

By far the most common means of obtaining free N-linked glycans is through the use of specific endoglycosidases. The most widely used endoglycosidase is peptide-*N*-glycosidase F (PNGase F). PNGase F is a glycoamidase that cleaves the linkage between the innermost or reducing terminal GlcNAc of the N-linked glycan and the asparagine residue at the site of protein attachment, freeing the glycan and converting the asparagine to an aspartic acid residue. One of the reasons for the wide use of the enzyme is its broad specificity. PNGase F is able to release all N-linked glycans except for those containing a Fuc linked α(1–3) to the reducing terminal GlcNAc residue. This structure is not observed in mammalian glycoproteins, but is commonly found in plant glycoproteins and can also be found in some insect glycoproteins.

When analyzing glycans derived from plant or insect expression systems, peptide-*N*-glycosidase A (PNGase A) is often used to release N-linked glycans. The specificity of PNGase A is similar to that of PNGase F, in that it cleaves between the reducing terminal GlcNAc and asparagine residue at the site of attachment. However, unlike PNGase F, PNGase A is able to release N-linked glycans that have a Fuc linked α(1–3) to the reducing terminal GlcNAc residue. An additional difference in enzymatic specificity is that PNGase A will only release glycans from glycopeptides, which means that the glycoprotein must first undergo protease digestion before PNGase A can act. Since both released glycans and peptides are present in the final digest, an additional cleanup step that removes the peptides is generally required prior to analysis or labeling of the glycans. One limitation of the use of PNGase A is that it is inefficient at releasing sialylated glycans, so if sialylated glycans are present, chemical release might be a better option.

In any endoglycosidase digestion there are a number of sample treatment concerns that should be considered. It is generally recommended that the glycoprotein be denatured, reduced, and alkylated prior to digestion with the endoglycosidase to ensure that the enzyme has unencumbered access to the glycosylation site. Although some proteins, in particular antibodies, can be completely deglycosylated in their native form, it is generally more prudent to unfold the glycoprotein to ensure complete deglycosylation. Confirmation of complete deglycosylation is also recommended. One way that this can be accomplished is by mass spectrometric analysis of the deglycosylated protein. An additional concern is selecting the proper pH for the enzymatic digestion. The reported optimum pH for PNGase F activity is 8.5 [43], so preparation of a digestion buffer that can maintain a pH of around 8.5 seems like a straightforward decision. However, at basic pH glycans can undergo epimerization in which the reducing terminal *N*-acetylglucosamine (GlcNAc) residue is converted to an *N*-acetylmannosamine (ManNAc) residue. Since GlcNAc and ManNAc are isomers and are therefore isobaric, this epimerization would not affect any mass spectrometric analysis. However, during chromatographic and electrophoretic analyses the two isomeric forms can be separated, which could alter the recovery and quantification of the glycans. Fortunately, PNGase F has a reported active pH range of 6–10, with efficient deglycosylation reported at pHs as low as 5.5 [43, 44]. Therefore, selection of a digest pH at the lower end of the active pH range of the enzyme might be preferable. After release of the glycans, it is recommended that the glycan solution be slightly acidified. This is usually accomplished by the addition of a dilute acetic acid solution. Acidification of the glycan solution converts the glycosylamine formed during the enzymatic release to a reducing form that can then be labeled by reductive amination. Conversion of the glycosylamine can also be important in some sample cleanup steps for mass spectrometric analysis that utilize ion exchange to remove unwanted cationic species. Since sialic acid residues are susceptible to acid hydrolysis, care should be taken to ensure that the treatment does not result in desialylation of the glycans. One additional consideration is the choice of buffering system used in the enzymatic release of glycans. Ammonium-containing buffers should be avoided if the glycans are to be labeled by reductive amination or analyzed by MS, since the ammonium ion tends to react with the reducing terminus of the glycan to produce a glycosylamine, which cannot be labeled by reductive amination and has a different mass than the native glycan in mass spectrometric analysis [45].

Historically, enzymatic deglycosylation protocols have called for incubation times of anywhere from several hours up to 24 h, depending on the nature of the glycoprotein. Recent technological developments have facilitated a more rapid release of glycans, thereby enabling the development of high-throughput assays with single-day analysis times. One such technological development was the use of 96-well plates with polyvinylidene floride (PVDF) membrane well bottoms to immobilize glycoproteins in preparation for enzymatic deglycosylation. The porous PVDF well bottoms allow the glycoproteins to be denatured, reduced, and alkylated in the wells of the plates, with the reagent removal and well washing between each step accomplished by applying a vacuum to the bottom of the plate. Using this method, the glycans from up to 60 glycoprotein samples can be released in 6 h [46]. The use of organic-aqueous solvent systems has also been demonstrated to improve the efficiency of enzymatic deglycosylation. Studies have shown that PNGase F remains active in solutions containing up to 60% acetonitrile and complete deglycosylation of most glycoproteins can be accomplished in as little as 1 h [47]. Microwave irradiation has also been demonstrated to accelerate the PNGase F-mediated release of N-linked glycans. With microwave-assisted enzymatic deglycosylation, most glycoproteins can be deglycosylated in less than 30 min and complete deglycosylation of antibodies can be accomplished in as little as 10 min. The microwave treatment does not adversely affect the protein moiety, so use of this technique also allows for the analysis of the deglycosylated protein [48]. Another promising development is the use of pressure-cycling technology (PCT) to increase the rate of enzymatic deglycosylation. In PCT an alternating cycle of atmospheric and high pressure is used during the enzymatic digestion. The pressure changes are believed to lead to a rapid folding and unfolding of the glycoprotein, which leads to better access of the enzyme to the glycosylation site, resulting in decreased digestion time and enzyme usage [49].

N-linked glycans can also be released by chemical means, the most common of which are alkaline β-elimination and hydrazinolysis. However, as discussed in the following section, these chemical treatments generally require additional sample cleanup and can result in both glycan degradation and incomplete release of N-linked glycans.

12.4.2 Release of O-linked Glycans

While enzymatic release is the preferred method for obtaining N-linked glycans, there are a few endoglycosidases that act on O-linked glycans and those that can act have very restrictive specificities. *Endo*-α-*N*-acetylgalactosaminidases (O-glycanases) are the only endoglycosidases that have proven useful in the analysis of mucin-type O-linked glycans. Most of the O-glycanases are entirely specific for the unsubstituted Galβ1-3GalNAc core 1 structure and any substitution to the core structure blocks enzymatic activity. However, two O-glycanase enzymes have been identified that not only cleave the core 1 structure, but also act on the unsubstituted GlcNAcβ1-3GalNAc core

3 disaccharide [50]. Since both core 1 and core 3 disaccharides are frequently substituted with other sugars, the glycans released by O-glycanase treatment might not be fully representative of all of the O-glycans present on a glycoprotein. Therefore, other methods must be used to release O-glycans in glycan characterization and profiling.

Alkaline β-elimination and hydrazinolysis are the two most common methods used to release O-linked glycans from glycoproteins. In alkaline β-elimination, glycans are released from glycoproteins by incubation in alkaline solutions at elevated temperatures. Typically solutions containing 0.05–0.1 M sodium hydroxide or potassium hydroxide are used with incubation at temperatures ranging from 45 to 60°C and at times ranging from 8 to 16 h. Incubation of glycans under alkaline conditions can lead to degradation of the glycan through a "peeling" reaction in which monosaccharides are lost from the reducing terminus of the glycan. Therefore, conditions must be selected that reduce or eliminate glycan degradation. Historically there have been two approaches to limiting the peeling reaction in alkaline β-elimination. One, termed reductive β-elimination, uses a reducing agent, typically sodium borohydride, to reduce the aldehyde group at the reducing terminus of the glycan thereby preventing glycan peeling. Unfortunately without the reducing terminal aldehyde group it is not possible to label the glycans by reductive amination. Since glycans do not have any inherent chromophores and are entirely hydrophilic in nature, this introduces substantial limitations to the types of analysis that can be performed, with analysis primarily restricted to the use of high-performance anion-exchange chromatography with pulsed amperometric detection (HPAEC-PAD), MS, and NMR methods. An alternative approach, termed nonreductive β-elimination, uses milder ammonia-based conditions to induce β-elimination in an attempt to limit the degradation of the glycans [51, 52]. The advantage of this approach is that the reducing terminus of the glycan remains intact, so labeling by reductive amination is possible. However, studies have shown that even with the use of milder alkaline conditions glycan peeling can still be an issue [53, 54] and the extent of peeling is directly related to the structure of the O-linked glycan [54]. Additionally, use of the milder alkaline conditions in nonreductive β-elimination has been associated with incomplete release of glycans [55]. One promising development that appears to have eliminated the peeling reaction associated with nonreductive β-elimination is the introduction of a method that simultaneously releases O-linked glycans and labels them with pyrazolone analogs [56]. The simultaneous release and labeling results in rapid derivatization of the reducing terminus of the glycan, thereby blocking the peeling reaction. The addition of microwave irradiation to the simultaneous release and labeling method has been shown to increase both reaction efficiency and glycan recovery [57]. Although recent developments in nonreductive β-elimination have led to a reduction in the levels of glycan degradation, questions about the completeness of the glycan release have not been fully addressed. Additionally, studies indicate that there could be a structural bias in the release efficiency of different kinds of O-linked glycans [58].

Hydrazinolysis is another method commonly used in the release of O-linked glycans from glycoproteins. In hydrazinolysis, glycoproteins are treated with anhydrous hydrazine, which releases O-linked glycans containing a free reducing terminus that can be labeled by reductive amination. Generally the release of O-linked glycans requires incubation of the glycoprotein in anhydrous hydrazine at 60°C for 6 h. N-linked glycans can also be released by hydrazinolysis; however, incubation at 95°C for 5 h is required [59]. During the hydrazinolysis process *N*-acetyl substitutions are lost from the amino sugars like GlcNAc and GalNAc, and *N*-acetyl, *N*-glycolyl and *O*-acyl groups are lost from sialic acids. After release of the glycans and evaporation of the anhydrous hydrazine, the N-acetylation of primary amines is accomplished by treatment with acetic anhydride. This treatment results in the regeneration of the amino sugars and the acetylation of the amine groups on sialic acid residues. The effect of the post-hydrazinolysis acetylation on the structure of the sialic acids can lead to incorrect structural assignments, since in the hydrazinolysis and subsequent acetylation of *N*-glycolylneuraminic acid the *N*-glycolyl functionality is replaced with an *N*-acetyl functionality resulting in the conversion of *N*-glycolylneuraminic acid to NeuAc. Additionally, all information about O-acylation is lost. Therefore, it is advisable to characterize the structure of sialic acids present in the glycoprotein sample by some other nondestructive means to determine whether the unsubstituted sialic acids observed in the released glycans are representative of the actual glycoprotein sialylation. Once the glycans have been released and re-N-acetylated, they must be separated from the protein-derived components. This can be accomplished through the use of paper chromatography, adsorptive chromatography, or gel filtration [60]. The use of hydrazinolysis has proven to be both more efficient and less structurally biased than the use of nonreductive β-elimination in the release of O-linked glycans [58]. Glycan peeling can result from hydrazinolysis, but can be minimized by reducing the water and cationic content of the reaction mixture [61, 62].

12.5 Determining Sites of Glycosylation

An essential part of characterizing protein glycosylation is establishing which sites within the protein are glycosylated. The ability to predict the sites of glycosylation, as well as the methods used to identify glycosylation sites, varies with the type of glycosylation. The identification of glycosylation sites is generally done at the glycopeptide level, with trypsin being the most common protease used in the generation of the peptides due to its predictable specificity for Lys and Arg residues and the high frequency of those residues within most protein sequences. Hence, the use of trypsin generally results in the generation of relatively small peptides that are easily sequenced to determine the sites of glycosylation.

In N-linked glycosylation, proteins are glycosylated both cotranslationally and posttranslationally through the actions of two oligosaccharyltransferase (OST) complex isoforms with different substrate specificities [63]. In the vast majority of cases these OST complexes specifically glycosylate asparagine residues at sites containing the amino acid consensus sequence NX(S/T), where X can be any amino acid except proline. If the amino acid sequence of the glycoprotein is known, then predicting potential N-linked glycosylation sites based upon the consensus sequence requirement might be expected to be relatively straightforward. However, the presence of a consensus sequence within the amino acid sequence of a protein is not a guarantee that the site will be glycosylated. The probability that a particular consensus sequence asparagine will be glycosylated is dependent upon a number of factors, including the type of hydroxyl amino acid in the consensus sequence [64–66] and the amino acid in the X position of the consensus sequence [67, 68], as well as the identity of the amino acids in the regions flanking the sequence, the location of the sequence within the polypeptide chain, and the conformation of the polypeptide chain in the regions around the sequence [69–72]. Predicting sites of N-linked glycosylation is further complicated by the fact that glycosylation has also been observed on asparagine residues that are not a part of the standard NX(S/T) coding sequence. In rare cases, N-linked glycosylation has been reported to occur on asparagine residues within the sequence NXC [73–75]. Still rarer cases have been reported in which N-glycosylation has been observed on asparagine residues in NSG, NQV, NEN, and NSQ sequences, as well as on glutamine residues in QGT sequences [76, 77].

O-linked glycosylation is a posttranslational protein modification in which one of six monosaccharides is transferred to the hydroxyl group of serine, threonine, or tyrosine residues of glycoproteins. Each of the sugars is transferred through the actions of enzymes that confer a particular anomerity to the sugar–protein linkage, with monosaccharide-specific linkages of α-GalNAc, β-GlcNAc, α-Man, α-Fuc, β-Xyl, β-Glc, and β-Gal. Predicting sites of O-linked glycosylation is far more complicated than predicting sites of N-linked glycosylation due to the absence of anything resembling a universal consensus sequence.

In the most common form of O-linked glycosylation, mucin-type glycosylation, synthesis is initiated by the attachment of GalNAc to Ser or Thr residues in an alpha linkage. There is no known consensus sequence for mucin-type glycosylation, although there is a correlation between sites of glycosylation and regions with a high frequency of Pro, Ser, Thr, and Ala residues [78]. The absence of a consensus sequence can be attributed to the fact that there are a number of different GalNAc transferases, each with its own substrate specificity and tissue distribution. The number of GalNAc transferases expressed has also been shown to be species specific, with humans expressing 20 different GalNAc transferase isoforms [79]. The identification of mucin glycosylation sites is further complicated by the tendency for mucin-type glycosylation to occur in regions that are rich in Thr, Ser, and Pro residues. Protein regions that contain a large proportion of Thr, Ser, and Pro residues tend to have few protease cleavage sites, due to not only the absence of amino acid residues recognized by the proteases, but also the high proportion of Pro residues that tend to inhibit protease activity when present at residues adjacent to a cleavage site. Glycosylation can also inhibit protease activity when it is present in the vicinity of a cleavage site. As a result, the peptides generated in regions with mucin-type glycosylation tend to be somewhat large, which makes the sequencing required to identify the sites of glycosylation somewhat challenging. Additional complexity results from the fact that mucin-type glycosylation tends to occur in regions with a high density of potential Ser and Thr glycosylation sites, thus increasing the challenge of assigning the glycosylation to a particular amino acid.

12.5.1 MS-Based Screening for Glycopeptides

Since predicting sites of glycosylation is not always straightforward and the identification of glycosylation sites requires some form of glycopeptide sequencing, identifying the glycopeptides within a protease digest is often the first step in pinpointing sites of glycosylation. MS-based methods have proven to be ideally suited for glycopeptide screening. Typically, peptides from a protease digest are separated chromatographically and tandem mass spectrometry (MS^2) is performed on the separated peptides to produce fragmentation that is characteristic of glycosylation. By screening for the presence of the glycopeptide-specific fragmentation within the mass spectrometric data acquired in the chromatographic separation, the glycopeptides in the peptide mixture can be identified. One strategy used to screen for the presence of N- and O-linked glycans involves analyzing materials that both have and have not been digested with PNGase F to remove the N-linked glycans. The O-linked glycopeptides can be identified in the materials treated with PNGase F, whereas the N-linked glycopeptides can be identified in the glycopeptide screening as additional glycopeptides that are only present in the analysis of the nontreated materials. Pretreatment of materials with sialidase can also be performed to increase the sensitivity of the MS-based glycopeptide screening methods.

One type of glycopeptide-specific fragmentation that can be used to identify glycopeptides in the liquid chromatography–tandem mass spectrometry (LC-MS^2) analysis of peptide mixtures is fragmentation of the glycan moiety by collision-induced dissociation (CID), which leads to the formation of positively charged sugar oxonium ions. Among the oxonium ions that can be used are *m*/*z* 204.09, 366.14, 147.07, 292.10, 308.10, and 163.06 ions. These ions result from the fragmentation of glycans containing *N*-acetylhexosamine ($HexNAc^+$), hexoylhexosamine ($HexNAc\text{-}Hex^+$), Fuc^+, $NeuAc^+$, *N*-glycolylneuraminic acid ($NeuGc^+$), and hexose (Hex^+) glycans, respectively. The two most commonly used oxonium reporter ions are the $HexNAc^+$ (*m*/*z* 204.09) and $HexNAc\text{-}Hex^+$ (*m*/*z* 366.14) ions, due to the common observance of these ions in spectra of both mucin-type O-linked glycopeptides and N-linked glycopeptides. Oxonium reporter ions can be generated either through "in-source" CID fragmentation or through MS^2 CID fragmentation [80]. Prior to initiating a screen for glycopeptides using sugar oxonium ions, it is good practice to optimize the collision energies used to induce the reporter ions using appropriate model glycopeptides. This not only increases the sensitivity of the acquisition, but it can reduce the potential of producing interfering peptide fragments that can be misidentified as carbohydrate oxonium ions. The use of mass spectrometers capable of high MS^2 mass accuracy can also reduce the number of false-positive glycopeptide identifications in precursor ion scan experiments [81, 82].

Another data-dependent mass spectrometric mode that can be used to identify glycopeptides in a chromatographic separation is neutral loss scanning. In this kind of MS^2 experiment, glycopeptides are identified by the loss of neutral sugars during the fragmentation of the glycopeptides. The neutral loss masses that can be used in screening for the presence of glycopeptides include 203.1, 365.2, 146.1, 291.1, 307.1, and 162.1 Da, which correspond to the loss of neutral *N*-acetylhexosamine (HexNAc), hexoylhexosamine (HexNAc-Hex), Fuc, *N*-NeuAc, *N*-glycolylneuraminic acid (NeuGc), and hexose (Hex) monosaccharides, respectively. The fragmentation that results in glycan neutral loss can be the result of either in-source fragmentation or standard MS^2 CID fragmentation [83, 84].

In both neutral loss and oxonium ion data-dependent experiments, the observation of glycan-specific fragmentation can be used to trigger further glycopeptide fragmentation that can, in some cases, be used to identify the sites of glycosylation. However, when the subsequent fragmentation fails to pinpoint the glycosylation site, other sequencing techniques must be used. Some of the methods used in glycosylation site identification will be discussed in subsequent sections.

12.5.2 Identification of Glycosylation Sites by Analysis of Native Glycopeptides

The most common mass spectrometric means of obtaining peptide sequence information is by fragmenting the peptide through energetic collisions with a nonreactive gas, such as argon or helium. This mode of fragmentation is referred to as collision-induced dissociation (CID). In CID fragmentation, the collision of the peptide with neutral gas molecules induces a vibrational

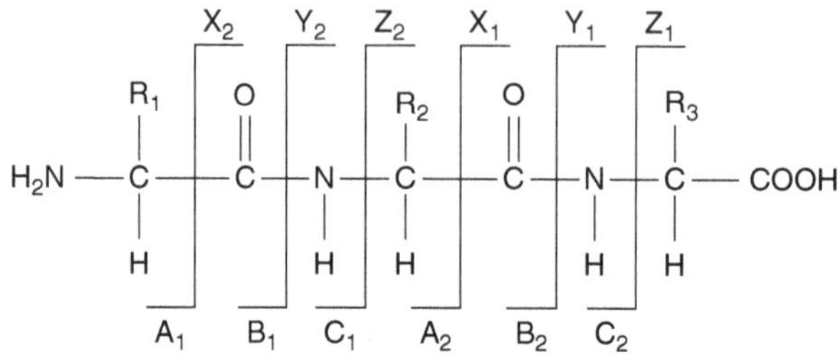

Figure 12.7 Common nomenclature for fragment ions in the mass spectra of peptides. In CID, peptide fragmentation occurs at the amide bond leading to the formation of b- and y-type ions. In ETD, peptide fragmentation occurs at the N–Cα bond resulting in the formation of c- and z-type ions. Source: Roepstorff and Fohlman [85]. Reproduced with permission of John Wiley & Sons.

activation which leads to peptide fragmentation that usually occurs at the amide bonds to produce characteristic b- and y-type fragment ions (Figure 12.7). Although CID is very useful in sequencing most peptides, it has proven to be generally ineffective in sequencing native glycopeptides due to preferential fragmentation of the glycan moiety over the peptide backbone. This glycan-dominated fragmentation leads to the production of spectra that are often not easily interpretable and frequently lack information about the site of glycosylation. Two general approaches have been used to overcome the limitations of CID fragmentation. One approach involves simultaneously removing the glycan moiety while labeling the site of glycosylation. The labeled peptide can then be sequenced using CID fragmentation to determine the site of glycosylation. The other approach makes use of another mode of mass spectrometric peptide fragmentation, referred to as electron-transfer dissociation (ETD), which preferentially fragments the peptide backbone while leaving the glycan moiety intact.

In ETD MS^2, peptide fragmentation is initiated by the transfer of electrons from gas phase radical anions (typically floranthene) to positively charged peptides, which induces fragmentation of the peptide backbone while leaving the glycan moiety attached to the peptide thereby allowing the assignment of the site of attachment. Specifically, the electron transfer promotes fragmentation of the peptide N–Cα bond leading to the formation of c- and z-type fragment ions (Figure 12.7). One potential limitation to the use of ETD fragmentation is that it tends to work best on peptides with higher charge states, so in general the fragmentation observed when fragmenting a +4 charged peptide is much better than the fragmentation observed when a +2 charged peptide is selected for fragmentation. In many cases the most abundant charge state of a peptide in a protease digest is the doubly protonated form, which does not fragment well in ETD fragmentation. Another limitation is that there is a general shift toward nondissociative electron transfer with increasing precursor m/z, which is observed as a decrease in peptide fragmentation and results in a corresponding decrease in the ability to assign the site of glycosylation.

Several approaches have been proposed for overcoming these potential limitations. One approach that has been used successfully with smaller peptides that typically lack abundant higher charge state ions is to introduce a "supercharging reagent," like *meta*-nitrobenzyl alcohol (*m*-NBA), either directly into the chromatographic mobile phase or into the postcolumn effluent. The addition of the supercharging reagent results in an increase in the average ion charge state of the peptides, which significantly improves the ETD fragmentation efficiency [86, 87]. Another approach that has been successfully used to overcome the nondissociative electron transfer limitation that is associated with larger peptides is to selectively fragment charge-reduced species observed in ETD spectra by low-energy CID. The supplemental activation of the ions generated in the ETD fragmentation with low-energy CID results in greater fragmentation of the peptide backbone while maintaining labile posttranslational modifications, like glycosylation. This method is referred to as charge-reduced collision-induced dissociation or CRCID [88].

12.5.3 Identification of N-linked Glycosylation Sites by Enzymatic Labeling of Glycosylation Sites

If the sequence of a glycopeptide is known, the mass of the glycopeptide is consistent with a peptide having N-linked glycans, and the glycopeptide has a single N-linked consensus sequence, then identification of the site of glycosylation can be considered a trivial matter. However, if a glycopeptide contains multiple consensus sequences, then further analysis is required to determine whether all potential sites have been glycosylated. Additionally, it is generally good practice to confirm the sites of glycosylation due to the absence of absolute fidelity to the N-linked consensus sequences by the glycosyltransferase complexes responsible for N-linked glycosylation.

The most widely used method for identifying sites of N-linked glycosylation is by enzymatically labeling the site of glycosylation through digestion with PNGase F, followed by LC-MS^2 analysis of a protease digest of the material. Treatment with PNGase F results in the release of N-linked glycans and the conversion of the Asn at the site of attachment to an Asp residue. This amino acid conversion results in an increase in the mass of the peptide by 1 Da, which is localized at the site of glycosylation. With the mass accuracy of many modern mass spectrometers, the site of glycosylation can be established by noting a 1 Da increase in peptide mass at an Asn residue that is within an N-linked consensus sequence. However, this approach can be misleading, because deamidation of a nonglycosylated Asn residue within an N-linked consensus sequence can also result in conversion of the Asn residue to an Asp residue. Since both deglycosylation and deamidation of an Asn residue can result in the conversion of the Asn residue to an Asp residue, an apparent increase in mass of 1 Da at an N-linked consensus sequence Asn residue following PNGase F treatment is not definitive proof of Asn glycosylation.

A widely used means of avoiding misassigning glycosylation sites due to deamidation is to perform the PNGase F digestion in the presence of $H_2{}^{18}O$, which results in the incorporation of an ^{18}O-label into the Asp residue that is formed during the digestion [89, 90]. The ^{18}O-label introduces a 3 Da increase in the overall mass of the peptide relative to the non-glycosylated peptide, with the mass shift localized to the Asn residue at the N-linked glycosylation site. Since Asn deamidation results in a 1 Da increase in mass at the site of deamidation and PNGase F treatment in the presence of $H_2{}^{18}O$ results in a 3 Da shift at the site of glycosylation, the two Asn modifications can be distinguished clearly. In generating ^{18}O-labeled glycopeptides, it is critical to ensure that no $H_2{}^{18}O$ is present during the protease digestion, as this will result in ^{18}O-labeling of all peptides at the protease cleavage site. One potential complication of the ^{18}O-labeling is the potential for deamidation to occur during the PNGase F treatment. However, since deamidation tends to take place at basic pH and PNGase F is enzymatically active between pH 6 and 10, labeling due to deamidation can be avoided by performing the PNGase F digestion at the lower end of the active pH range.

Even when the generation of ^{18}O-labeled glycopeptides is fully optimized, there are factors that can complicate the interpretation of the data. Of particular concern is the fact that the deamidated and ^{18}O-labeled forms of a peptide coelute and that the isotopic distribution of an ^{18}O-labeled peptide lies within the isotopic distribution of the deamidated form of a peptide. Therefore, glycosylation sites with low levels of occupancy can be overlooked if there is deamidation at a potential glycosylation site. One potential method for circumventing this problem is by enzymatically labeling the site of glycosylation through digestion with a combination of *endo*-β-*N*-acetylglucosaminidases, rather than with PNGase F. Digestion with *endo*-β-*N*-acetylglucosaminidases results in glycan cleavage between the two GlcNAc residues that make up the chitobiose core, leaving either a GlcNAc or a GlcNAc-Fuc glycan at the site of attachment. As a result, the mass of the peptide is increased by either 203.1 or 349.1 Da relative to the non-glycosylated peptide and the increase in mass is localized to the Asn residue at the glycan attachment site. The endoglycosidase cocktail that has been shown to be effective in glycan site labeling contains *endo*-β-*N*-acetylglucosaminidase D (Endo D) and *endo*-β-*N*-acetylglucosaminidase H (Endo H). Endo H cleaves high Man and most hybrid glycans, whereas Endo D is specific for complex glycans. By combining the specificities of the two enzymes, all N-linked glycan classes can be enzymatically cleaved. In addition to the two endoglycosidases, the enzymatic cocktail must contain neuraminidase, β-galactosidase, and β-*N*-acetylglucosaminidase.

The presence of these exoglycosidases is required because Endo D only recognizes complex glycans after the monosaccharides that make up the antennae have been removed [91].

Although enzymatic labeling of O-linked glycosylation sites is possible using labeling strategies similar to the ones described earlier, since O-linked glycans do not have a common core structure and there are no enzymes that recognize all classes of O-linked glycans, the enzymatic methods used would need to be specific to the types of O-linked glycan present on the glycoprotein. Since designing an enzymatic method for labeling of O-linked glycosylation sites requires prior knowledge of the structures of all O-linked glycans present on a glycoprotein, chemical methods are generally preferred in the labeling of O-linked glycosylation sites.

12.5.4 Identification of O-linked Glycosylation Sites by Chemical Labeling of Glycosylation Sites

The identification of O-linked glycosylation sites using MS most frequently involves labeling the O-linked glycopeptides by a chemical reaction involving β-elimination followed by Michael addition (Figure 12.8). In this reaction the O-linked glycans are removed by base-catalyzed β-elimination to form a

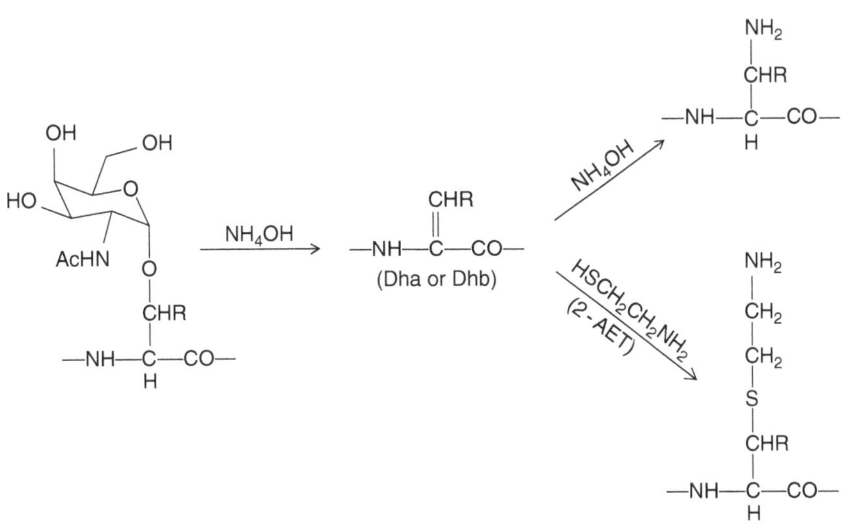

Figure 12.8 β-Elimination/Michael addition reactions. O-linked glycans are removed from Ser or Thr residues by base-catalyzed β-elimination to form dehydroalanine (Dha) or dehydrobutyric acid (Dhb), respectively. Nucleophiles then react with the dehydroamino acids to form stable derivatives. Shown are Michael additions using NH_4OH and 2-aminoethanethiol (2-AET). In the chemical structure of the peptide, R = H for Ser and R = CH_3 for Thr.

dehydroamino acid at the site of attachment. In the case of serine-linked glycans the product is dehydroalanine (Dha), and in the case of threonine-linked glycans the product is dehydrobutyric acid (Dhb). The dehydroamino acid (either Dha or Dhb) then reacts with a nucleophilic moiety to produce a stable derivative with a characteristic mass that differs from the native amino acid. The β-elimination of the glycan moiety has been carried out using a number of different bases, including NaOH, $Ba(OH)_2$, and NH_4OH. The use of NH_4OH is generally recommended, since the use of stronger bases is more likely to lead to peptide degradation. Additionally, since NH_4OH can be removed from the sample by vacuum centrifugation, its use eliminates the need for sample cleanup and desalting that can result in significant loss of materials.

When NH_4OH alone is used in the reaction, the β-elimination is accompanied by the Michael addition of NH_3 at the site of Ser or Thr glycosylation to produce an amino acid containing a primary amine in place of the hydroxyl group of the respective Ser or Thr residue. As a result, the mass of the peptide is increased by 1 Da relative to the non-glycosylated peptide and that increase in mass is localized to the site of attachment. This difference in mass can then be used to identify sites of O-linked glycosylation in MS^2 experiments. The effective yield of the reaction can vary depending on the structure of the peptide and the reaction conditions. Studies have shown that O-glycosylated Thr residues are more resistant to β-elimination than Ser residues, presumably because of steric hindrance by the methyl group in the Thr side chain [92]. Additionally, O-linked glycosylation sites that are adjacent to Pro residues have a tendency to lose glycosylation, but are resistant to the addition of the NH_3 moiety, so unless reaction conditions are adjusted properly they are observed in MS^2 experiments as dehydroamino acids (Dha or Dhb). Generally speaking, incubation of glycopeptides in 25% NH_4OH at 45°C for 15–16 h provides effective labeling of O-linked glycosylation sites [92]. One caveat to assigning sites of O-linked glycosylation based upon β-elimination is that, under some reaction conditions, Ser and Thr phosphorylation can produce the same reaction products as Ser and Thr glycosylation, making it impossible to distinguish sites of glycosylation from sites of phosphorylation. If peptide phosphorylation is suspected, it is best to eliminate phosphorylation by digestion of the peptide with alkaline phosphatase prior to β-elimination to eliminate any chance of misassignment.

A number of different nucleophilic reagents have been used in labeling by β-elimination/Michael addition, including alkyl amines and various thiol-containing nucleophiles [93–95]. The primary impetus for incorporating the use of these nucleophiles is to increase the ionization efficiencies of the derivatized peptides and thereby increase assay sensitivity. One nucleophile whose use has resulted in a completely novel approach to assigning glycosylation sites is 2-aminoethanethiol (2-AET). When 2-AET is used as the nucleophile in a β-elimination/Michael addition reaction with O-glycosylated Ser or Thr, the reaction products are structural analogs of lysine. Treatment of 2-AET-labeled O-linked glyocopeptides with endoproteinase Lys-C results in cleavage of the peptides at the sites of O-linked glycan attachment, which not only labels the sites of glycosylation but also shortens the peptides, thereby facilitating their sequencing by tandem mass spectrometric analysis [96].

12.5.5 Identification of Glycosylation Sites by Edman Degradation

Another means of identifying sites of both N- and O-linked glycosylation is by sequencing glycopeptides *via* Edman degradation (Figure 12.9). In Edman degradation isolated peptides are immobilized, typically on PVDF membranes or polybrene-coated glass fiber filters, and then reacted with phenyl isothiocyanate (PITC) under mildly basic conditions. The PITC reacts with the α-amino group of the N-terminal amino acid of the peptide to produce a cyclic phenylthiocarbamoyl (PTC) derivative of the terminal amino acid. Treatment of the derivatized peptide with a strong acid under anhydrous conditions results in cleavage of the N-terminal peptide bond, thereby releasing the derivatized amino acid residue to produce an anilinothiazolinone (ATZ) form of the amino acid. The ATZ derivative is separated from the remainder of the peptide by extraction with an organic solvent, typically ethyl acetate or chlorobutane, and is then converted to a more stable phenylthiohydantoin (PTH) form by acid treatment. Identification and quantification of the released PTH derivative is generally accomplished by reversed-phase high-performance liquid chromatography (rp-HPLC). Since the remainder of the peptide remains intact and

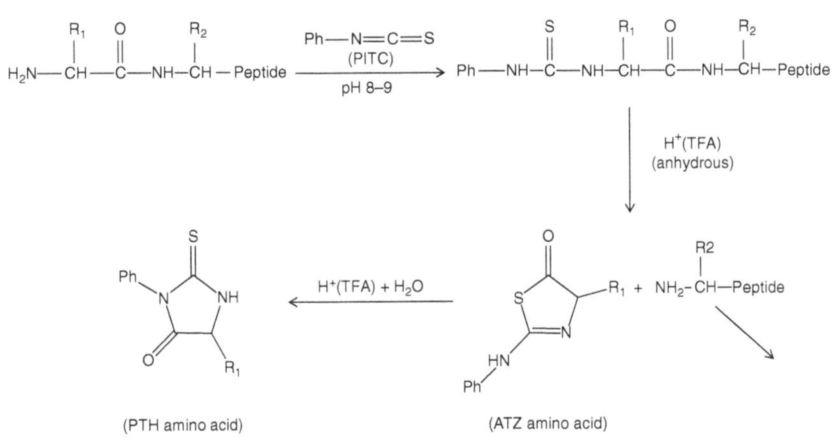

Figure 12.9 Edman degradation reaction. Phenyl isothiocyanate (PITC) reacts with the amine group of the N-terminal amino acid of a peptide. Treatment with a strong acid, like TFA, under anhydrous conditions releases the N-terminal amino acid as an anilinothiazolinone (ATZ) amino acid, which is then converted to a phenylthiohydantoin (PTH) amino acid by treatment with an aqueous acid solution.

unaltered by the derivatization process, the peptide can undergo sequential treatments to determine the amino acid sequence of the peptide.

In practice, when sequencing glycopeptides a "blank" cycle is observed at points in the amino acid sequence where glycosylation occurs. This absence of signal is attributed to the hydrophilic nature of glycoamino acids, which are poorly soluble in the organic solvents used to extract the amino acid ATZ derivatives from the membranes used to immobilize the peptide. Although a blank cycle can be indicative of the presence of a glycosylated amino acid, if the glycosylation site has a low percent occupancy it can be difficult to identify the site as being glycosylated, since there would be partial recovery of the non-glycosylated amino acid at the glycosylation site. One approach used to identify O-linked glycosylation sites that largely circumvents this issue is pre-treatment of the glycopeptide with trifluoromethanesulfonic acid (TFMSA). Mild TFMSA treatment eliminates all of the monosaccharide constituents of O-linked glycans except for the GalNAc residue that is directly attached to the Ser or Thr residue. The resultant glycoamino acid is less hydrophilic than the fully glycosylated amino acid, which allows for the extraction of the ATZ derivative by the organic solvents used in the sequencing process. As a result, both the glycoamino acid and the non-glycosylated amino acid that are present at partially occupied glycosylation sites can be chromatographically separated, identified, and relatively quantified [97].

An alternative approach to recovering glycosylated amino acids during sequencing by Edman degradation is to covalently attach the glycopeptide to a solid support so that polar solvents can be used to extract the glycosylated amino acids. The support matrix generally used is an arylamine-derivatized PVDF membrane that preferentially reacts with the C-terminus of peptides after treatment of the peptide-coated membranes with *N*-ethyl-*N*′-dimethylaminopropyl carbodiimide (EDC). Prior to attaching the glycopeptide to the arylamine-derivatized membrane, the sialic acid residues must be removed from the glycan. Desialylation is necessary because the carboxyl group of the sialic acid can react with amine groups on the solid support forming an amide bond that prevents the recovery of the glycoamino acid. Using this method and a modified chromatographic separation it is possible not only to identify sites of glycosylation, but to relatively quantify different O-linked glycans present at a particular glycosylation site [98].

12.6 Determining N-linked Glycan Distribution

In the development and production of biotherapeutics, it is important to monitor the types and distribution of the glycans expressed on the therapeutic in order to ensure consistent product quality. Even in glycoproteins with a relatively simple glycan profile, like mAbs, monitoring glycan distribution can be somewhat challenging due to the number of different glycans expressed and the structural similarities between some of the glycans, which include glycan structural isomers. An example of the number and types of glycans that can be observed in a CHO cell–derived recombinant mAb is provided in Table 12.1. Techniques for monitoring glycan distribution include a wide array of mass spectrometric, electrophoretic, and chromatographic methods. This section will provide an overview of the most commonly used techniques.

12.6.1 Assessing Glycan Distribution by MS

MS is a very powerful tool for assessing glycan distribution that has found wide use in the biopharmaceutical industry. One factor that contributes to its widespread use is that mass spectrometric analysis can simultaneously provide information about both glycan structure and glycan distribution. An additional advantage to the use of MS over other analytical techniques is that mass spectrometric methods do not generally require significant methods development in order to be able to assess glycan distribution, since glycans are resolved from each other based upon their inherent mass. The two mass spectrometric techniques most widely used in the analysis of glycan distribution are electrospray-ionization MS and matrix-assisted time-of-flight mass spectrometry.

12.6.1.1 Assessing Glycan Distribution by Mass Spectrometric Analysis of Glycoproteins

Electrospray ionization–mass spectrometric (ESI-MS) analysis of intact glycoprotein and glycoprotein fragments is a commonly used means of assessing N-linked glycan distribution. The utility of the analysis is largely dependent upon the size of the glycoprotein or glycoprotein fragment, as well as the number of different glycosylation sites on the glycoform being analyzed. The analysis of antibodies and antibody fragments can serve as an example of the relationship between glycoform size, glycoform occupancy, and the information that can be gleaned from ESI-MS analysis.

The ESI-MS analysis of intact glycoproteins is a very rapid means of assessing glycan distribution that generally consists of injecting the protein onto a reversed-phase column to remove unwanted salts, followed by elution of the glycoprotein into the electrospray source of a mass spectrometer. The resultant spectrum which contains clusters of ions that are associated with multiple charge states of the glycoprotein is then deconvoluted by mass spectrometric software to display the spectrum of the intact glycoprotein with its associated glycosylation. An example of the deconvoluted mass spectrum of an intact antibody produced in CHO cells is provided in Figure 12.10.

In the mass spectrometric analysis of most intact antibodies, the glycoforms observed result from a combination of the glycans that are present at the two N-linked sites in the Fc region of the antibody. In the intact antibody spectrum

Table 12.1 Glycans observed in a recombinant mAb produced in CHO cells.

Glycan	Structure	Glycan	Structure
G0F-GlcNAc	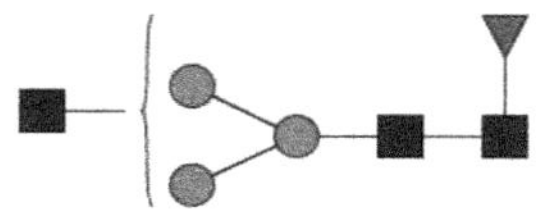	G2FS	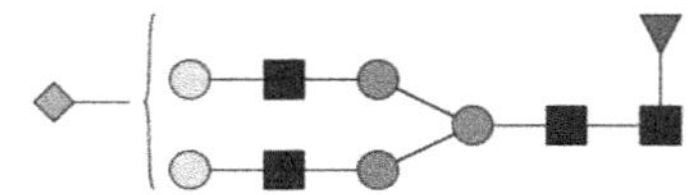
G0	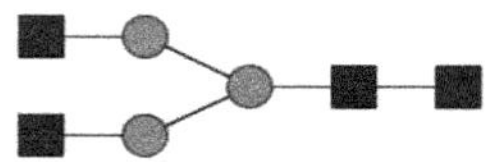	G2FS2	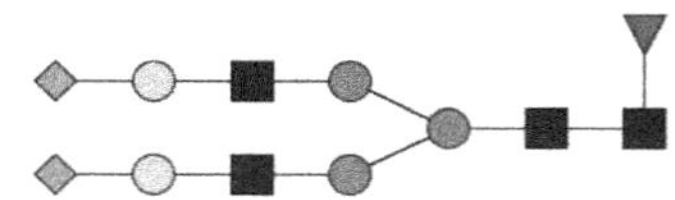
G0F	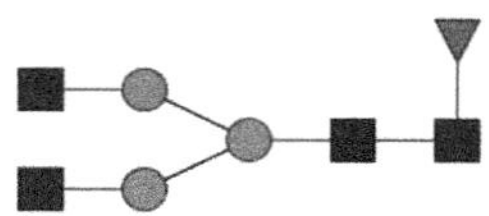	G0F + GlcNAc	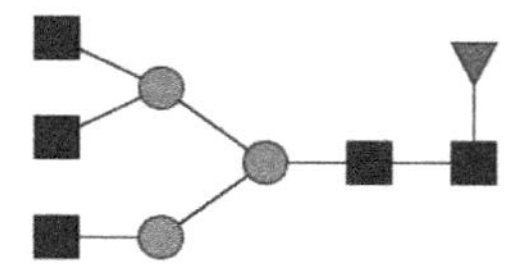
G1-GlcNAc	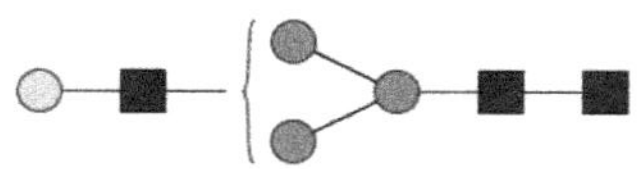	G1F + GlcNAc	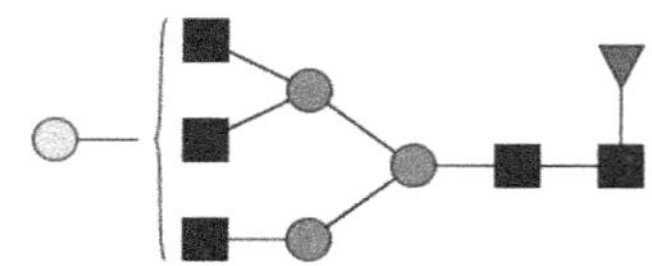
G1	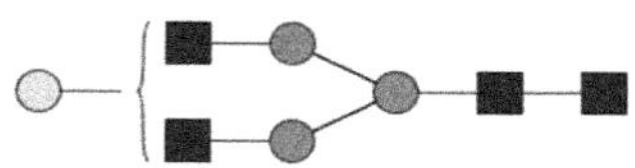	G2F + GlcNAc	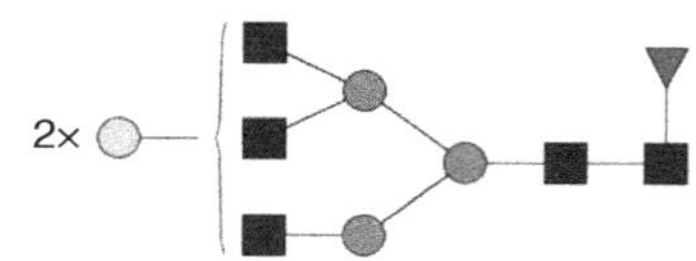

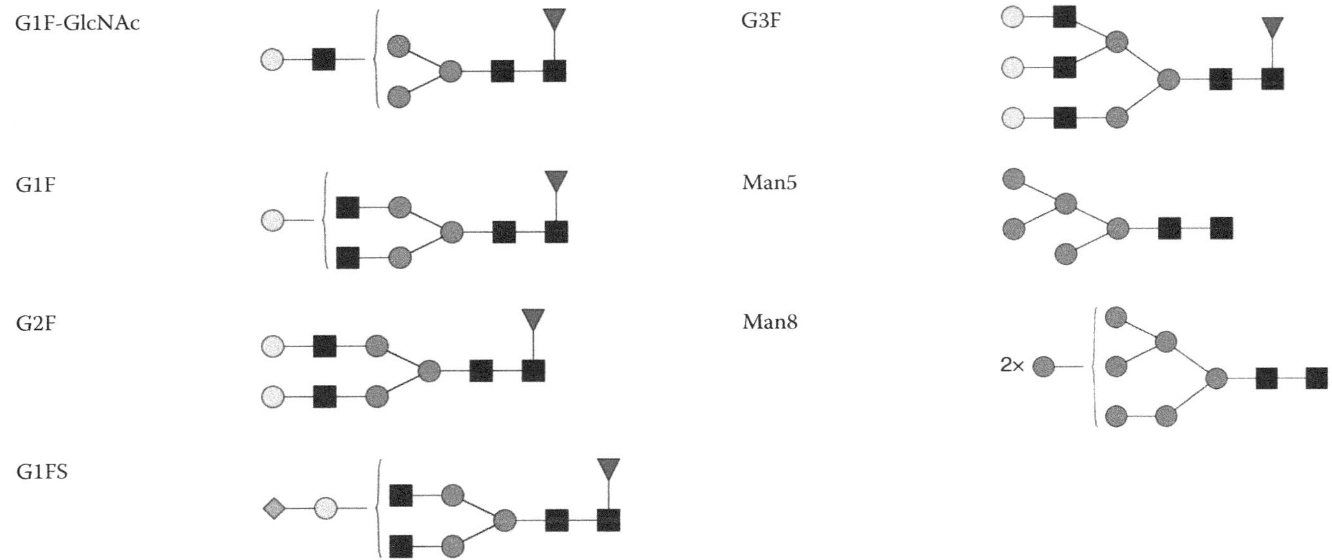

, *N*-Acetylglucosamine; , mannose; , galactose; , fucose; , *N*-acetylneuraminic acid.

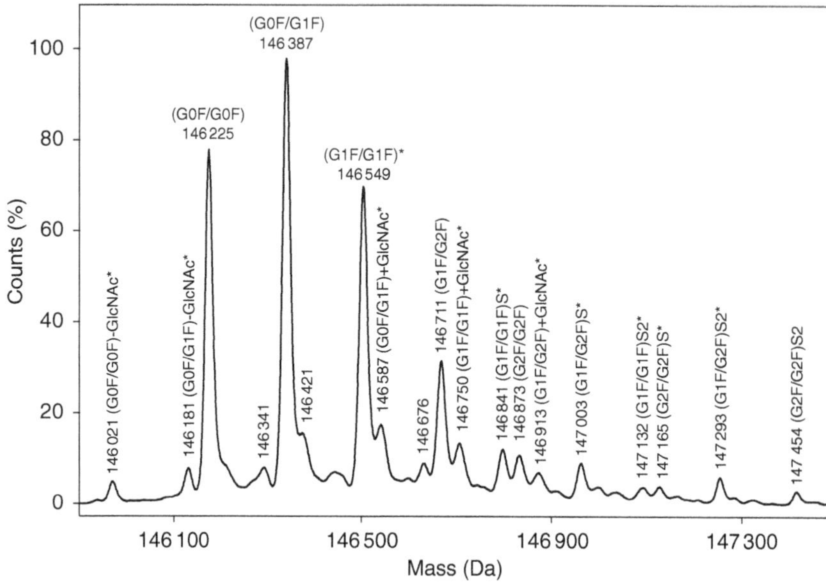

Figure 12.10 ESI-MS spectrum of an intact antibody. Both antibody Fc glycoforms associated with a particular mass are provided. For example, the antibody glycovariant associated with a mass of 146 225 Da contains a G0F glycan at both Fc glycosylation sites (G0F/G0F). *Masses associated with isobaric glycoforms. All (G1F/G1F) glycoforms are isobaric with (G0F/G2F) glycoforms. In all glycoforms containing "+GlcNAc," "−GlcNAc," or "S" in their nomenclature, isomeric forms are possible in which the GlcNAc or NeuAc residue in question can be associated with either of the two glycans that comprise the antibody Fc glycosylation. The structures of the glycans assigned to each glycoform are provided in Table 12.1. Glycan structural assignments are inferred based on a correlation between the observed mass of the glycan moiety and the mass of glycans known to be expressed on CHO-derived glycoproteins.

provided, the most abundant peaks observed are glycoforms of the antibody in which the glycosylation at both Fc region N-linked sites consists of G0F, G1F, and G2F glycans (refer to Table 12.1 for glycan structures). The most abundant glycoform observed in the spectrum is the G0F/G1F glycoform in which one of the N-linked sites in the Fc domain is the G0F glycan and the glycan at the other N-linked site is a G1F glycan. In the ESI-MS analysis of proteins, the width of the peaks observed in deconvoluted spectra tend to increase with increasing molecular mass, which results in an inverse relationship between the size of the protein and spectral resolution. With a glycoprotein the size of an intact antibody, the spectral peaks can be broad enough that it is not possible to resolve some antibody variants. For example, protein oxidation which results in a 16 Da increase in the mass of the antibody is commonly not resolved in the intact antibody spectra. Some glycovariants are also not resolved in intact antibody spectra, like the G0F and G1 glycoforms which also differ in mass by 16 Da. The broad spectral peak widths in intact spectra can also preclude the resolution of some combinations of glycovariants and protein variants, such as C-terminal lysine variants and fucosylated glycoforms. The lack of resolution between C-terminal lysine and fucosylated variants arises from the fact that there is only an 18 Da difference in mass between lysine and Fuc, which cannot be resolved in the deconvoluted spectra of intact antibodies. This lack of resolution is evident in the deconvoluted spectrum of the intact antibody provided in that the peak with a deconvoluted mass of 146 421 Da is not completely resolved from the peak with a deconvoluted mass of 146 387 Da, even though there is a 34 Da difference in mass between the two species.

In addition to the inability to resolve certain glycoforms, the ability to determine glycan distribution using ESI-MS analysis of intact glycoproteins is hindered by the inability to distinguish between glycoforms having the same mass, termed isobaric glycoforms. In particular, distinguishing between glycoforms that contain Man and Gal can be difficult, since both are hexose sugars found in N-linked glycans that have the same mass. In the intact antibody spectrum provided, an example of the uncertainty in structural assignment that results from the possibility of isobaric glycans is provided by the peak labeled G1F/G1F*. The mass associated with this peak is consistent with either the antibody containing G1F glycans at both N-linked sites in the Fc region or the antibody containing a G0F glycan at one N-linked site and a G2F glycan at the other site. The same uncertainty arises in the structural assignment of glycan pairs that contain high Man glycans. For example, since Man and Gal are isobaric, it is not possible to assign structure to a glycoform with a mass that is consistent with a Man6/G0F glycan pair, since this mass is also consistent with a Man5/G1F glycan pair. The uncertainty in glycan distribution introduced by the possibility of isobaric glycoforms is an issue in any analysis in which multiple N-linked glycosylation sites are present in the mass spectrometric analysis of a glycoprotein. Adding to the complexity is the possibility of glycoprotein glycation, which involves the nonenzymatic reaction of Glc with amine groups on the protein, primarily at the N-terminus of the protein and at lysine residues. Since Glc is also a hexose sugar and is therefore isobaric with Man and Gal, it is not possible to distinguish between some protein N-linked glycovariants containing Man or Gal and glycovariants that are attached to glycated proteins. The uncertainty in glycan structure introduced by the possibility of protein glycation is an issue that affects any mass spectrometric determination of glycan distribution in which the glycan remains attached to a protein or protein fragment. This structural uncertainty is present even when the protein moiety contains only a single glycosylation site. One means of assessing whether protein glycation is contributing significantly to the perceived glycan distribution is to determine the level of protein glycation by treating with PNGase F to remove the N-linked glycans and then analyzing the deglycosylated protein by

ESI-MS to determine the relative proportions of glycated and non-glycated protein.

Another consideration in assessing antibody glycan distribution through the analysis of the intact antibody is that many of the low-abundance glycoforms are not detected. When the glycans from the antibody analyzed in Figure 12.10 were released, labeled with 2-aminobenzamide (2-AB), and then analyzed by hydrophilic interaction liquid chromatography (HILIC), 25 different glycans were observed, whereas in Figure 12.10 only 16 glycoforms are observed, each with two constituent glycans. This can partly be explained by the inability of the mass spectrometric analysis to resolve glycoforms, but not entirely, since some of the low-abundance glycans observed in the HILIC analysis have masses that are not isobaric with the glycoforms observe in Figure 12.10. Thus, it appears that the determination of glycan distribution by mass spectrometric analysis of intact protein suffers from a lack of sensitivity relative to some other forms of analysis. This lack of sensitivity not only leads to an inability to detect some glycoforms, but also results in the underrepresentation of some glycoforms in the glycan distribution.

Although there are limitations to assessing glycan distribution through the analysis of intact antibodies, there are some advantages. Studies have shown that the distribution of the major glycoforms derived from the mass spectrometric analysis of intact antibodies provides is consistent and can provide a rapid and effective means of monitoring antibody glycosylation during the manufacturing process [99–101]. Additionally, unlike methods that analyze released glycans, the analysis of intact proteins and protein fragments allows for the assessment of the proportion of the protein that is not glycosylated. Moreover, in the analysis of intact antibodies, the mass spectrometric analysis allows for the assessment of glycan pairing in the Fc domain of the antibody. For example, in the intact antibody mass spectrum provided in Figure 12.10, the dominant glycan pairing observed is the G0F/G1F pairing. This glycan pairing is not what would be predicted based upon the relative abundance of the individual glycans present in the antibody. In a statistically random pairing of glycans, the dominant antibody glycoform would be one in which the most abundant glycan was paired with itself, which in this case would result in a G0F/G0F configuration.

When either the size of the protein is decreased or the number of glycosylation sites present is decreased, the assessment of glycan distribution becomes more complete and informative. In the analysis of antibody glycosylation, this can be accomplished through reduction of the antibody, which not only reduces the size of the glycoprotein fragment, but in most cases produces a glycoprotein fragment containing a single glycosylation site. As an example, Figure 12.11 provides the deconvoluted heavy chain spectrum that was acquired following reduction of the same antibody used to produce the intact spectrum in Figure 12.10. Since following reduction the pairing of antibody glycoforms is

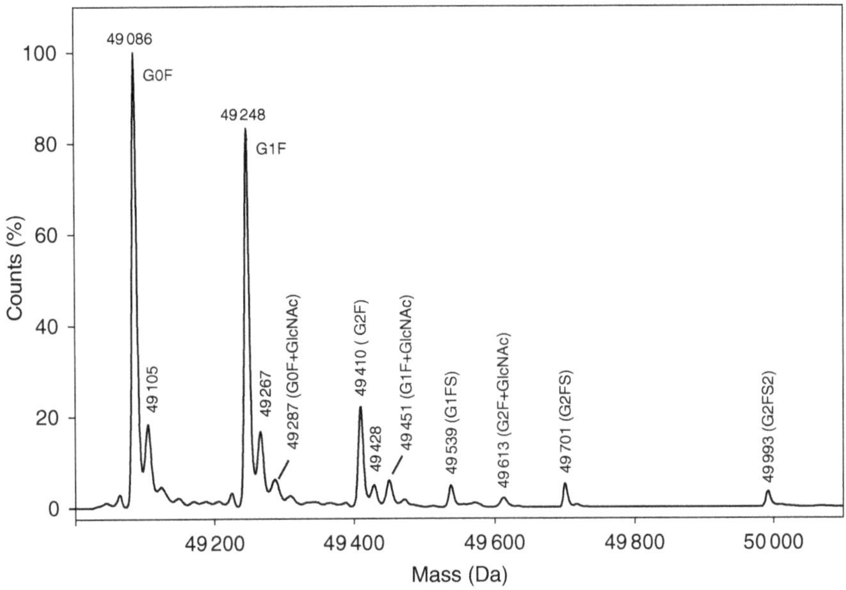

Figure 12.11 ESI-MS spectrum of the heavy chain from the analysis of reduced antibody. The structures of the glycans assigned in the spectrum are provided in Table 12.1. Glycan structural assignments are inferred based on a correlation between the observed mass of the glycan moiety and the mass of glycans known to be expressed on CHO-derived glycoproteins.

removed, the analysis of the heavy chain of reduced antibodies can be used to quantify individual glycoforms. Antibody reduction not only eliminates glycoform structural ambiguity by producing a protein fragment containing a single glycosylation site, the decrease in the size of the glycosylated polypeptide increases the mass spectrometric resolution. In the mass spectrometric analysis of the reduced antibody in Figure 12.11, this can be seen as a complete resolution of the G1F + GlcNAc and G2F + GlcNAc glycoforms and the near baseline resolution of the G0F + GlcNAc glycoform. In the mass spectrum of the intact antibody, none of the "+GlcNAc" glycoforms are completely resolved.

Further evidence for the effect of decreasing size on the resolution and sensitivity of the ESI-MS analysis is provided by a study that was conducted to assess the relative utility of 11 mass spectrometric methods for assessing the glycan distribution in a model antibody [99]. As part of this study ESI-MS was used to assess the glycan distribution of antibody fragments produced either by reduction or by a combination of reduction and proteolytic cleavage. The results of the study demonstrated that the ESI-MS analysis of the reduced antibody heavy chain was able to detect 4 out of the 14 glycoforms present in the antibody, whereas ESI-MS analysis of a proteolytic fragment of the heavy chain produced by treatment with reduction and proteolytic cleavage by IdeS (FabRICATOR) was able to detect 11 glycoforms. Although more glycoforms are resolved in the ESI-MS analysis of the smaller protein fragments, the analysis of protein fragments by MS is generally less sensitive in detecting low-abundance glycoforms than methods in which released glycans are analyzed [99, 101]. Furthermore, as in any mass spectrometric analysis in which the glycans remain attached to the protein, some protein modifications, like glycation, can affect the assessment of glycan distribution.

12.6.1.2 Assessing Glycan Distribution by Mass Spectrometric Analysis of Glycopeptides

One means of assessing the glycan distribution at a particular glycosylation site is through the ESI-MS analysis of glycopeptides. In this analysis, the glycoprotein is digested with a specific protease and then the resultant peptides are separated chromatographically and analyzed by online MS. Assessment of glycan distribution is accomplished by integration of extracted ion chromatograms generated using the ions expected from the various glycoforms. Although this technique can be very effective in elucidating the types of glycans present at a particular site, caution should be used in using the data to assess site-specific glycan distribution. Glycosylation markedly decreases the hydrophobicity of the glycopeptide, which in turn decreases ionization efficiency. Therefore, glycopeptide ionization efficiency is expected to decrease with an increase in the size of the glycan moiety. Based upon this differential ionization, it is not surprising to find that studies have shown that the assessment of glycan distribution based upon the ESI-MS analysis of glycopeptides is not fully quantitative [73, 101, 102]. However, the data can be used as a quick, semiquantitative means of assessing glycan distribution. If a more precise evaluation of site-specific glycan distribution is required, then glycopeptide isolation, followed by the analysis of released glycans, might provide the most accurate assessment.

12.6.1.3 Determining Glycan Distribution by Mass Spectrometric Analysis of Native Glycans

The most commonly used mass spectrometric method for determining glycan distribution is through the analysis of released glycans by matrix-assisted laser desorption/ionization time-of-flight mass spectrometry (MALDI-TOF MS). In preparation for MALDI-TOF MS analysis, glycan samples are mixed with an appropriate matrix, allowed to crystallize on the surface of a MALDI target, and then placed into the source of the mass spectrometer where the sample is irradiated with light from a laser. The pulsed laser irradiation both vaporizes/desorbs the sample and transfers energy to the light absorbing matrix, which in turn transfers energy to the analyte resulting in analyte ionization. In most cases the laser used is a nitrogen laser which emits UV light at a wavelength of

337 nm; however, infrared lasers can also be used. The matrix used in the analysis is dependent upon the laser in the mass spectrometer, since one of the properties of an appropriate matrix is that it absorbs the light emitted by the laser. The most common matrix used for glycan analysis in the MALDI instruments equipped with UV lasers is 2,5-dihydroxybenzoic acid (DHB); however, a substantial number of other compounds have been demonstrated to be effective matrices in MALDI mass spectrometric analysis of carbohydrates [103–105].

Sample cleanup following glycan release is an important part of MALDI-TOF MS analysis of glycans. The signal strength and quality of the glycan spectrum can be adversely affected by the presence of buffer salts and other contaminants. Therefore, most sample cleanup strategies are designed to remove both protein and salts from the glycan sample. A number of high-throughput methods utilizing 96-well plate formats have been proposed for sample cleanup, including a method that utilizes 96-well PVDF membranes for protein removal followed by spin-column cation-exchange chromatography to remove salts [46] and a method that uses 96-well HILIC membranes to isolate released glycans [106].

One important consideration in determining glycan distribution by MALDI-TOF mass spectrometry is the propensity for glycans to lose sialic acid if the acquisition parameters are not properly set. In determining the glycan distribution by MALDI-TOF mass spectrometry it is important that the spectral acquisition is performed in linear mode rather than reflector mode. The rationale for performing the analysis in the linear mode is that it is not uncommon for sialylated glycans to lose sialic acid in the field-free flight tube after the glycans have accelerated in the MALDI source. Since in the linear mode the glycans do not experience any postsource electrical field to separate the freed sialic acid from the rest of the glycan, the two components hit the detector at the same time and are therefore observed as a single ion. However, in reflector mode the two components experience a second electromagnetic field which separates the sialic acid from the rest of the sialylated glycan, so the observed glycan ion has an associated mass that is consistent with the desialylated form of the glycan. Another factor that can affect the observed glycan distribution is the laser power setting used in the spectral acquisition. If the laser power is set too high, this can lead to in-source loss of sialic acid and a consequent shift in the observed glycan distribution toward less sialylated glycans. One means of avoiding laser-induced desialylation is to titrate the laser power in the negative ion mode using a doubly or triply sialylated glycan standard. In titrating the laser power, a series of spectra are acquired using the sialylated glycan standard, with the laser power increasing with each acquisition in the series. The laser power limit is determined as the first spectrum in the series in which a shift in the spectrum toward less sialylated glycans is first observed. The laser power can then be reduced relative to this limit to ensure that no desialylation takes place during spectral acquisition.

Unlabeled neutral glycans are generally analyzed in the positive ion mode using a DHB matrix spiked with sodium salts. The addition of sodium to the matrix results in the formation of glycan-Na^+ adducts, which ionize very well in the positive ion mode. Since glycans can form adducts with a number of alkali metal salts, the addition of sodium to the matrix also prevents the formation of other alkali metal adducts, thereby ensuring that only glycan-Na^+ adducts will be observed in the spectrum. Inhibiting glycan adduct formation with non-sodium alkali metals not only simplifies glycan quantification and increases the glycan signal-to-noise ratio, but it can also prevent misidentification of glycan-related peaks in the spectrum. Of particular concern with respect to misidentifications is the formation of glycan-K^+ adducts. The possibility of misidentification arises from the fact that there is a 16 Da difference in mass between Na^+ and K^+ glycan adducts and there is also a 16 Da difference in mass between Gal and Fuc. An example of how this mass difference can lead to glycan misidentification is illustrated in Figure 12.12 with the *m/z* 1501.5 peak. There is a 16 Da difference between this ion and the ion that has been

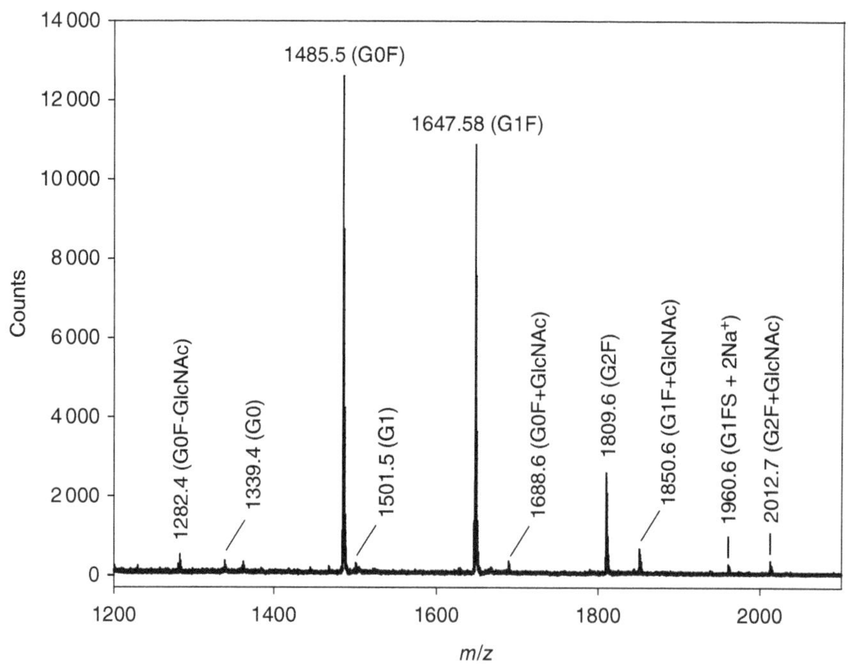

Figure 12.12 Positive ion mode MALDI-TOF MS analysis of glycans released from a recombinant antibody. The structures of the glycans assigned in the spectrum are provided in Table 12.1. Glycan structural assignments are inferred based on a correlation between the observed mass of the glycan moiety and the mass of glycans known to be expressed on CHO-derived glycoproteins.

assigned to the G0F glycan, which is actually a G0F-Na^+ adduct. The 16 Da difference in mass between the G0F-Na^+ ion and the m/z 1501.5 ion is consistent with the identification of this peak as a G0F-K^+ adduct of the G0F glycan that is present in the spectrum because the sample had a small amount of potassium in it. However, this m/z value is also consistent with a G1F-Na^+ glycan adduct that lacks a Fuc residue and would therefore be identified as a G1 glycan. Unless all of the potassium is removed from a sample prior to analysis, many of the most abundant glycans in the spectrum could become associated with potassium and the resultant ions would be misidentified as afucosylated forms of existing glycans. In the case of the spectrum acquired in Figure 12.12, the PNGase F release of the glycans was performed in a salt-free Tris buffer and prior to analysis the sample underwent spin-column cation-exchange chromatography to remove any trace cationic salts, so confident identification of the glycan associated with the m/z 1501.5 ion could be made.

Unlabeled negatively charged glycans, like glycans containing sialic acid, are analyzed in the negative ion mode. Two matrices that have proven to be very effective in the analysis of charged glycans are DHB and 2,4,6-trihydroxyacetophenone (THAP). These matrices provide good ionization in the negative ion mode and because they tend not to catalyze the loss of sialic acid from the glycans, they are considered "cool" matrices. Between the two matrices, THAP is the "cooler" matrix, as it has been demonstrated to be less likely to induce loss of sialic acid [107]. Removal of cationic salts is also important in the acquisition of sialylated glycans, since cationic salts can neutralize the negative charge on sialic acids, thereby decreasing ionization efficiency.

In assigning glycan ions in the negative ion mode, care must be taken when both NeuAc and *N*-glycolylneuraminic acid (NeuGc) are present in the glycan sample due to the possible presence of isobaric glycan species. The difference in mass between these two forms of sialic acid is 16 Da, which is due to the presence of an additional oxygen in the structure of NeuGc relative to NeuAc. As discussed previously, there is also a 16 Da difference in the mass of Gal and Fuc. This correlation in mass difference can lead to uncertainty in the identification of some glycans in negative ion spectra due to the fact that some afucosylated forms of NeuAc-containing glycans are isobaric with fucosylated forms of NeuGc-containing glycans. For example, a G2S1 non-fucosylated glycan with a NeuAc sialic acid is isobaric with a G1FS1 fucosylated glycan with a NeuGc sialic acid.

Some advantages that the mass spectrometric analysis of released glycans has over the mass spectrometric analysis of glycoproteins and glycoprotein fragments include an increased ability to detect glycans that are present at low abundance and the ability to completely resolve all glycan ions in a glycan mixture. Shown in Figure 12.12 is the positive ion spectrum of glycans released from the same model antibody that was analyzed to produce Figures 12.10 and 12.11. In the MALDI-TOF spectrum of the released glycans all of the glycan ions are baseline resolved and there are no signals observed in the spectrum that are not attributable to glycans, whereas in the ESI-MS spectrum of the reduced heavy chain not all glycoform related signals are baseline resolved, and there are signals in the spectrum that are not related to glycans that could interfere with the assessment of glycan distribution. Additionally, there are some low-abundance glycans, like the G0, G1, and G0F-GlcNAc glycans, that are observed in the MALDI-TOF spectrum but are not observed in the ESI-MS spectrum.

One disadvantage to analyzing native glycans by MALDI-TOF mass spectrometry is that not all glycans can be analyzed in a single mass spectrometric mode. The positive ion mode is very effective in assessing the distribution of neutral glycans; however, the sodium in the matrices used to enhance the signal from neutral glycans tends to suppress the ionization of sialylated glycans. The effects of this suppression can be seen in the example MALDI-TOF spectrum (Figure 12.12), where only one out of the three sialylated glycans observed in the ESI-MS analysis of the heavy chain is also observed in the MALDI-TOF spectrum of the released glycans. The signal from the sialylated G1FS1 glycan that is observed in the MALDI spectrum is the doubly sodiated ion which has a much lower relative signal in the MALDI spectrum relative to the signal in the ESI-MS spectrum. Just as the signals from the sialylated glycans are suppressed in the positive ion mode, the signals from the neutral glycans are suppressed in the negative ion mode, since neutral glycans tend not to take on a negative charge. Therefore, MALDI-TOF MS is an effective means of separately determining the glycan distribution of native neutral glycans and sialylated glycans, but it cannot be used to determine the overall glycan distribution. In order to assess the overall glycan distribution using MALDI-TOF mass spectrometry, the glycans must be chemically altered.

12.6.1.4 Determining Glycan Distribution by Mass Spectrometric Analysis of Derivatized Glycans

One approach to enabling the analysis of both neutral and sialylated glycans in a single mass spectrometric mode is to neutralize the charge on the sialic acid, thereby allowing the assessment of the overall glycan distribution in the positive ion mode. Two derivatization methods that have proven to be effective in both stabilizing sialic acids and neutralizing sialic acid charge are glycan permethylation and sialic acid esterification.

In glycan permethylation, all hydrogens on hydroxyl groups, amine groups, and carbonyl groups are replaced by methyl groups to produce a hydrophobic glycan derivative in which the charge on sialic acid residues has been neutralized. The permethylated glycans can then be analyzed in the positive ion mode, typically using a DHB matrix spiked with sodium. Permethylation not only neutralizes and stabilizes sialic acid residues thereby allowing both sialylated and neutral glycans to be observed in a single spectrum, it also significantly improves ionization efficiency and, due to the absence of hydroxyl groups, stabilizes other glycosidic bonds. The disadvantages of permethylation include the possibility of side reactions or incomplete permethylation, the requirement for sample cleanup, and the inability to use enzymatic methods in the structural characterization of the permethylated glycans.

In preparing permethylated glycans, the glycan mixture is first evaporated to dryness and then deprotonated by incubation in a dimethylsulfoxide (DMSO) solution containing finely powdered sodium hydroxide. Following complete deprotonation, the glycan is permethylated by the addition of methyl iodide to the reaction mixture. It is important to allow enough time for complete deprotonation to take place prior to the addition of the methyl iodide in order to prevent the reaction of the methyl iodide with the DMSO solvent to form a dimethylmethoxysulfonium salt that can induce oxidation of the glycan [108]. The use of a DMSO/NaOH solution containing trace amounts of water, as well as incorporating the addition of a second aliquot of DMSO/NaOH solution after the addition of the methyl iodide, has also proven to reduce the chances of glycan oxidation [109]. Once permethylated, the glycans are purified through organic extraction by adding equal volumes of water and an immiscible organic solvent, usually chloroform or methylene chloride. The organic phase containing the permethylated glycans is then evaporated to dryness, after which the purified glycans can be dissolved in methanol/water (1 : 1, v/v) in preparation for analysis.

Sialic acids can also be neutralized and stabilized through linkage-specific esterification. In this form of derivatization, released glycans are derivatized by incubation in an ethanol solution containing 1-ethyl-3-(3-dimethylaminopropyl)carbodiimide hydrochloride and hydroxybenzotriazole. This treatment results in a differential derivatization of sialic acid residues that is dependent upon the linkage of the sialic acid to the adjacent monosaccharide. If the sialic acid is attached to an adjacent Gal residue through an α2,6-linkage, then the proton of the sialic acid carboxyl group is replaced by an ethyl group to form an ethyl ester group. If the sialic acid is attached to an adjacent Gal residue through an α2,3-linkage, then the sialic acid carboxyl group interacts with the hydroxyl group in position 2 of the adjacent galactose to form a lactone [106]. Since there is a 14 Da difference in mass between the two sialic acid derivatives, sialic acid esterification not only stabilizes and neutralizes the sialic acid, thereby allowing the quantification of both neutral and sialylated glycans in a single spectrum, but it allows separate quantification of glycans containing α2,3- and α2,6-linked sialic acid residues.

Neutral and sialylated glycans can also be analyzed in a single mass spectrometric mode by derivatizing the glycans with a charged label. One commonly used label is 2-AA, which has a primary amine functional group that enables the labeling of glycans by reductive amination, and a carboxylic acid functional group that imparts a negative charge to all derivatized glycans. Once labeled, the negative charge glycans can undergo mass spectrometric analysis in the negative ion mode to quantitatively assess the total glycan distribution. One of the advantages to using the 2-AA label is that there are commercially available kits that provide for both labeling and cleanup of the labeled glycans. In addition, high-throughput methods have been developed that allow for the derivatization and cleanup of samples using a 96-well plate format [110]. Although mass spectrometric analysis is able to provide excellent resolution of glycans in complex mixtures, it is not able to resolve isomeric species with the same monosaccharide composition. When resolution of structural isomers is required, either electrophoretic or chromatographic methods can be employed.

12.6.2 Assessing Glycan Distribution by Chromatography and CE

The methods most commonly used in the biopharmaceutical industry for assessing relative glycan distribution are normal phase liquid chromatography or HILIC, high-performance anion-exchange chromatography (HPAEC), and CE. All three methods have been demonstrated to be highly sensitive and able to resolve glycan structural isomers.

12.6.2.1 Analysis of N-linked Glycans by CE

CE is a widely used means of assessing glycan distribution in recombinant glycoproteins. The separation of glycans by CE is dependent on the differential migration of charged glycans in an applied electric field. Since neutral glycans contain no inherent charge, it is necessary to alter them in a manner that will introduce a charged functional group, preferably a stable functional group that remains ionized over a wide pH range. Glycans also contain no inherent chromophore, so it is necessary to introduce a functionality that will make detection possible, preferably a fluorophore with a relatively high fluorescence quantum yield that absorbs at a wavelength close to the emission wavelength of commercially available lasers.

The fluorescent label most widely used in the CE analysis of glycans is 8-aminopyrene-1,3,6-trisulfonic acid (APTS). The APTS label contains three sulfonate groups and a single primary amine group. The sulfonate groups impart three negative charges to the labeled glycans that remain ionized over a wide pH range, and the primary amine group makes it possible to efficiently label glycans by reductive amination. In the electrophoretic analysis, labeled glycans move rapidly through the capillary, which results in high-resolution separations that require very little analysis time. With respect to the APTS fluorescent properties, APTS-labeled glycans have a broad absorbance maximum at 455 nm, with an associated emission maximum at 512 nm. Due to the broad 455 nm APTS absorbance maximum, an argon-ion laser with emission at 488 nm can be used to enable sensitive fluorescent detection of the labeled glycans. Interestingly, the APTS fluorophore itself has an absorption maximum at 424 nm with very little absorbance at 488 nm and an emission maximum at 501 nm. Based upon the differences in absorbance and emission characteristics of the APTS and APTS-labeled glycans, a typical detection system for APTS-labeled glycans contains an argon-ion laser and an emission band-pass filter of 520 ± 10 nm, which enables the preferential detection of the APTS-labeled glycans over the underivatized APTS fluorophore.

In any CE analysis of N-linked glycans, the typical workflow includes enzymatic release of the glycans using PNGase F, removal of the deglycosylated protein by precipitation or centrifugal filtration, evaporation of the supernatant or filtrate using a centrifugal vacuum evaporator, labeling the released glycans with the fluorophore by reductive amination, and finally CE analysis of the labeled glycans [111, 112]. The CE separation is performed using coated capillaries that reduce electroosmotic flow, which results in a reverse polarity electrophoretic separation in which the negatively charged labeled glycans migrate toward the anode. The separation of the labeled glycans is based upon relative electrophoretic mobility, which is a function of both hydrodynamic size and molecular charge. Thus, glycans with the same charge will be separated by differences in their hydrodynamic size. An example of this kind of separation is provided in Figure 12.13, where there is a direct correlation

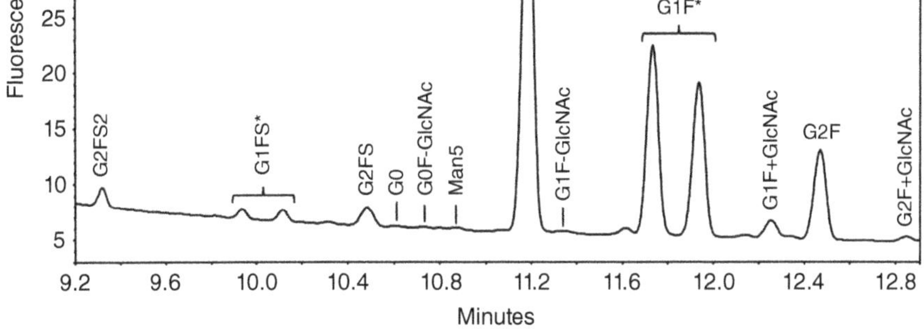

Figure 12.13 CE analysis of APTS-labeled glycans released from a recombinant antibody. The structures of the glycans assigned in the electropherogram are provided in Table 12.1. Glycan structural assignments are inferred based on a combination of experimental data, including coelution with glycan standards, enzymatic studies, and a correlation between the observed mass of the glycan moiety and the mass of glycans known to be expressed on CHO-derived glycoproteins or by coelution with glycan standards in CE-MS experiments. Refer to Section 12.8 for a general description of these methodologies. Glycan assignments were provided by Lynn A. Gennaro. *Resolved glycan structural isomers.

between the migration time of the G0F, G1F, and G2F glycans and their relative size, with the G0F reaching the detector first, followed by the larger G1F glycan, and finally followed by the largest glycan, the G2F glycan. An example of the resolving power of the CE analysis can be seen in the electropherogram as the baseline resolution between the two possible G1F isomers. These two isomers have the same monosaccharide composition and mass, but differ in the antenna that contains the terminal Gal residue. This difference in terminal galactosylation evidently confers enough difference in hydrodynamic size to allow for separation in the CE analysis.

Shown in Figure 12.13 is the CE electropherogram from the analysis of the APTS-labeled glycans released from the same model antibody that was analyzed by ESI-MS in Figures 12.10 and 12.11 and by MALDI-TOF MS in Figure 12.12. As can be seen from a comparison of the analytical profiles, the CE analysis is able to detect a greater number of glycans than was detected by either the MALDI-TOF MS of released glycans or the ESI-MS of the reduced heavy chain, with the CE analysis detecting 14 different glycans and the MALDI-TOF MS and ESI-MS analyses detecting 10 and 9 different glycans, respectively. Not only is the CE method able to detect more glycans, but unlike any of the mass spectrometric methods, the CE method is able to resolve glycan structural isomers.

Although the APTS label has proven to be very effective in the CE analysis of the simple glycan mixtures observed in the analysis of recombinant antibodies, the analysis of complex glycan mixtures containing tri- and tetra-antennary glycans can be problematic. In particular, the sialylated tri- and tetra-antennary glycans are not well resolved using the APTS label because of the number of negative charges that are present on the labeled glycans. In a CE separation of this kind, the resolution between two labeled glycans is proportional to the ratio of the charge differences to the average charge on the labeled glycans ($R \propto \Delta Z / Z_{Avg}$); therefore, the resolution between highly sialylated glycans should be better when a label containing fewer negative charges is used. One label that has proven to be superior to APTS in the analysis of sialylated glycans is 5-amino-2-naphthalenesulfonic acid (ANSA) [113]. The ANSA label contains a sulfonate group and a primary amine group. The primary amine permits glycan labeling through reductive amination and the sulfonate group imparts a single negative charge which enables the electrophoretic separation of all glycans. Since the fluorescent moiety used is naphthalene-based rather than pyrene-based, a helium–cadmium laser with an emission at 325 nm is used in place of the argon-ion laser to excite the fluorophore. In addition to increasing the resolution of sialylated glycans in CE analysis, use of the ANSA label permits both the analysis of labeled glycans by MALDI-TOF mass spectrometry and resolution of labeled glycans in high-performance liquid chromatography (HPLC), which the APTS label does not allow. Since the ANSA label imparts a negative charge to all labeled glycans, all glycans can be observed in a negative ion mode mass spectrum. By combining the ability of CE and HPLC to resolve structural isomers with the structural information obtained in MALDI-TOF MS analysis, use of the ANSA label makes it possible to fully characterize complex glycan mixtures.

12.6.2.2 Analysis of N-linked Glycans by HILIC

HILIC is the most widely used chromatographic method for determining glycan distribution in the biopharmaceutical industry. HILIC is a form of normal phase chromatography, which is a term used for chromatographic methods in which a hydrophilic stationary phase is used in combination with a hydrophobic mobile phase to separate hydrophilic analytes. Once retained, the polar analytes are then eluted from the stationary phase by gradually increasing the hydrophilic nature of the mobile phase, which results in the elution of the analytes in order of their hydrophilicity, with the least hydrophilic analytes eluting first. HILIC differs from other normal phase chromatographic methods in that the hydrophilic component used in the elution of the polar analytes is aqueous in nature, whereas in other normal phase separations nonaqueous solvents are used.

The HILIC separation mechanism is a little more complex than a simple direct interaction between the analyte and the stationary phase. The separation involves the partitioning of analytes between the hydrophobic mobile phase and a hydrophilic aqueous layer that forms at the surface of the hydrophilic stationary phase. In addition to retention based upon aqueous partitioning, direct interaction between analytes and the stationary phase can lead to retention that stems from hydrogen bonding or electrostatic interaction, depending on the nature of the stationary phase [114]. Common HILIC stationary phases include silica-based columns containing amine, amide, diol, or sulfobetaine zwitterionic functionalities [115].

The determination of glycan distribution by HILIC is most often accomplished through the separation of 2-AB-labeled glycans using an amide-bonded stationary phase. The primary amine group of the 2-AB label enables derivatization of released glycans by reductive amination, and the inherent fluorescence of the label allows the sensitive detection of glycans using excitation and emission wavelengths of 330 and 420 nm, respectively. As with CE analysis, sample preparation generally proceeds by releasing N-linked glycans by PNGase F treatment, removing the deglycosylated protein by precipitation or centrifugal filtration, evaporation of the supernatant or filtrate using a centrifugal vacuum evaporator, and labeling the released glycans with the fluorophore by reductive amination. An additional purification step is generally performed prior to analysis to remove excess labeling reagents. This purification step usually involves some form of HILIC solid phase extraction or gel filtration; 2-AB labeling and cleanup kits are commercially available.

The use of buffers in the HILIC separation can substantially affect the separation of 2-AB-labeled glycans using an amide stationary phase. If the

HILIC separation is carried out below a pH of ~4.4, then the separation on the amide column is based upon both hydrogen bonding and a weak anion-exchange mechanism. If the separation is performed above ~ pH 4.4, then there is decreased retention of sialylated glycans, which can lead to a decrease in overall glycan resolution. The best resolution of glycans is obtained when the HILIC separation relies solely upon hydrophilicity/hydrogen bonding and when the resolution of sialylated forms is maximized; therefore, use of a buffer that can maintain a pH of about 4.4 is optimal [116]. The concentration and type of buffer salts used in the separation can also affect chromatographic resolution. Since it is preferable to use buffers that are amenable to mass spectrometric analysis to aid in glycan identification, volatile salts are typically used in the preparation of HILIC mobile phases. The salts used in the buffer must also be soluble in the hydrophobic mobile phase component used in the separation. Two volatile buffer salts that have the necessary solubility and have pK_a values that enable maintenance of pH in the desired range are ammonium formate and ammonium acetate. Of the two buffer systems, use of the ammonium formate buffer provides better resolution of sialylated glycans than the ammonium acetate buffer, so the use of the ammonium formate buffer is preferable when the glycan mixture contains sialylated glycans. Additionally, formate is the more volatile of the two ammonium counterions and so is better suited for mass spectrometric analysis using ESI. Buffer concentration also affects glycan resolution in HILIC-based separations, with greater suppression of sialic acid ionic interactions associated with increasing buffer concentration [116]. A common mobile phase combination is 100 mM ammonium formate, pH 4.5 (solvent A), and acetonitrile (solvent B).

In a typical HILIC separation of 2-AB-labeled glycans, the initial mobile phase contains 70–75% acetonitrile. The high organic content of the mobile phase promotes the partitioning of the labeled glycans into the hydrophilic aqueous layer surrounding the stationary phase. As the gradient proceeds and the mobile phase becomes more hydrophilic, the glycans elute from the column in order of their polarity, which in glycan analysis is related to the number and type of monosaccharides present. As evidenced by the ability of HILIC to resolve glycan isoforms, the separation is also influenced by the orientation of the component monosaccharides with respect to each other. The introduction of ultra-high-performance liquid chromatography (UHPLC) systems and UHPLC HILIC columns has both substantially increased chromatographic resolution of glycans and significantly decreased analysis time [117].

The chromatographic profile from the HILIC analysis of 2-AB-labeled glycans released from the same model antibody used in the ESI-MS, MALDI-TOF MS, and CE analyses is provided in Figure 12.14. The chromatographic analysis was performed using an ACQUITY UHPLC system equipped with a 2.1 mm × 150 mm ACQUITY UHPLC BEH column packed with a 1.7 μm amide-based stationary phase using mobile phase and separation conditions

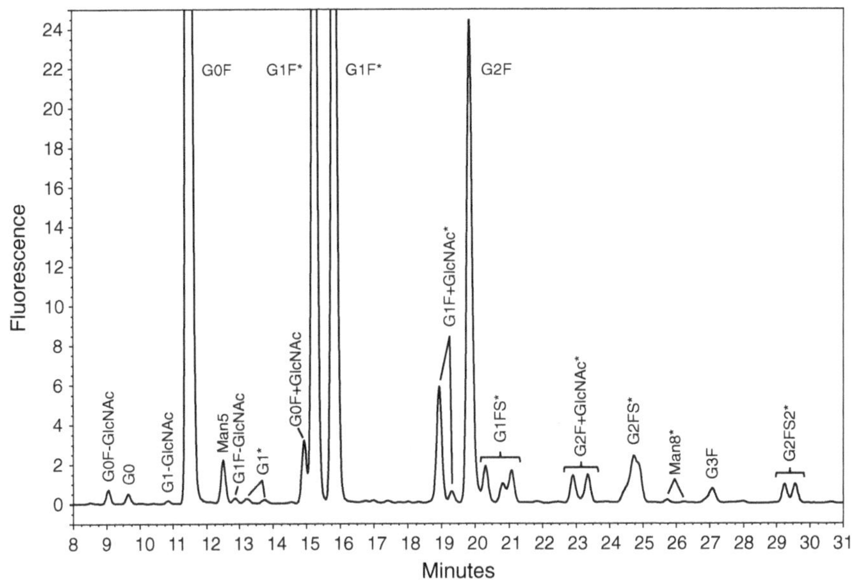

Figure 12.14 HILIC analysis of 2-AB-labeled glycans released from a recombinant antibody (expanded view). The structures of the glycans assigned in the chromatogram are provided in Table 12.1. Glycan structural assignments are inferred based on a combination of experimental data, including coelution with glycan standards, enzymatic studies, and a correlation between the observed mass of the glycan moiety and the mass of glycans known to be expressed on CHO-derived glycoproteins or by coelution with glycan standards in CE-MS experiments. Refer to Section 12.8 for a general description of these methodologies. Glycan profile and structural assignment provided by Tomasz K. Baginski. *Resolved glycan structural isomers.

similar to those described elsewhere [117]. As indicated by the analytical profiles, the HILIC separation has the greatest glycan resolution, with 25 different glycans resolved in the chromatography as compared with the 14 glycans resolved in the CE analysis, the 9 glycans resolved in the ESI-MS analysis, and the 10 glycans resolved in the positive mode MALDI-TOF MS analysis. These results are consistent with other studies that indicate that between ESI-MS analysis of the antibody heavy chain, MALDI-TOF analysis of released glycans, CE analysis of APTS-labeled glycans, and HILIC analysis of 2-AB-labeled glycans, the method that provides the greatest sensitivity and resolution of glycans is HILIC analysis [99, 118].

12.6.2.3 Determining Glycan Distribution by HPAEC

Another sensitive, high-resolution chromatographic method that is commonly used for the relative quantification of glycoprotein glycans is HPAEC-PAD of released, unlabeled glycans. One of the advantages of using HPAEC-PAD in determining glycan distribution is that it does not require the labeling of released glycans prior to analysis. As was described in Section 12.3.1, HPAEC is an anion-exchange chromatographic separation that is performed using sodium hydroxide-based mobile phases at pH ≥ 12. The hydroxyl groups of monosaccharides have pK_a values that range between 12 and 14, so by conducting the separation in mobile phases with a pH of 12 or higher, the hydroxyl groups of glycans become ionized and interact with the anion-exchange stationary phase. A typical mobile phase used in the HPAEC separation of glycans contains 100 mM sodium hydroxide. Glycans are generally eluted from the HPAEC column by applying a sodium acetate gradient.

The stationary phases used in HPAEC must be made of materials that have a sufficient chemical stability to be able to support the use of high-pH mobile phases. HPAEC stationary phases are polymer-based pellicular columns with high mechanical and chemical stability that use quaternary ammonium functionalities in the separation of ionized glycans. The most commonly used columns are the Dionex CarboPac PA100 and CarboPac PA200 columns. The use of high-pH mobile phases also calls for a specialized chromatographic system. These specialized chromatographic systems have completely metal-free polyether ether ketone (PEEK) flowpaths that eliminate the possibility of the metal contamination that can be observed in typical stainless-steel chromatographic systems. The absence of metals in the mobile phase is important because metals can contaminate both the column and the PAD working electrode, which usually results in the need to replace the contaminated component. The high pH of the mobile phase is not only required for the ionization of the glycan hydroxyl groups, but it is essential for the detection of the glycans by PAD. In pulsed amperometric detection, the electrocatalytic oxidation of weakly acidic sugars at a gold electrode is measured. Efficient electrocatalytic oxidation of glycans requires that the pH be maintained at or above pH 12.

In HPAEC-PAD analysis, glycans are separated based upon their charge, with neutral glycans eluting first followed by sialylated glycans. Sialylated glycans elute in order of the number of sialic acid residues present. Within glycan groups containing the same number of sialic acids, the glycans tend to elute in order of size, with the larger glycans having greater retention times. As in HILIC separations, the separation is also influenced by the orientation of the component monosaccharides with respect to each other, so HPAEC-PAD is capable of separating glycan isomers that have the same size and composition, but different monosaccharide linkage(s) [119, 120]. With respect to glycan resolution, one study on the analysis of antibody glycans suggests that the resolution observed in HPAEC-PAD analysis is on par with that observed in HILIC analysis of 2-AB-labeled glycans [118]. Although the resolution obtained in HPAEC-PAD analysis of glycans is comparable to that observed in HILIC analysis, HPAEC-PAD analysis requires specialized instrumentation, so HILIC analysis tends to be preferred over HPAEC-PAD analysis. Additionally,

HILIC analysis is more amenable to online mass spectrometric characterization of the separated glycans than HPAEC-PAD analysis.

When determining glycan distribution using HPAEC-PAD there are a number of caveats to be aware of. One factor to be cognizant of is the potential for glycan epimerization. Glycan epimerization is a phenomenon in which the conformation at the C2 position of the reducing terminal GlcNAc residue undergoes rearrangement in the presence of a base to form ManNAc. As with other glycan isomeric forms, the ManNAc-containing glycans are separated from the native GlcNAc-containing glycans in HPAEC analysis. Thus, there is the potential to have two peaks in an HPEAC analysis representing a single glycoprotein glycan. If the peak associated with the ManNAc epimer coelutes with another glycan in the chromatographic separation, then this would introduce error into the determination of glycan distribution. In HPAEC glycan separations the mobile phase typically contains 100 mM NaOH. Studies have shown that within 1 h at room temperature the equilibrium for epimerization of GlcNAc to ManNAc is reached in 100 mM NaOH, so it is beneficial to keep retention times to a minimum in the HPAEC analysis of glycans [30]. Another factor to consider in HPAEC-PAD analysis are reports that indicate that electrocatalytic oxidation at the gold electrode in pulsed amperometric detection is dependent upon the structure of the glycan [120, 121]. In the report by Townsend, the molar responses observed for different glycans varied from 0.2 to 3.2 depending on the structure of the glycan. This kind of variability could lead to the introduction of significant error into the determination of glycan distribution. Although it is known that glycan structure can affect PAD response, some studies suggest that in the analysis of structurally similar N-linked glycans the differences in response factors can be negligible [118, 122].

12.7 Comparison of Methods Used in Determining Glycan Distribution

The methods for determining glycan distribution that have been described in this chapter each have their strengths and weaknesses, some of which are outlined in Table 12.2. The "best method" to use will depend on the purpose of the analysis. If the purpose of the analysis is bioprocess monitoring for protein glycosylation, then the preferred method would be one with little sample manipulation and a rapid turnaround time that would provide sufficient information about the glycan distribution to be able to evaluate product quality. In such cases the preferred method would likely involve mass spectrometric analysis of the intact glycoprotein or a glycoprotein fragment. Although this mass spectrometric analysis would not provide the sensitivity necessary to detect some minor glycoforms that might be present, and some protein variants can interfere with the accurate quantification of glycoforms due to lack of

Table 12.2 Comparison of common methods used in determining glycan distribution.

Attribute	ESI-MS	MALDI-TOF MS	MALDI-TOF MS	APTS CE	2-AB HILIC	HPAEC-PAD
Analyte	Glycoprotein or glycoprotein fragment	Glycans	Derivatized glycans	APTS-labeled glycans	2AB-labeled glycans	Glycans
Turnaround time (9 samples)	1 day	2 days	2 days	2 days	2 days	2 days
Detection of nonglycosylated glycoforms	Yes	No	No	No	No	No
Interference by protein variants	Yes	No	No	No	No	No
Quantification of both neutral and charged glycans	Yes	No	Yes	Yes/no[a]	Yes	Yes
Resolution	Moderate	Very high	Very high	High	High	High
Isomer separation	None	None	None	Very good	Excellent	Excellent
Detection sensitivity	Moderate	High[b]	High	High	Very high	Very high
Often implemented in quality control	No	No	No	Yes	Yes	No

[a] The APTS CE method cannot be used in the quantification of highly sialylated glycans due to the inability to resolve highly sialylated glycans.

[b] Neutral glycans can be quantified in the positive ion mode and sialylated glycans can be quantified in the negative ion mode, but the total distribution of both neutral and sialylated glycans cannot be determined when analyzing underivatized glycans.

resolution, the information gleaned from the analysis is sufficient to be able to determine whether the process is delivering materials with the desired glycosylation pattern. On the other hand if the purpose of the analysis is to get an accurate assessment of the total glycan distribution with as much structural detail as possible, then a chromatographic or electrophoretic method that is capable of resolving structural isomers would be the method of choice. In cases where extended characterization is required, the best option is to use a combination of orthogonal methods to get a complete picture of glycoprotein glycosylation with confirmatory and complementary data from various sources.

12.8 Assessing N-linked Glycan Structure

12.8.1 Characterization of Glycan Structure Using Standards and Enzymatic Studies

In the quantification of glycans using chromatographic and electrophoretic techniques, the quantification is inextricably linked to structural characterization of the glycans being quantified. One means of assessing the structural identity of a glycan that is separated in a chromatographic or electrophoretic analysis is through the use of glycan standards whose structures have been fully characterized. If it can be established that a glycan in a glycan mixture coelutes or comigrates with a glycan standard, then there is evidence that the coeluting/comigrating glycan in the glycan mixture has the same structure as the glycan standard. One limitation to the use of standards in identifying glycan structure is that there are a limited number of commercially available standards. Another limitation is that although a correlation between retention time or migration time is suggestive of the structure of the glycan in question, it is not confirmatory of the glycan structure. The reliability of this kind of structural assignment will depend upon the resolution of the chromatographic or electrophoretic method used in the analysis. To confirm the purported identity of the glycan, further structural analysis is required. One means of confirming glycan structure is through the use of structure-specific enzymatic studies.

In the enzymatic characterization of glycans, an aliquot of a glycan mixture is digested with an exoglycosidase that is specific for a particular monosaccharide that is present at the nonreducing terminus of glycans in a particular anomeric linkage. Aliquots of both the digested and nondigested glycan mixture are then analyzed chromatographically or electrophoretically. A change in the chromatographic profile or electrophoretic profile indicates that the glycosidase-specific linkage was present at the nonreducing terminus of all glycans whose retention time was altered. Through sequential treatment with a series of exoglycosidases, the sequence and linkage of the constituent monosaccharides can be elucidated. Alternatively, multiple combinations of exoglycosidases can be added to separate aliquots of the glycan mixture to establish glycan structure [112, 123, 124].

Enzymatic studies can also be used in the mass spectrometric characterization of glycan structure, where confirmation of monosaccharide sequence and linkage is derived from a shift in glycan mass following enzymatic treatment. Exoglycosidase digestion can be a very powerful means of assessing glycan linkage in that, other than NMR, it is the only means of determining both the position and anomericity of monosaccharide linkages. However, the utility of the method is limited by the relatively few highly specific exoglycosidases that are commercially available.

12.8.2 Characterization of Glycan Linkage by Methylation Analysis

Another means of characterizing monosaccharide linkages is through the analysis of partially methylated alditol acetates (PMAAs) by gas chromatography–mass spectrometry (GC-MS). This type of analysis is commonly referred to as methylation analysis. Due to protein interference, glycans must first be released and separated from the protein before derivatization. In preparing glycan PMAAs, the isolated glycan is first evaporated to dryness. The dried glycan preparation is then deprotonated by incubation in a DMSO solution containing finely powdered sodium hydroxide. Once the deprotonation is complete, the glycan is methylated by addition of methyl iodide to the reaction mixture. In the methylation procedure it is important to allow enough time for complete deprotonation of the hydroxyl groups of the glycan prior to adding the methyl iodide. Complete deprotonation of the hydroxyl groups promotes a rapid reaction of the glycan with the methyl iodide, which not only enhances full methylation, but prevents the reaction of the methyl iodide with DMSO to form a dimethylmethoxysulfonium salt that can alter the structure of the carbohydrate by oxidation [108]. The use of a NaOH/DMSO solution containing trace amounts of water as well as the addition of a second aliquot of NaOH/DMSO solution following the methyl iodide addition has also been recommended to eliminate oxidative degradation [109]. After extracting the permethylated glycans into an organic phase by partitioning between an aqueous and an immiscible organic solvent, usually chloroform or methylene chloride, the sample is evaporated to dryness and then subjected to acid hydrolysis. Alternatively, the permethylated samples can be purified by solid phase extraction using a reversed-phase column or cartridge [125]. The partially methylated monosaccharides are then reduced to alditols by incubation with sodium borodeuteride and acetylated by incubation with acetic anhydride. The PMAAs that are produced by this series of reactions contain monosaccharides that are O-methylated at all hydroxyl groups that were not involved in linkage to another monosaccharide, deuterated at the C-1 or anomeric carbon, and O-acetylated at all hydroxyl groups that were involved in linkage to another monosaccharide and at the hydroxyl groups formed at the C-1 and C-5 positions following reduction. GC-MS is used to characterize the PMAAs with each PMAA derivative giving a characteristic elution time and fragmentation pattern that allows unambiguous identification of the derivatized monosaccharides resolved in the gas chromatographic separation. The gas chromatographic separation is generally conducted using fused silica capillary columns, a helium carrier gas, and a temperature gradient. Fragmentation of the PMAA derivatives is accomplished either by bombarding the derivatives with energetic electrons using electron impact ionization (EI) or by energetic collision with an inert reagent gas, such as methane, using chemical ionization (CI). Figure 12.15 provides an overview of the preparation of PMAAs and examples

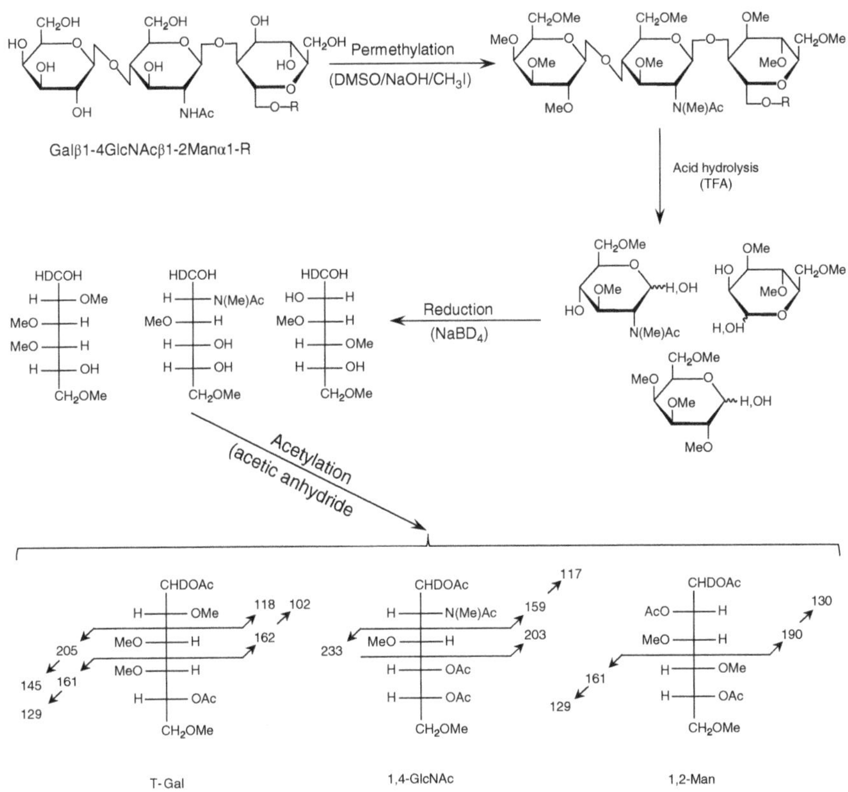

Figure 12.15 Preparation and GC–MS analysis of glycan partially methylated alditol acetates (PMAAs). Glycan PMAAs are produced by sequential permethylation, acid hydrolysis, reduction with sodium borodeuteride ($NaBD_4$), and acetylation. Provided are the representations of the derivatives produced in the preparation of PMAAs from the Galβ1-4GlcNAcβ1-2Manα1 sequence commonly observed in mammalian glycans. Also shown are the masses of some of the characteristic fragment ions observed in the GC–MS analysis of the PMAAs (bottom panel). Ions produced by secondary fragmentation are often observed which result from the loss of formaldehyde ($m/z = 30$), methanol ($m/z = 32$), or acetic acid ($m/z = 60$) from the primary fragment ions.

of the types of fragmentation observed in the GC-MS analysis. Confirmation of the identities of the separated PMAA derivatives is typically accomplished by comparison of the observed elution times and fragmentation patterns to those of a PMAA standard containing the relevant derivatives.

12.8.3 Characterization of Glycan Structure by MS^2

Mass spectrometry is a very powerful and sensitive means of assessing glycan structure. From the mass of a glycan and a knowledge of glycan core structures and structural motifs, it is possible to deduce glycan monosaccharide composition and, to some extent, the overall structure of the glycan. An understanding of the kinds of glycosytransferases expressed by the host cell used in the production of a recombinant glycoprotein can also inform the assessment of glycan structure. Although simply knowing the mass of the glycan and the glycosyltransferase repertoire of the host cell can greatly assist in suggesting potential glycan structures, due to the existence of isobaric structural isomers, confirmation of the structure a glycan requires further analysis. One means of confirming the monosaccharide sequence and linkage of a glycan is through the use of MS^2.

The most common mass spectrometric means of obtaining glycan structural information is by electrospray ionization–tandem mass spectrometry (ESI-MS^2) in which glycans are fragmented through energetic collisions with a non-reactive gas, such as argon, helium, or nitrogen. This mode of fragmentation is referred to as collision-induced dissociation (CID). The primary kinds of cleavage observed in CID spectra are glycosidic cleavages, in which glycosidic bonds between the monosaccharides in a glycan are cleaved, and cross-ring cleavages, in which bonds within a monosaccharide are cleaved. Glycosidic cleavages provide information about monosaccharide composition and sequence, whereas cross-ring cleavages provide information about the linkage positions of constituent monosaccharides. Of the two primary kinds of fragmentation, cross-ring cleavages are typically much less abundant due to the high energy needed to cleave two covalent bonds. Internal fragments in which a fragment is generated from the cleavage of residues from both the reducing and nonreducing terminus of the glycan can also be observed. Examples of the types of fragmentation observed in CID spectra and the nomenclature associated with glycan fragmentation are provided in Figure 12.16.

MS^2 analysis of glycans has most commonly been performed in the positive ion mode using either native or permethylated glycans. One advantage to analyzing permethylated glycans is that permethylation decreases the intermolecular hydrogen bonding between glycans, which increases sample volatility and thereby increases ionization efficiency. Furthermore, permethylation both eliminates the negative charge associated with sialic acids and stabilizes sialic acid residues, which further enhances the signal of sialylated glycans in

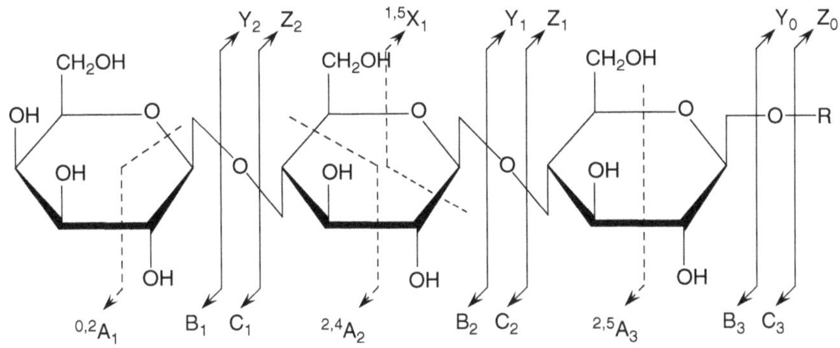

Figure 12.16 Glycan fragmentation nomenclature. Glycan fragments are labeled using the nomenclature proposed by Domon and Costello [126]. A, B, and C ions are fragment ions that retain the nonreducing terminus of the glycan, whereas X, Y, and Z ions are fragment ions that retain the reducing terminus of the glycan. A and X ions are ions produced by cross-ring cleavage of two bonds within a constituent monosaccharide. Subscript numbers denote the glycosidic bond cleaved or the monosaccharide involved in cross-ring cleavage, with reducing terminal fragments (A, B, and C ions) starting with the number 0 and nonreducing terminal fragments (X, Y, and Z ions) starting with the number 1. Superscript numbers on A and X fragment nomenclature denote the bonds within the monosaccharide ring that are cleaved to generate cross-ring cleavage, with the bonds labeled 0 through 5 starting with the bond between the oxygen and the anomeric carbon.

the positive ion mode. Permethylated glycans have also been demonstrated to produce more diagnostic fragment ions in MS^2 experiments than underivatized glycans, including a greater number of ions resulting from cross-ring cleavages. Some disadvantages to analyzing permethylated glycans include increased sample requirements due to potential losses during sample cleanup, the potential for glycan degradation during the derivatization process, and the inability to use highly resolving chromatographic or electrophoretic methods to separate permethylated glycans for online MS^2 analysis.

The MS^2 analysis of $[M+H]^+$ protonated glycans by positive ion ESI-MS results in a very simple fragmentation pattern that tends to consist primarily of glycosidic cleavages with few, if any, diagnostic cross-ring cleavages, so typically only the monosaccharide composition and sequence can be determined. However, care must be taken in interpreting $[M+H]^+$ glycan fragmentation, because monosaccharide rearrangement and internal residue loss have been reported [127, 128]. These monosaccharide rearrangements and losses can be avoided by analyzing the $[M+Na]^+$ sodium adducts of glycans rather than the protonated forms. An additional advantage to analyzing the sodiated forms is that there is an increased likelihood of observing cross-ring cleavages [129–131]. Glycan adducts formed using divalent cations, like $[M+Ca]^{2+}$, have been shown to provide better glycan ionization than the corresponding sodium adducts, while maintaining the glycan sequence integrity and the ability to induce cross-ring cleavages observed with the sodium adducts [132]. One limitation to the mass spectrometric analysis of glycan cation adducts is that, since adduct formation is required prior to introduction of the glycans into the source, the technique is not conducive to online analysis coupled to a chromatographic or electrophoretic separation. Additionally, even under the best conditions, cross-ring fragmentation is rarely observed in positive ion spectra outside of the reducing terminal GlcNAc residue, so little monosaccharide linkage information is obtained.

An appealing alternative to positive ion tandem ESI-MS analysis is negative ion tandem ESI-MS analysis. Negative ion tandem ESI-MS has several advantages over positive ion tandem ESI-MS including the observation of more cross-ring cleavages in negative ion CID spectra, the observation of simpler fragmentation patterns with more diagnostic ions than in positive ion spectra, and the absence of the monosaccharide rearrangements and internal residue losses associated with the analysis of $[M+H]^+$ protonated glycans in the positive ion mode. As in the positive ion tandem ESI-MS analysis of glycans, the best results are obtained when glycan adducts are analyzed. Negative ion ESI-MS analysis of deprotonated $[M-H]^-$ glycans results in substantial in-source fragmentation with accompanying loss of signal. However, studies have shown that when certain anionic glycan adducts are analyzed, the in-source fragmentation is eliminated resulting in a substantial increase in signal intensity. Among the anionic species shown to produce stable glycan adducts are chloride, bromide, iodide, nitrate, sulfate, and phosphate, with nitrate adducts providing the greatest overall signal intensity. Unlike positive ion CID spectra where glycosidic B- and Y-type fragment ions predominate, the dominant ions in negative ion CID spectra are A-type cross-ring and C-type glycosidic ion fragments [133, 134]. As in the positive ion mode analysis of glycan adducts, the analysis of glycan anion adducts in the negative ion mode is not conducive to online chromatographic or electrophoretic mass spectrometric analysis, because adduct formation is required prior to the introduction of the glycans into the source. The increased cross-ring cleavage and heightened presence of fragment ions that can be unambiguously assigned to glycan structural elements make negative ion CID preferable to positive ion CID in the mass spectrometric characterization of glycan structure.

As in the analysis of glycans in the positive ion mode, it is recommended that glycans containing sialic acid should be derivatized to stabilize sialic acid residues. However, permethylation is not recommended for negative ion mode analysis, since glycan hydroxyl groups can no longer be deprotonated to produce negatively charged glycans following permethylation. One means of stabilizing sialic acid residues while allowing hydroxyl group ionization is to selectively methylate the carboxyl group of sialic acids to form methyl esters [135, 136]. Once derivatized in this manner, the sialylated glycans fragment in much the same manner as neutral glycans. One caveat to the analysis of sialylated glycan methyl esters is that some of the diagnostic fragment ions observed in CID spectra lose methanol during the fragmentation process. Another method for stabilizing sialic acids is by incubating them with 4-(4,6-dimethoxy-1,3,5-triazin-2-yl)-4-methylmorpholinium chloride (DMT-MM) in methanol. In reacting sialic acids with DMT-MM in methanol, α2,6-linked sialic acids are methylated at their carboxyl group to form methyl esters, whereas the carboxyl groups of α2,3-linked sialic acids react with hydroxyl groups on the adjacent galactose residues to form lactones. As a result of the differential derivatization of α2,3- and α2,6-linked sialic acids, glycans with α2,6-linked sialic acids increase in mass by 14 Da for each α2,6-linked sialic acid and glycans with α2,3-linked sialic acids decrease in mass by 18 Da for each α2,3-linked sialic acid, which results in a 32 Da difference in mass between glycans containing the two forms of derivatized sialic acid. Thus the derivatization not only stabilizes sialic acid residues, thereby increasing the abundance of some of the diagnostic fragments in negative ion mode spectra, but it also makes it possible to differentiate between glycans containing α2,3- and α2,6-linked sialic acids. Another advantage that DMT-MM derivatization has over the selective methylation method is that it is a cleaner derivatization with fewer side products. One caveat to the use of the DMT-MM derivatization is that while the methyl ester derivatives are stable, the lactones slowly decompose when dissolved in water [137].

Of course potentially the most informative means of characterizing glycan structure by MS^2 are methods that couple a high-resolution separation to mass spectrometric analysis. In hyphenated mass spectrometric techniques, like LC-MS and capillary electrophoresis–mass spectrometry (CE-MS), the ability of the chromatographic or electrophoretic separation to resolve structural isomers is combined with the ability of the mass spectrometer to provide structural information.

The mass spectrometric assignment of glycan structure to labeled glycans separated in a CE-MS analysis is complicated by a number of factors. One complicating factor is the relatively low concentration of the glycans entering the source of the mass spectrometer. The glycan concentration is limited by the low solute concentrations and volumes allowed by the CE separation, as well as the requirement for a postcolumn addition of a sheath liquid to supplement the CE flow rate in order to support mass spectrometric analysis. Another complicating factor is the requirement to modify the electrophoretic system in order to accommodate mass spectrometric analysis. Typical modifications include increasing the length of the capillary and changing the composition of the running buffer, both of which can affect the overall glycan separation in the CE analysis. To ensure that changes in the glycan separation do not lead to misidentification of the glycans in the standard CE separation, it is recommended that a detector be introduced into the capillary system just prior to the mass spectrometric interface so that glycan relative peak areas can be used to correlate peak assignments between the CE and CE-MS separations [138]. Due to the low sample concentration, sample dilution, and ionization efficiency considerations inherent in the CE-MS analysis, acquisition of CID spectra is typically not possible. As a result, in most cases glycan identification relies solely on the mass of the glycan derived from the observed mass spectrum. To confirm glycan structure and differentiate glycan isomers, supplemental characterization, including enzymatic studies, is typically performed.

The two most commonly used chromatographic methods for the separation of glycans are HPAEC and HILIC. In HPAEC analysis, a typical mobile phase contains 50–100 mM sodium hydroxide, which is often supplemented with a sodium acetate gradient to elute the glycans. The high salt content of the mobile phase is not compatible with the ESI used in ESI-MS analysis. Consequently, separated glycans must be desalted prior to mass spectrometric analysis. Online HPAEC-MS can be performed by the introduction of a carbohydrate membrane desalter between the column and the ESI source of the mass spectrometer. The carbohydrate membrane desalter exchanges sodium ions in the column eluent with hydronium ions, which results in the conversion of the sodium hydroxide and sodium acetate in the mobile phase to water and acetic acid, both of which are compatible with ESI-MS analysis. The incorporation of the desalter into the HPAEC system has in some cases been reported to enable the determination of both glycan mass and sequence through the acquisition of CID spectra [139]. However, supplemental characterization is generally required to determine glycan monosaccharide linkages.

In the analysis of glycans by hydrophilic interaction chromatography, fluorescently labeled glycans are separated based upon their relative polarity, with more hydrophilic glycans being retained to a greater extent and eluting later than less hydrophilic glycans. HILIC mobile phase components typically consist of volatile salts and organic solvents, which are amenable to ESI used in online LC-MS analysis. As a result, no desalting or alteration of the chromatographic system is required to enable mass spectrometric analysis. The relative compatibility of HILIC separations with mass spectrometric analysis has enabled not only the generation of MS^2 spectra containing characteristic glycan fragment ions, but it has also allowed the generation of MS^3 spectra resulting from CID fragmentation of the initial glycan fragment ions [140]. Together the MS^2 and MS^3 spectra provide a more complete picture of glycan monosaccharide sequence and linkage.

Assigning glycan structure based upon tandem mass spectra can be challenging due to the relatively large number of theoretical fragment ions that can be generated in glycan CID fragmentation. Fortunately, a number of different software tools have been developed to help interpret glycan fragmentation patterns. These programs can provide proposed glycan structures based upon the input of fragment ion m/z values, or they can provide proposed fragment ion *m*/*z* values based upon the input of glycan structure. GlycoFragment, GlycoSearchMS, and GlycoWorkbench are examples of free programs that are readily available online. The GlycoFragment (http://www.glycosciences.de/tools/) and GlycoSearchMS (http://www.glycosciences.de/database/) programs are online tools [141], while the GlycoWorkbench software (http://download.glycoworkbench.org/) can be downloaded for use on any computer [142]. Other commercial and free online software is also available to help with glycan mass spectrometric analysis [143, 144].

Although mass spectrometric analysis can be used to establish monosaccharide sequence and linkage positions, it cannot provide information about linkage anomericity. The two means of establishing the anomeric configuration of linkages are through enzymatic studies using exoglycosidases of known specificity or through the analysis of the glycan by NMR spectroscopy.

12.8.4 Characterization of Glycan Structure by NMR

NMR spectroscopy is a very powerful tool for assessing glycan structure. It is the only analytical technology that is capable of full structural characterization of glycans, including the constituent monosaccharide sequence, linkage positions, and anomericity. When used, NMR analysis is usually the first analysis performed on glycans, since it is nondestructive and therefore allows the sample to be used in other analyses after the NMR analysis is complete. While NMR analysis is a very powerful tool for glycan structural characterization, it also has sample limitations that restrict its widespread routine use, including the requirement for complete glycan isolation and the requirement of milligram quantities of glycan for complete structural assignment.

Glycan structural characterization by NMR is accomplished using both one-dimensional (1D) and two-dimensional (2D) ^{1}H and ^{13}C NMR experiments, with glycans dissolved in D_2O. In the 1D ^{1}H spectra of N-linked glycans, the protons have characteristic resonances that can be used to determine the number, type, and linkage of the monosaccharides that constitute the glycan. An example of a glycan 1D ^{1}H spectrum is provided in Figure 12.17. Most of the ring protons of the constituent monosaccharides resonate in a crowded region of the ^{1}H spectrum between 3.3 and 4.0 ppm. Due to the close proximity of the resonances in this "bulk" region, it is difficult to assign structural elements to glycans using these signals. However, in other regions of the ^{1}H NMR spectrum proton signals are better resolved and key information about glycan structure can be derived. The resonances within these less crowded regions that provide critical information about glycan structure have been termed "structural-reporter group" signals [146]. The structural-reporter group signals of N-linked glycans include the signals from the H-1 anomeric, Man H-2 and H-3, Gal H-3, GlcNAc and NeuAc *N*-acetyl groups, NeuAc H-3 axial and H-3 equatorial, Fuc H-5, and Fuc methyl group protons. The Man H-2 and

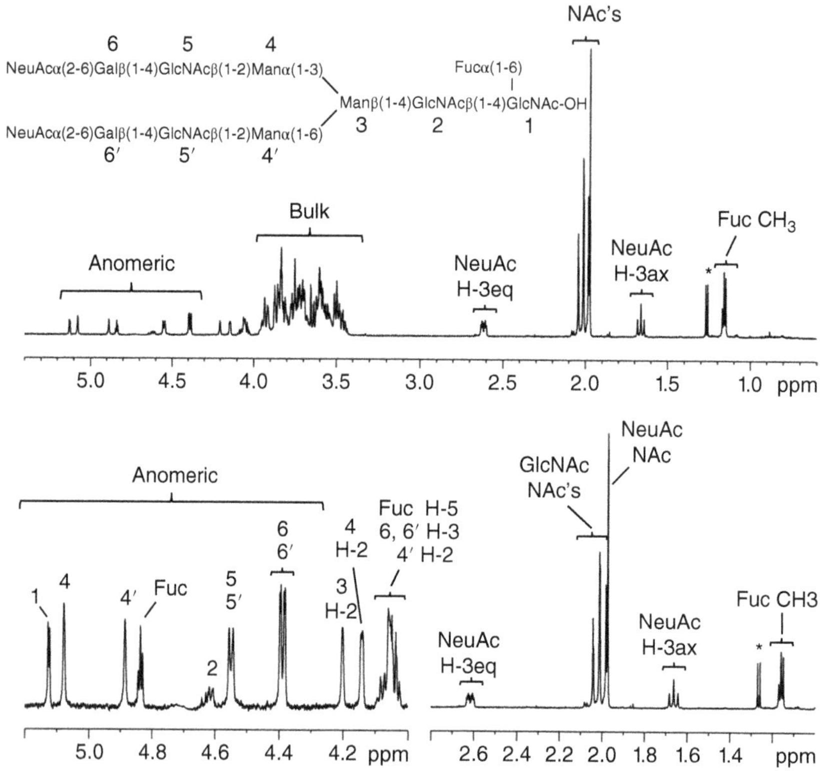

Figure 12.17 ^{1}H NMR spectrum of a fucosylated fully sialylated biantennary glycan. Provided are the full spectrum (top) and expanded views of the anomeric/H-2 region (bottom left) and upfield region (bottom right) of the spectrum. Anomeric and Man H-2 proton signals are labeled based upon the numbering system provided in the glycan structure provided. The Fuc H-5 and Gal H-3 proton signals are also observed in the Man H-2 region and overlap with the signal from the α1,6-linked Man H-2 signal. * Methyl protons from isopropanol in the glycan sample. Spectral assignments are based upon published assignments from the ^{1}H NMR analysis of fully sialylated biantennary glycans [16, 145, 146]. Spectral acquisition and consultation provided by Ken Skidmore.

H-3, Gal H-3, and Fuc H-5 proton resonances can be found in a relatively uncrowded part of the ^{1}H NMR spectrum between 4.0 and 4.3 ppm and can be used to assess the Man, Gal, and Fuc content of the glycan. Further assessment of the Fuc content of the glycan can be obtained by the observation of Fuc methyl protons with resonances between 1.1 and 1.3 ppm.

All of the anomeric proton signals are observed in the ^{1}H NMR spectrum in the chemical shift range between 4.3 and 5.5 ppm, with α-anomeric signals observed between 4.8 and 5.5 ppm and β-anomeric protons observed between 4.3 and 4.8 ppm. The signals from α- and β-anomeric protons can also be distinguished based upon their coupling constants, which are measured as the difference, in Hz, between the doublet peaks that are characteristically observed in the signals of anomeric protons. The coupling constant value is dependent on the orientation of the anomeric proton relative to the H-2 proton of the monosaccharide. For GlcNAc and Gal residues with H-2 protons in an axial orientation, the axial–axial orientation of the anomeric and H-2 protons that results from a β-anomeric linkage is associated with a $J_{1,2}$ coupling constant of 7–9 Hz, whereas an α-anomeric linkage with an equatorial–axial orientation between the anomeric and H-2 protons has $J_{1,2}$ values of 2–4 Hz. Fuc residues, which in N-linked glycans are invariably in α-anomeric linkages, have coupling constants that fall within the range observed for α-linked Gal or GlcNAc. Man residues that contain an H-2 proton that is in an equatorial configuration have $J_{1,2}$ coupling values of around 1.6 Hz in α-anomeric linkages and $J_{1,2}$ coupling values of around 0.8 Hz for β-anomeric linkages. The relatively small $J_{1,2}$ coupling constants associated with Man residues can at times result in not resolving the doublet peaks commonly observed in anomeric proton resonances [147]. Integration of the signals in the anomeric region can provide an estimate of the number of neutral monosaccharides in a glycan, as well as the number of α- and β-linked neutral sugars in a glycan. No signals related to sialic acid are observed in the anomeric region, since sialic acids do not have anomeric protons.

The primary structural-reporter group signals that are used to assess the sialic acid content of glycans are the signals from the sialic acid H-3 axial and equatorial protons. The H-3 axial proton signals are centered in the ^{1}H NMR spectrum at around 1.6–1.8 ppm and the H-3 equatorial signals are centered at about 2.6–2.8 ppm. Identifying the kinds of sialic acid that are present can be accomplished by determining whether signals associated with the methyl protons of the *N*-acetyl group of NeuAc are present or the methylene protons of the *N*-glycolyl group of NeuGc are present. The *N*-acetyl group proton signals of NeuAc are found along with the *N*-acetyl group signals of GlcNAc at 1.9–2.1 ppm, whereas the *N*-glycolyl proton signals are present in the ^{1}H NMR spectrum at 3.9–4.1 ppm.

Historically the assessment of glycan structure using structural-reporter groups was based upon a comparison of the chemical shifts observed in the ^{1}H NMR spectrum of the glycan sample with chemical shifts observed in the analysis of glycan standards whose structure was known. However, online searchable databases now exist that contain glycan ^{1}H and ^{13}C chemical shift information based upon the reported analysis of glycans over the past 35 years. These databases can provide proposed glycan structures based upon the input chemical shift information, or they can provide proposed chemical shifts based upon the input of glycan structure. CASPER (http://www.casper.organ.su.se/casper/) and GLYCOSCIENCES.de (http://www.glycosciences.de/index.php) are examples of these kinds of online databases [148–150]. Supplementary structural data derived from orthogonal techniques, such as enzymatic studies, methylation analysis, or mass spectrometric analysis, is generally used to confirm the glycan structures that are proposed based upon 1D NMR studies [16].

To fully characterize the structure of a glycan using only NMR analysis, the acquisition of a combination of different two-dimensional homo- and heteronuclear NMR spectra is required. These 2D experiments generally require a milligram or more of isolated glycan and significant technical expertise to acquire and interpret the spectral data. Among the homonuclear 2D experiments that are used are ^{1}H–^{1}H correlation spectroscopy (COSY) and ^{1}H–^{1}H total correlation spectroscopy (TOCSY). The ^{1}H–^{1}H COSY experiment is used to establish the connectivity between protons on neighboring carbon atoms in a particular monosaccharide, while the ^{1}H–^{1}H TOCSY experiment can be used to establish the connectivity between all, or at least most, of the protons in each of the glycan monosaccharides. A variation of the 2D ^{1}H–^{1}H TOCSY experiment is the ^{1}H–^{1}H selective 1D TOCSY experiment in which protons within a constituent monosaccharide are selectively excited to establish the connectivity between the protons in the monosaccharide. Once the proton connectivities within the glycan monosaccharides have been established, the carbons that are directly attached to the protons in the monosaccharides can be determined using the 2D ^{1}H–^{13}C heteronuclear single quantum correlation (HSQC) experiment. To establish the connectivity between the different monosaccharides that comprise the glycan, the ^{1}H–^{13}C heteronuclear multiple bond correlation (HMBC) experiment is generally used, which identifies the carbons that are separated from a particular proton by two or three bonds. Thus the carbons and hydrogens that are separated by a glycosidic bond can be identified. Another 2D experiment that can be used to establish glycan linkage positions is the ^{1}H–^{1}H nuclear Overhauser effect spectroscopy or NOESY experiment, which identifies protons that are in close spatial proximity to each other (<5 Å). The development of new selective pulse sequences has led to significant improvements in the efficiency of glycan NMR characterization, like the pulse sequences used in selective-TOCSY-NOESY, selective-TOCSY-HSQC, and selective-TOCSY-DQFCOSY experiments [151, 152].

References

1 Apweiler, R., Hermjakob, H., & Sharon, N. (1999). On the frequency of protein glycosylation, as deduced from analysis of the SWISS-PROT database. *Biochim Biophys Acta*, 1473(1), 4–8.

2 Krapp, S., Mimura, Y., Jefferis, R., Huber, R., & Sondermann, P. (2003). Structural analysis of human IgG-Fc glycoforms reveals a correlation between glycosylation and structural integrity. *J Mol Biol*, 325(5), 979–989.

3 Parodi, A. J. (2000). Role of N-oligosaccharide endoplasmic reticulum processing reactions in glycoprotein folding and degradation. *Biochem J*, 348(Pt 1), 1–13.
4 Karpusas, M., Whitty, A., Runkel, L., & Hochman, P. (1998). The structure of human interferon-beta: implications for activity. *Cell Mol Life Sci*, 54(11), 1203–1216.
5 Mimura, Y., Church, S., Ghirlando, R., Ashton, P. R., Dong, S., Goodall, M., *et al.* (2000). The influence of glycosylation on the thermal stability and effector function expression of human IgG1-Fc: properties of a series of truncated glycoforms. *Mol Immunol*, 37(12–13), 697–706.
6 Chung, C. H., Mirakhur, B., Chan, E., Le, Q. T., Berlin, J., Morse, M., *et al.* (2008). Cetuximab-induced anaphylaxis and IgE specific for galactose-alpha-1,3-galactose. *N Engl J Med*, 358(11), 1109–1117.
7 Galili, U., Shohet, S. B., Kobrin, E., Stults, C. L., & Macher, B. A. (1988). Man, apes, and Old World monkeys differ from other mammals in the expression of alpha-galactosyl epitopes on nucleated cells. *J Biol Chem*, 263(33), 17755–17762.
8 Gribben, J. G., Devereux, S., Thomas, N. S., Keim, M., Jones, H. M., Goldstone, A. H., & Linch, D. C. (1990). Development of antibodies to unprotected glycosylation sites on recombinant human GM-CSF. *Lancet*, 335(8687), 434–437.
9 Baynes, J. W., & Wold, F. (1976). Effect of glycosylation on the in vivo circulating half-life of ribonuclease. *J Biol Chem*, 251(19), 6016–6024.
10 Goetze, A. M., Liu, Y. D., Zhang, Z., Shah, B., Lee, E., Bondarenko, P. V., & Flynn, G. C. (2011). High-mannose glycans on the Fc region of therapeutic IgG antibodies increase serum clearance in humans. *Glycobiology*, 21(7), 949–959.
11 Szkudlinski, M. W., Thotakura, N. R., Tropea, J. E., Grossmann, M., & Weintraub, B. D. (1995). Asparagine-linked oligosaccharide structures determine clearance and organ distribution of pituitary and recombinant thyrotropin. *Endocrinology*, 136(8), 3325–3330.
12 Boyd, P. N., Lines, A. C., & Patel, A. K. (1995). The effect of the removal of sialic acid, galactose and total carbohydrate on the functional activity of Campath-1H. *Mol Immunol*, 32(17–18), 1311–1318.
13 Shields, R. L., Lai, J., Keck, R., O'Connell, L. Y., Hong, K., Meng, Y. G., *et al.* (2002). Lack of fucose on human IgG1 N-linked oligosaccharide improves binding to human Fcgamma RIII and antibody-dependent cellular toxicity. *J Biol Chem*, 277(30), 26733–26740.
14 Wasley, L. C., Timony, G., Murtha, P., Stoudemire, J., Dorner, A. J., Caro, J., *et al.* (1991). The importance of N- and O-linked oligosaccharides for the biosynthesis and in vitro and in vivo biologic activities of erythropoietin. *Blood*, 77(12), 2624–2632.

15 Hossler, P., Khattak, S. F., & Li, Z. J. (2009). Optimal and consistent protein glycosylation in mammalian cell culture. *Glycobiology*, 19(9), 936–949.
16 Spellman, M. W., Basa, L. J., Leonard, C. K., Chakel, J. A., O'Connor, J. V., Wilson, S., & van Halbeek, H. (1989). Carbohydrate structures of human tissue plasminogen activator expressed in Chinese hamster ovary cells. *J Biol Chem*, 264(24), 14100–14111.
17 Kornfeld, R., & Kornfeld, S. (1985). Assembly of asparagine-linked oligosaccharides. *Annu Rev Biochem*, 54, 631–664.
18 Raju, T. S., Briggs, J. B., Borge, S. M., & Jones, A. J. (2000). Species-specific variation in glycosylation of IgG: evidence for the species-specific sialylation and branch-specific galactosylation and importance for engineering recombinant glycoprotein therapeutics. *Glycobiology*, 10(5), 477–486.
19 Varki, A. (2001). Loss of N-glycolylneuraminic acid in humans: mechanisms, consequences, and implications for hominid evolution. *Am J Phys Anthropol, Suppl* 33, 54–69.
20 Teranishi, K., Manez, R., Awwad, M., & Cooper, D. K. (2002). Anti-Gal alpha 1-3Gal IgM and IgG antibody levels in sera of humans and old world non-human primates. *Xenotransplantation*, 9(2), 148–154.
21 Qian, J., Liu, T., Yang, L., Daus, A., Crowley, R., & Zhou, Q. (2007). Structural characterization of N-linked oligosaccharides on monoclonal antibody cetuximab by the combination of orthogonal matrix-assisted laser desorption/ionization hybrid quadrupole-quadrupole time-of-flight tandem mass spectrometry and sequential enzymatic digestion. *Anal Biochem*, 364(1), 8–18.
22 Hossler, P. (2012). Protein glycosylation control in mammalian cell culture: past precedents and contemporary prospects. *Adv Biochem Eng Biotechnol*, 127, 187–219.
23 Borch, R. F., Bernstein, M. D., & Durst, H. D. (1971). The cyanoborohydride anion as a selective reducing agent. *J Am Chem Soc*, 93(12), 2897–2904.
24 Anumula, K. R. (2006). Advances in fluorescence derivatization methods for high-performance liquid chromatographic analysis of glycoprotein carbohydrates. *Anal Biochem*, 350(1), 1–23.
25 Lamari, F. N., Kuhn, R., & Karamanos, N. K. (2003). Derivatization of carbohydrates for chromatographic, electrophoretic and mass spectrometric structure analysis. *J Chromatogr B Analyt Technol Biomed Life Sci*, 793(1), 15–36.
26 Abdel-Magid, A. F., Carson, K. G., Harris, B. D., Maryanoff, C. A., & Shah, R. D. (1996). Reductive amination of aldehydes and ketones with sodium triacetoxyborohydride. Studies on direct and indirect reductive amination procedures (1). *J Org Chem*, 61(11), 3849–3862.
27 Dalpathado, D. S., Jiang, H., Kater, M. A., & Desaire, H. (2005). Reductive amination of carbohydrates using NaBH(OAc)3. *Anal Bioanal Chem*, 381(6), 1130–1137.

28 Sato, S., Sakamoto, T., Miyazawa, E., & Kikugawa, Y. (2004). One-pot reductive amination of aldehydes and ketones with α-picoline-borane in methanol, in water, and in neat conditions. *Tetrahedron*, 60, 7899–7906.
29 Hardy, M. R., Townsend, R. R., & Lee, Y. C. (1988). Monosaccharide analysis of glycoconjugates by anion exchange chromatography with pulsed amperometric detection. *Anal Biochem*, 170(1), 54–62.
30 Lee, Y. C. (1990). High-performance anion-exchange chromatography for carbohydrate analysis. *Anal Biochem*, 189(2), 151–162.
31 Behan, J. L., & Smith, K. D. (2011). The analysis of glycosylation: a continued need for high pH anion exchange chromatography. *Biomed Chromatogr*, 25(1–2), 39–46.
32 Weitzhandler, M., Pohl, C., Rohrer, J., Narayanan, L., Slingsby, R., & Avdalovic, N. (1996). Eliminating amino acid and peptide interference in high-performance anion-exchange pulsed amperometric detection glycoprotein monosaccharide analysis. *Anal Biochem*, 241(1), 128–134.
33 Anumula, K. R. (1994). Quantitative determination of monosaccharides in glycoproteins by high-performance liquid chromatography with highly sensitive fluorescence detection. *Anal Biochem*, 220(2), 275–283.
34 Rustighi, I., Campa, C., Rossi, M., Semeraro, S., Vetere, A., & Gamini, A. (2009). Analysis of N-acetylaminosugars by CE: a comparative derivatization study. *Electrophoresis*, 30(15), 2632–2639.
35 Racaityte, K., Kiessig, S., & Kalman, F. (2005). Application of capillary zone electrophoresis and reversed-phase high-performance liquid chromatography in the biopharmaceutical industry for the quantitative analysis of the monosaccharides released from a highly glycosylated therapeutic protein. *J Chromatogr A*, 1079(1–2), 354–365.
36 Corfield, A. P., Sander-Wewer, M., Veh, R. W., Wember, M., & Schauer, R. (1986). The action of sialidases on substrates containing O-acetylsialic acids. *Biol Chem Hoppe Seyler*, 367(5), 433–439.
37 Schauer, R. (2000). Achievements and challenges of sialic acid research. *Glycoconj J*, 17(7–9), 485–499.
38 Varki, A., & Diaz, S. (1984). The release and purification of sialic acids from glycoconjugates: methods to minimize the loss and migration of O-acetyl groups. *Anal Biochem*, 137(1), 236–247.
39 Hara, S., Yamaguchi, M., Takemori, Y., Furuhata, K., Ogura, H., & Nakamura, M. (1989). Determination of mono-O-acetylated N-acetylneuraminic acids in human and rat sera by fluorometric high-performance liquid chromatography. *Anal Biochem*, 179(1), 162–166.
40 Manzi, A. E., Diaz, S., & Varki, A. (1990). High-pressure liquid chromatography of sialic acids on a pellicular resin anion-exchange column with pulsed amperometric detection: a comparison with six other systems. *Anal Biochem*, 188(1), 20–32.

41 Anumula, K. R. (1995). Rapid quantitative determination of sialic acids in glycoproteins by high-performance liquid chromatography with a sensitive fluorescence detection. *Anal Biochem*, 230(1), 24–30.

42 Klein, A., Diaz, S., Ferreira, I., Lamblin, G., Roussel, P., & Manzi, A. E. (1997). New sialic acids from biological sources identified by a comprehensive and sensitive approach: liquid chromatography-electrospray ionization-mass spectrometry (LC-ESI-MS) of SIA quinoxalinones. *Glycobiology*, 7(3), 421–432.

43 Plummer, T. H., Jr., Elder, J. H., Alexander, S., Phelan, A. W., & Tarentino, A. L. (1984). Demonstration of peptide:N-glycosidase F activity in endo-beta-N-acetylglucosaminidase F preparations. *J Biol Chem*, 259(17), 10700–10704.

44 Liu, Y., Salas-Solano, O., & Gennaro, L. A. (2009). Investigation of sample preparation artifacts formed during the enzymatic release of N-linked glycans prior to analysis by capillary electrophoresis. *Anal Chem*, 81(16), 6823–6829.

45 Kuster, B., & Harvey, D. J. (1997). Ammonium containing buffers should be avoided during enzymatic release of glycans from glycoproteins when followed by reducing terminal derivatization. *Glycobiology*, 7(2), vii–ix.

46 Papac, D. I., Briggs, J. B., Chin, E. T., & Jones, A. J. (1998). A high-throughput microscale method to release N-linked oligosaccharides from glycoproteins for matrix-assisted laser desorption/ionization time-of-flight mass spectrometric analysis. *Glycobiology*, 8(5), 445–454.

47 Fenaille, F., Le Mignon, M., Groseil, C., Siret, L., & Bihoreau, N. (2007). Combined use of 2,4,6-trihydroxyacetophenone as matrix and enzymatic deglycosylation in organic-aqueous solvent systems for the simultaneous characterization of complex glycoproteins and N-glycans by matrix-assisted laser desorption/ionization time-of-flight mass spectrometry. *Rapid Commun Mass Spectrom*, 21(5), 812–816.

48 Sandoval, W. N., Arellano, F., Arnott, D., Raab, H., Vandlen, R., & Lill, J. R. (2007). Rapid removal of N-linked oligosaccharides using microwave assisted enzyme catalyzed deglycosylation. *Int J Mass Spectrom*, 259(1–3), 117–123.

49 Szabo, Z., Guttman, A., & Karger, B. L. (2010). Rapid release of N-linked glycans from glycoproteins by pressure-cycling technology. *Anal Chem*, 82(6), 2588–2593.

50 Koutsioulis, D., Landry, D., & Guthrie, E. P. (2008). Novel endo-alpha-N-acetylgalactosaminidases with broader substrate specificity. *Glycobiology*, 18(10), 799–805.

51 Huang, Y., Mechref, Y., & Novotny, M. V. (2001). Microscale nonreductive release of O-linked glycans for subsequent analysis through MALDI mass spectrometry and capillary electrophoresis. *Anal Chem*, 73(24), 6063–6069.

52 Miura, Y., Kato, K., Takegawa, Y., Kurogochi, M., Furukawa, J., Shinohara, Y., *et al.* (2010). Glycoblotting-assisted O-glycomics: ammonium carbamate allows for highly efficient o-glycan release from glycoproteins. *Anal Chem*, 82(24), 10021–10029.

53 Houel, S., Hilliard, M., Yu, Y. Q., McLoughlin, N., Martin, S. M., Rudd, P. M., *et al.* (2014). N- and O-glycosylation analysis of etanercept using liquid chromatography and quadrupole time-of-flight mass spectrometry equipped with electron-transfer dissociation functionality. *Anal Chem*, 86(1), 576–584.
54 Yu, G., Zhang, Y., Zhang, Z., Song, L., Wang, P., & Chai, W. (2010). Effect and limitation of excess ammonium on the release of O-glycans in reducing forms from glycoproteins under mild alkaline conditions for glycomic and functional analysis. *Anal Chem*, 82(22), 9534–9542.
55 Tarelli, E. (2007). Resistance to deglycosylation by ammonia of IgA1 O-glycopeptides: implications for the beta-elimination of O-glycans linked to serine and threonine. *Carbohydr Res*, 342(15), 2322–2325.
56 Zauner, G., Koeleman, C. A., Deelder, A. M., & Wuhrer, M. (2012). Mass spectrometric O-glycan analysis after combined O-glycan release by beta-elimination and 1-phenyl-3-methyl-5-pyrazolone labeling. *Biochim Biophys Acta*, 1820(9), 1420–1428.
57 Furukawa, J., Piao, J., Yoshida, Y., Okada, K., Yokota, I., Higashino, K., *et al.* (2015). Quantitative O-glycomics by microwave-assisted beta-elimination in the presence of pyrazolone analogues. *Anal Chem*, 87(15), 7524–7528.
58 Turyan, I., Hronowski, X., Sosic, Z., & Lyubarskaya, Y. (2014). Comparison of two approaches for quantitative O-linked glycan analysis used in characterization of recombinant proteins. *Anal Biochem*, 446, 28–36.
59 Merry, T., & Astrautsova, S. (2003). Chemical and enzymatic release of glycans from glycoproteins. *Methods Mol Biol*, 213, 27–40.
60 Patel, T. P., & Parekh, R. B. (1994). Release of oligosaccharides from glycoproteins by hydrazinolysis. *Methods Enzymol*, 230, 57–66.
61 Kozak, R. P., Royle, L., Gardner, R. A., Fernandes, D. L., & Wuhrer, M. (2012). Suppression of peeling during the release of O-glycans by hydrazinolysis. *Anal Biochem*, 423(1), 119–128.
62 Merry, A. H., Neville, D. C., Royle, L., Matthews, B., Harvey, D. J., Dwek, R. A., & Rudd, P. M. (2002). Recovery of intact 2-aminobenzamide-labeled O-glycans released from glycoproteins by hydrazinolysis. *Anal Biochem*, 304(1), 91–99.
63 Ruiz-Canada, C., Kelleher, D. J., & Gilmore, R. (2009). Cotranslational and posttranslational N-glycosylation of polypeptides by distinct mammalian OST isoforms. *Cell*, 136(2), 272–283.
64 Bause, E., & Legler, G. (1981). The role of the hydroxy amino acid in the triplet sequence Asn-Xaa-Thr(Ser) for the N-glycosylation step during glycoprotein biosynthesis. *Biochem J*, 195(3), 639–644.
65 Breuer, W., & Bause, E. (1995). Oligosaccharyl transferase is a constitutive component of an oligomeric protein complex from pig liver endoplasmic reticulum. *Eur J Biochem*, 228(3), 689–696.
66 Kasturi, L., Eshleman, J. R., Wunner, W. H., & Shakin-Eshleman, S. H. (1995). The hydroxy amino acid in an Asn-X-Ser/Thr sequon can influence N-linked

core glycosylation efficiency and the level of expression of a cell surface glycoprotein. *J Biol Chem*, 270(24), 14756–14761.

67 Malaby, H. L., & Kobertz, W. R. (2014). The middle X residue influences cotranslational N-glycosylation consensus site skipping. *Biochemistry*, 53(30), 4884–4893.

68 Shakin-Eshleman, S. H., Spitalnik, S. L., & Kasturi, L. (1996). The amino acid at the X position of an Asn-X-Ser sequon is an important determinant of N-linked core-glycosylation efficiency. *J Biol Chem*, 271(11), 6363–6366.

69 Ben-Dor, S., Esterman, N., Rubin, E., & Sharon, N. (2004). Biases and complex patterns in the residues flanking protein N-glycosylation sites. *Glycobiology*, 14(2), 95–101.

70 Gavel, Y., & von Heijne, G. (1990). Sequence differences between glycosylated and non-glycosylated Asn-X-Thr/Ser acceptor sites: implications for protein engineering. *Protein Eng*, 3(5), 433–442.

71 Petrescu, A. J., Milac, A. L., Petrescu, S. M., Dwek, R. A., & Wormald, M. R. (2004). Statistical analysis of the protein environment of N-glycosylation sites: implications for occupancy, structure, and folding. *Glycobiology*, 14(2), 103–114.

72 Walmsley, A. R., & Hooper, N. M. (2003). Glycosylation efficiency of Asn-Xaa-Thr sequons is independent of distance from the C-terminus in membrane dipeptidase. *Glycobiology*, 13(9), 641–646.

73 Borisov, O. V., Field, M., Ling, V. T., & Harris, R. J. (2009). Characterization of oligosaccharides in recombinant tissue plasminogen activator produced in Chinese hamster ovary cells: two decades of analytical technology development. *Anal Chem*, 81(23), 9744–9754.

74 Grinnell, B. W., Walls, J. D., & Gerlitz, B. (1991). Glycosylation of human protein C affects its secretion, processing, functional activities, and activation by thrombin. *J Biol Chem*, 266(15), 9778–9785.

75 Titani, K., Kumar, S., Takio, K., Ericsson, L. H., Wade, R. D., Ashida, K., *et al.* (1986). Amino acid sequence of human von Willebrand factor. *Biochemistry*, 25(11), 3171–3184.

76 Valliere-Douglass, J. F., Eakin, C. M., Wallace, A., Ketchem, R. R., Wang, W., Treuheit, M. J., & Balland, A. (2010). Glutamine-linked and non-consensus asparagine-linked oligosaccharides present in human recombinant antibodies define novel protein glycosylation motifs. *J Biol Chem*, 285(21), 16012–16022.

77 Valliere-Douglass, J. F., Kodama, P., Mujacic, M., Brady, L. J., Wang, W., Wallace, A., *et al.* (2009). Asparagine-linked oligosaccharides present on a non-consensus amino acid sequence in the CH1 domain of human antibodies. *J Biol Chem*, 284(47), 32493–32506.

78 Wilson, I. B., Gavel, Y., & von Heijne, G. (1991). Amino acid distributions around O-linked glycosylation sites. *Biochem J*, 275(Pt 2), 529–534.

79 Gill, D. J., Clausen, H., & Bard, F. (2011). Location, location, location: new insights into O-GalNAc protein glycosylation. *Trends Cell Biol*, 21(3), 149–158.
80 Carr, S. A., Huddleston, M. J., & Bean, M. F. (1993). Selective identification and differentiation of N- and O-linked oligosaccharides in glycoproteins by liquid chromatography-mass spectrometry. *Protein Sci*, 2(2), 183–196.
81 Jebanathirajah, J., Steen, H., & Roepstorff, P. (2003). Using optimized collision energies and high resolution, high accuracy fragment ion selection to improve glycopeptide detection by precursor ion scanning. *J Am Soc Mass Spectrom*, 14(7), 777–784.
82 Sullivan, B., Addona, T. A., & Carr, S. A. (2004). Selective detection of glycopeptides on ion trap mass spectrometers. *Anal Chem*, 76(11), 3112–3118.
83 Hsiao, H. H., & Urlaub, H. (2010). Pseudo-neutral-loss scan for selective detection of phosphopeptides and N-glycopeptides using liquid chromatography coupled with a hybrid linear ion-trap/orbitrap mass spectrometer. *Proteomics*, 10(21), 3916–3921.
84 Stalnaker, S. H., Hashmi, S., Lim, J. M., Aoki, K., Porterfield, M., Gutierrez-Sanchez, G., *et al.* (2010). Site mapping and characterization of O-glycan structures on alpha-dystroglycan isolated from rabbit skeletal muscle. *J Biol Chem*, 285(32), 24882–24891.
85 Roepstorff, P., & Fohlman, J. (1984). Proposal for a common nomenclature for sequence ions in mass spectra of peptides. *Biomed Mass Spectrom*, 11(11), 601.
86 Kjeldsen, F., Giessing, A. M., Ingrell, C. R., & Jensen, O. N. (2007). Peptide sequencing and characterization of post-translational modifications by enhanced ion-charging and liquid chromatography electron-transfer dissociation tandem mass spectrometry. *Anal Chem*, 79(24), 9243–9252.
87 Williams, J. P., Pringle, S., Richardson, K., Gethings, L., Vissers, J. P., De Cecco, M., *et al.* (2013). Characterisation of glycoproteins using a quadrupole time-of-flight mass spectrometer configured for electron transfer dissociation. *Rapid Commun Mass Spectrom*, 27(21), 2383–2390.
88 Wu, S. L., Huhmer, A. F., Hao, Z., & Karger, B. L. (2007). On-line LC-MS approach combining collision-induced dissociation (CID), electron-transfer dissociation (ETD), and CID of an isolated charge-reduced species for the trace-level characterization of proteins with post-translational modifications. *J Proteome Res*, 6(11), 4230–4244.
89 Angel, P. M., Lim, J. M., Wells, L., Bergmann, C., & Orlando, R. (2007). A potential pitfall in 18O-based N-linked glycosylation site mapping. *Rapid Commun Mass Spectrom*, 21(5), 674–682.
90 Kaji, H., Saito, H., Yamauchi, Y., Shinkawa, T., Taoka, M., Hirabayashi, J., *et al.* (2003). Lectin affinity capture, isotope-coded tagging and mass spectrometry to identify N-linked glycoproteins. *Nat Biotechnol*, 21(6), 667–672.

91 Hagglund, P., Bunkenborg, J., Elortza, F., Jensen, O. N., & Roepstorff, P. (2004). A new strategy for identification of N-glycosylated proteins and unambiguous assignment of their glycosylation sites using HILIC enrichment and partial deglycosylation. *J Proteome Res*, 3(3), 556–566.

92 Rademaker, G. J., Pergantis, S. A., Blok-Tip, L., Langridge, J. I., Kleen, A., & Thomas-Oates, J. E. (1998). Mass spectrometric determination of the sites of O-glycan attachment with low picomolar sensitivity. *Anal Biochem*, 257(2), 149–160.

93 Hanisch, F. G., Jovanovic, M., & Peter-Katalinic, J. (2001). Glycoprotein identification and localization of O-glycosylation sites by mass spectrometric analysis of deglycosylated/alkylaminylated peptide fragments. *Anal Biochem*, 290(1), 47–59.

94 Klemm, C., Schroder, S., Gluckmann, M., Beyermann, M., & Krause, E. (2004). Derivatization of phosphorylated peptides with S- and N-nucleophiles for enhanced ionization efficiency in matrix-assisted laser desorption/ionization mass spectrometry. *Rapid Commun Mass Spectrom*, 18(22), 2697–2705.

95 Wells, L., Vosseller, K., Cole, R. N., Cronshaw, J. M., Matunis, M. J., & Hart, G. W. (2002). Mapping sites of O-GlcNAc modification using affinity tags for serine and threonine post-translational modifications. *Mol Cell Proteomics*, 1(10), 791–804.

96 Rusnak, F., Zhou, J., & Hathaway, G. M. (2002). Identification of phosphorylated and glycosylated sites in peptides by chemically targeted proteolysis. *J Biomol Tech*, 13(4), 228–237.

97 Gerken, T. A., Owens, C. L., & Pasumarthy, M. (1997). Determination of the site-specific O-glycosylation pattern of the porcine submaxillary mucin tandem repeat glycopeptide. Model proposed for the polypeptide:galnac transferase peptide binding site. *J Biol Chem*, 272(15), 9709–9719.

98 Muller, S., Alving, K., Peter-Katalinic, J., Zachara, N., Gooley, A. A., & Hanisch, F. G. (1999). High density O-glycosylation on tandem repeat peptide from secretory MUC1 of T47D breast cancer cells. *J Biol Chem*, 274(26), 18165–18172.

99 Reusch, D., Haberger, M., Falck, D., Peter, B., Maier, B., Gassner, J., *et al.* (2015a). Comparison of methods for the analysis of therapeutic immunoglobulin G Fc-glycosylation profiles-Part 2: Mass spectrometric methods. *MAbs*, 7(4), 732–742.

100 Siemiatkoski, J., Lyubarskaya, Y., Houde, D., Tep, S., & Mhatre, R. (2006). A comparison of three techniques for quantitative carbohydrate analysis used in characterization of therapeutic antibodies. *Carbohydr Res*, 341(3), 410–419.

101 Sinha, S., Pipes, G., Topp, E. M., Bondarenko, P. V., Treuheit, M. J., & Gadgil, H. S. (2008). Comparison of LC and LC/MS methods for quantifying N-glycosylation in recombinant IgGs. *J Am Soc Mass Spectrom*, 19(11), 1643–1654.

102 Leymarie, N., Griffin, P. J., Jonscher, K., Kolarich, D., Orlando, R., McComb, M., *et al.* (2013). Interlaboratory study on differential analysis of protein glycosylation by mass spectrometry: the ABRF glycoprotein research multi-institutional study 2012. *Mol Cell Proteomics*, 12(10), 2935–2951.
103 Harvey, D. J. (2012). Analysis of carbohydrates and glycoconjugates by matrix-assisted laser desorption/ionization mass spectrometry: an update for 2007-2008. *Mass Spectrom Rev*, 31(2), 183–311.
104 Harvey, D. J. (2015a). Analysis of carbohydrates and glycoconjugates by matrix-assisted laser desorption/ionization mass spectrometry: an update for 2009-2010. *Mass Spectrom Rev*, 34(3), 268–422.
105 Harvey, D. J. (2015b). Analysis of carbohydrates and glycoconjugates by matrix-assisted laser desorption/ionization mass spectrometry: an update for 2011-2012. *Mass Spectrom Rev*, 36(3), 255–422.
106 Reiding, K. R., Blank, D., Kuijper, D. M., Deelder, A. M., & Wuhrer, M. (2014). High-throughput profiling of protein N-glycosylation by MALDI-TOF-MS employing linkage-specific sialic acid esterification. *Anal Chem*, 86(12), 5784–5793.
107 Hsu, N. Y., Yang, W. B., Wong, C. H., Lee, Y. C., Lee, R. T., Wang, Y. S., & Chen, C. H. (2007). Matrix-assisted laser desorption/ionization mass spectrometry of polysaccharides with 2′,4′,6′-trihydroxyacetophenone as matrix. *Rapid Commun Mass Spectrom*, 21(13), 2137–2146.
108 Needs, P. W., & Selvendran, R. R. (1993). Avoiding oxidative-degradation during sodium-hydroxide methyl iodide-mediated carbohydrate methylation in dimethyl-sulfoxide. *Carbohydr Res*, 245(1), 1–10.
109 Ciucanu, I., & Costello, C. E. (2003). Elimination of oxidative degradation during the per-O-methylation of carbohydrates. *J Am Chem Soc*, 125(52), 16213–16219.
110 Ruhaak, L. R., Huhn, C., Waterreus, W. J., de Boer, A. R., Neususs, C., Hokke, C. H., *et al.* (2008). Hydrophilic interaction chromatography-based high-throughput sample preparation method for N-glycan analysis from total human plasma glycoproteins. *Anal Chem*, 80(15), 6119–6126.
111 Chen, F. T., & Evangelista, R. A. (1998). Profiling glycoprotein n-linked oligosaccharide by capillary electrophoresis. *Electrophoresis*, 19(15), 2639–2644.
112 Ma, S., & Nashabeh, W. (1999). Carbohydrate analysis of a chimeric recombinant monoclonal antibody by capillary electrophoresis with laser-induced fluorescence detection. *Anal Chem*, 71(22), 5185–5192.
113 Briggs, J. B., Keck, R. G., Ma, S., Lau, W., & Jones, A. J. (2009). An analytical system for the characterization of highly heterogeneous mixtures of N-linked oligosaccharides. *Anal Biochem*, 389(1), 40–51.
114 Boersema, P. J., Mohammed, S., & Heck, A. J. (2008). Hydrophilic interaction liquid chromatography (HILIC) in proteomics. *Anal Bioanal Chem*, 391(1), 151–159.

115 Ruhaak, L. R., Zauner, G., Huhn, C., Bruggink, C., Deelder, A. M., & Wuhrer, M. (2010). Glycan labeling strategies and their use in identification and quantification. *Anal Bioanal Chem*, 397(8), 3457–3481.

116 Guile, G. R., Rudd, P. M., Wing, D. R., Prime, S. B., & Dwek, R. A. (1996). A rapid high-resolution high-performance liquid chromatographic method for separating glycan mixtures and analyzing oligosaccharide profiles. *Anal Biochem*, 240(2), 210–226.

117 Ahn, J., Bones, J., Yu, Y. Q., Rudd, P. M., & Gilar, M. (2010). Separation of 2-aminobenzamide labeled glycans using hydrophilic interaction chromatography columns packed with 1.7 microm sorbent. *J Chromatogr B Analyt Technol Biomed Life Sci*, 878(3–4), 403–408.

118 Reusch, D., Haberger, M., Maier, B., Maier, M., Kloseck, R., Zimmermann, B., *et al.* (2015b). Comparison of methods for the analysis of therapeutic immunoglobulin G Fc-glycosylation profiles—part 1: separation-based methods. *MAbs*, 7(1), 167–179.

119 Hardy, M. R., & Townsend, R. R. (1988). Separation of positional isomers of oligosaccharides and glycopeptides by high-performance anion-exchange chromatography with pulsed amperometric detection. *Proc Natl Acad Sci U S A*, 85(10), 3289–3293.

120 Townsend, R. R., Hardy, M. R., Hindsgaul, O., & Lee, Y. C. (1988). High-performance anion-exchange chromatography of oligosaccharides using pellicular resins and pulsed amperometric detection. *Anal Biochem*, 174(2), 459–470.

121 Kunz, C., Rudloff, S., Hintelmann, A., Pohlentz, G., & Egge, H. (1996). High-pH anion-exchange chromatography with pulsed amperometric detection and molar response factors of human milk oligosaccharides. *J Chromatogr B Biomed Appl*, 685(2), 211–221.

122 Rice, K. G., Takahashi, N., Namiki, Y., Tran, A. D., Lisi, P. J., & Lee, Y. C. (1992). Quantitative mapping of the N-linked sialyloligosaccharides of recombinant erythropoietin: combination of direct high-performance anion-exchange chromatography and 2-aminopyridine derivatization. *Anal Biochem*, 206(2), 278–287.

123 Prime, S., & Merry, T. (1998). Exoglycosidase sequencing of N-linked glycans by the reagent array analysis method (RAAM). *Methods Mol Biol*, 76, 53–69.

124 Rudd, P. M., & Dwek, R. A. (1997). Rapid, sensitive sequencing of oligosaccharides from glycoproteins. *Curr Opin Biotechnol*, 8(4), 488–497.

125 Morelle, W., & Michalski, J. C. (2007). Analysis of protein glycosylation by mass spectrometry. *Nat Protoc*, 2(7), 1585–1602.

126 Domon, B., & Costello, C. E. (1988). A systematic nomenclature for carbohydrate fragments in FAB-MS/MS spectra of glycoconjugates. *Glycoconj J*, 5, 397–409.

127 Brull, L. P., Heerma, W., Thomas-Oates, J., & Haverkamp, J. (1997). Loss of Internal 1-6 substituted monosaccharide residues from underivatized and per-O-methylated trisaccharides. *J Am Soc Mass Spectrom*, 8, 43–49.

128 Ernst, E., Muller, D. R., & Richter, W. J. (1997). False sugar sequence ions in electrospray tandem mass spectrometry of underivatized sialyl-Lewis-type oligosaccharides. *Int J Mass Spectrom Ion Processes*, 160, 283–290.

129 Brull, L. P., Kovacik, V., Thomas-Oates, J. E., Heerma, W., & Haverkamp, J. (1998). Sodium-cationized oligosaccharides do not appear to undergo 'internal residue loss' rearrangement processes on tandem mass spectrometry. *Rapid Commun Mass Spectrom*, 12(20), 1520–1532.

130 Franz, A. H., & Lebrilla, C. B. (2002). Evidence for long-range glycosyl transfer reactions in the gas phase. *J Am Soc Mass Spectrom*, 13(4), 325–337.

131 Harvey, D. J., Mattu, T. S., Wormald, M. R., Royle, L., Dwek, R. A., & Rudd, P. M. (2002). "Internal residue loss": rearrangements occurring during the fragmentation of carbohydrates derivatized at the reducing terminus. *Anal Chem*, 74(4), 734–740.

132 Harvey, D. J. (2001). Ionization and collision-induced fragmentation of N-linked and related carbohydrates using divalent cations. *J Am Soc Mass Spectrom*, 12(8), 926–937.

133 Harvey, D. J. (2005a). Fragmentation of negative ions from carbohydrates: part 1. Use of nitrate and other anionic adducts for the production of negative ion electrospray spectra from N-linked carbohydrates. *J Am Soc Mass Spectrom*, 16(5), 622–630.

134 Harvey, D. J., Royle, L., Radcliffe, C. M., Rudd, P. M., & Dwek, R. A. (2008). Structural and quantitative analysis of N-linked glycans by matrix-assisted laser desorption ionization and negative ion nanospray mass spectrometry. *Anal Biochem*, 376(1), 44–60.

135 Harvey, D. J. (2005b). Fragmentation of negative ions from carbohydrates: part 3. Fragmentation of hybrid and complex N-linked glycans. *J Am Soc Mass Spectrom*, 16(5), 647–659.

136 Powell, A. K., & Harvey, D. J. (1996). Stabilization of sialic acids in N-linked oligosaccharides and gangliosides for analysis by positive ion matrix-assisted laser desorption/ionization mass spectrometry. *Rapid Commun Mass Spectrom*, 10(9), 1027–1032.

137 Wheeler, S. F., Domann, P., & Harvey, D. J. (2009). Derivatization of sialic acids for stabilization in matrix-assisted laser desorption/ionization mass spectrometry and concomitant differentiation of alpha(2 --> 3)- and alpha(2 --> 6)-isomers. *Rapid Commun Mass Spectrom*, 23(2), 303–312.

138 Gennaro, L. A., & Salas-Solano, O. (2008). On-line CE-LIF-MS technology for the direct characterization of N-linked glycans from therapeutic antibodies. *Anal Chem*, 80(10), 3838–3845.

139 Maier, M., Reusch, D., Bruggink, C., Bulau, P., Wuhrer, M., & Molhoj, M. (2016). Applying mini-bore HPAEC-MS/MS for the characterization and quantification of Fc N-glycans from heterogeneously glycosylated IgGs. *J Chromatogr B Analyt Technol Biomed Life Sci*, 1033–1034, 342–352.

140 Zhao, J., Li, S., Li, C., Wu, S. L., Xu, W., Chen, Y., *et al.* (2016). Identification of low abundant isomeric N-glycan structures in biological therapeutics by LC/MS. *Anal Chem*, 88(14), 7049–7059.

141 Lohmann, K. K., & von der Lieth, C. W. (2004). GlycoFragment and GlycoSearchMS: web tools to support the interpretation of mass spectra of complex carbohydrates. *Nucleic Acids Res*, 32(Web Server issue), W261–W266.

142 Ceroni, A., Maass, K., Geyer, H., Geyer, R., Dell, A., & Haslam, S. M. (2008). GlycoWorkbench: a tool for the computer-assisted annotation of mass spectra of glycans. *J Proteome Res*, 7(4), 1650–1659.

143 Woodin, C. L., Maxon, M., & Desaire, H. (2013). Software for automated interpretation of mass spectrometry data from glycans and glycopeptides. *Analyst*, 138(10), 2793–2803.

144 Yu, C. Y., Mayampurath, A., & Tang, H. (2013). Software tools for glycan profiling. *Methods Mol Biol*, 951, 269–276.

145 Hard, K., Mekking, A., Damm, J. B., Kamerling, J. P., de Boer, W., Wijnands, R. A., & Vliegenthart, J. F. (1990). Isolation and structure determination of the intact sialylated N-linked carbohydrate chains of recombinant human follitropin expressed in Chinese hamster ovary cells. *Eur J Biochem*, 193(1), 263–271.

146 Vliegenthart, J. F. G., Van Halbeek, H., & Lambertus, D. (1981). The applicability of 500-MHz high-resolution 1H-NMR spectroscopy for the structure determination of carbohydrates derived from glycoproteins. *Pure Appl Chem*, 53, 45–77.

147 Bubb, W. A. (2003). NMR spectroscopy in the study of carbohydrates: characterizing the structural complexity. *Concepts Magn Reson Part A*, 19A(1), 1–19.

148 Lundborg, M., & Widmalm, G. (2015). NMR chemical shift prediction of glycans: application of the computer program CASPER in structural analysis. *Methods Mol Biol*, 1273, 29–40.

149 Lutteke, T., Bohne-Lang, A., Loss, A., Goetz, T., Frank, M., & von der Lieth, C. W. (2006). GLYCOSCIENCES.de: an Internet portal to support glycomics and glycobiology research. *Glycobiology*, 16(5), 71R–81R.

150 Stenutz, R., Jansson, P. E., & Widmalm, G. (1998). Computer-assisted structural analysis of oligo- and polysaccharides: an extension of CASPER to multibranched structures. *Carbohydr Res*, 306(1–2), 11–17.

151 Sato, H., Fukae, K., & Kajihara, Y. (2008). 2D selective-TOCSY-DQFCOSY and HSQC-TOCSY NMR experiments for assignment of a homogeneous asparagine-linked triantennary complex type undecasaccharide. *Carbohydr Res*, 343(8), 1333–1345.

152 Sato, H., & Kajihara, Y. (2005). An unambiguous assignment method by 2D selective-TOCSY-HSQC and selective-TOCSY-DQFCOSY and structural analysis by selective-TOCSY-NOESY experiments of a biantennary undecasaccharide. *Carbohydr Res*, 340(3), 469–479.

Index

Analytical Characterization of Biotherapeutics, First Edition. Edited by Jennie R. Lill and Wendy Sandoval.

d

e

www.ingramcontent.com/pod-product-compliance
Lightning Source LLC
Chambersburg PA
CBHW071350220926
51613CB00008B/5
9781119053101